The Quantum and the Cosmic Realms

The Quantum and the Cosmic Realms

The Great Divide Between General Relativity and Quantum Theory

By
Les Short

The Quantum and the Cosmic Realms

ISBN: 979-8-9970229-0-7 (Paperback)
ISBN: 979-8-9970229-5-2 (Hardback)
ISBN: 979-8-9970229-1-4 (Hardback KDP)
ISBN: 979-8-9970229-6-9 (eBook)

Acknowledgments

Ψ

Thanks to Professor Carlo Rovelli for his works on quantum gravity and other topics. His natural flair for poetic and elegant writing has renewed my enthusiasm.

Ψ

I am indebted to Andy Richter, a friend and colleague who graciously reviewed this book and offered helpful suggestions and corrections.

Ψ

Many thanks to Ville Hirvonen, Abir Ghosh, and Cameron Bunney for allowing me to use some of their work from their excellent *Profound Physics* website.

Table of Contents

Preface

This book is for those excited by subatomic physics, astronomy, and science fiction. I'm not a theoretical physicist, but I wanted to put together something about the two main theories that fascinate me as a layperson.

Quantum Mechanics and General Relativity can frustrate some theoretical physicists. Both are incredibly successful but incompatible. They describe the universe at vastly different scales with conflicting rules. Einstein's General Relativity describes gravity as the curvature of a smooth, continuous spacetime caused by mass and energy. It governs the motion of planets, stars, and galaxies. Quantum Mechanics explains the strange, probabilistic world of subatomic particles, and three other forces: electromagnetism and the strong and weak nuclear forces.

General Relativity combines space and time (spacetime) and predicts the motions of apples, planets, and galaxies under definite physical laws that describe a smooth, continuous universe. Quantum Mechanics tells us that reality is blurry and quantized at the smallest scales. It emphasizes that probabilities and inherent uncertainty are at the root of our physical, matter-based, world. This makes finding a unified "*Theory of Everything*" very difficult. The two theories thus form *The Great Divide.*

If you want to start with something lighter to get you started, read *Quantum Physics for Beginners* by Carl J. Pratt or *Schrödinger's Rabbits* by Colin Bruce for quantum physics; for general relativity, consider *Relativity For Beginners* by John Stoddard or *General Relativity: The Essentials* by Carlo Rovelli.

In this book, I wanted to convey as much information as possible on both subjects. Some related topics could fill a book on their own. My intent was to provide enough detail to pique your interest in the topics covered. You may have heard that quantum theory is *weird, bizarre,* and *nonintuitive* – it is. Don't worry, just enjoy the ride! Physicists who understand the underlying mathematics often do not dwell on the oddities – "*Shut up and calculate*" – as the saying goes. That doesn't stop some scientists and writers from trying to translate it into English or develop new theories. It may never make total sense because the theory doesn't fit our classical way of thinking. (don't ask, Schrödinger's cat)

Instead of getting frustrated with any of the gobbledygook you don't understand, make the rest of it an engaging and ongoing experience. Lacking the mathematics and physics shouldn't diminish the excitement of learning about *the quantum and the cosmic realms.*

Reader's Guide

With apologies to Frank Wilczek and his wonderful book, *The Lightness of Being*, the idea of a reader's guide was so good that I wanted one too. I'm aiming for a comprehensive, concise view of quantum physics and general relativity. You can read the book from start to finish, but if you want to look at topics out of order, I also provide:

- An extensive *Background Materials* section that can help you with the main chapters. For example, Gauge Theory, Hilbert Space, Planck Units, etc.
- At the start of the Alphabetical Index is a list of all the topics covered in the Background Materials chapter.
- The *Glossary* contains words that may be unfamiliar, and some additional information on terms such as what a Landau level is.
- Even though I don't really like them, the Endnotes contain some key references to discoveries and events I thought might be relevant.[†]
- Several significant equations and use-case examples are included for completeness, but it won't hurt if you skip over them.

I recommend using the Background Materials and Glossary whenever something comes up that will help you understand the main topic.

Chapter One: The Standard Model of Particle Physics

In which we learn that reality is not what it seems,
that empty space is actually full of foam; particles are waves,
waves can be particles, both might be strings, and that photons
and electrons seem to know when we are looking at them.

Ψ

I want to begin with the current theory of particle physics because it provides our most accurate description of the subatomic world. The Standard Model of Particle Physics is a comprehensive framework based on Quantum Field Theory, encompassing Quantum Electrodynamics, Quantum Chromodynamics, and the weak nuclear force. It also incorporates the Higgs mechanism. This theory describes the fundamental constituents of matter and the forces governing their interactions. While it explains nearly all known forces in the universe, gravity (and its hypothetical graviton) remains outside its scope.

You may have encountered the phrase "A Theory of Everything"; however, the Standard Model of Particle Physics is often referred to as the "Theory of *Almost* Everything." It stands as the most successful and rigorously tested scientific framework to date, serving as an "instruction manual" for the visible universe. Analogous to a periodic table for the universe, it classifies the fundamental constituents of matter and the particles responsible for mediating forces. The Standard Model reduces the universe to 17 fundamental particles. It accounts for three of the four fundamental forces: electromagnetism, the strong nuclear force, and the weak nuclear force. Its mathematical predictions have been confirmed with remarkable precision; for instance, it predicted the electron's magnetic moment to within one part in a trillion, comparable to measuring the Earth-Moon distance with the accuracy of a human hair. Additionally, it provides the mechanism – via the Higgs field – by which particles acquire mass.

The Standard Model did not arise from a single breakthrough; rather, it developed over several decades through a series of experimental discoveries and theoretical advancements.

A Timeline of Discovery

In the 1860s, James Clerk Maxwell unified electricity and magnetism into a single force. In 1900, Max Planck introduced the idea of energy quantization to explain blackbody radiation. Einstein built on this idea with the photoelectric effect – showing that light consisted of particles – photons.

The landscape of physics changed dramatically in 1905, when Albert Einstein, then a patent clerk, introduced his theory of Special Relativity. This theory established the speed of light as a universal speed limit, prohibiting any object with mass from surpassing it. Special Relativity posed significant challenges for physicists, who now had to develop theoretical frameworks for particles that could approach light speed while also exhibiting the unpredictable behavior characteristic of quantum objects.

By the 1920s, the cast of characters was growing. J.J. Thomson proposed the plum pudding model. Ernest Rutherford then showed that atoms were mostly empty space (though, as we will see, not really) with a dense nucleus. Niels Bohr arranged electrons into neat, quantized orbits, and Schrödinger's wave mechanics theory improved Bohr's model by describing electron orbitals. But it wasn't until 1928 that Paul Dirac brought Einstein's relativity and quantum mechanics together. His equation didn't just describe the electron; it also predicted the *positron* (antimatter).

In 1932, physicist Carl Anderson, studying cosmic rays using a cloud chamber, captured clear evidence of the positron. This discovery confirmed that the universe has a dual structure. For every fundamental particle of regular matter that builds the physical world, a corresponding antiparticle of antimatter can exist.

The Golden Age of Fields (1940s–1950s)

By the 1940s, physicists hit a wall in their attempts to calculate the mass of an electron using Quantum Field Theory. The calculations returned "infinities" – it seemed that the theory had run into an insurmountable problem. Along came Richard Feynman, Julian Schwinger, and Shin'ichirō Tomonaga, who independently developed Renormalization, a mathematical method that allowed infinities to be subtracted, leaving behind finite, testable predictions. This gave rise to Quantum Electrodynamics (QED).

Although infinities suggest that the theory is missing *something* fundamental, renormalization makes quantum field theory finite. It redefines "bare" charge and mass (initially infinite) as physical observables with finite, experimentally measured values.

Taming the Particle Zoo (1960s–2012)

During the 1950s and 1960s, advancements in particle accelerator technology enabled physicists to collide protons and electrons at unprecedented energies. These experiments revealed a multitude of new particles, leading to what became known as the "Particle Zoo." Initially, this abundance suggested that nature was inherently disordered. However, further investigation by Murray Gell-Mann and George Zweig in 1964 led to the proposal that protons and neutrons are not indivisible spheres, but are instead composed of more fundamental constituents called quarks (also referred to as aces).[1]

Gell-Mann named the new particle a "quark", borrowed from the novel Finnegans Wake by James Joyce. With quarks providing the building blocks, the forces themselves needed to be organized. In the late 1960s, Steven Weinberg, Abdus Salam, and Sheldon Glashow achieved a remarkable unification. They proved that the Weak Nuclear Force (responsible for radioactive decay) and electromagnetism were actually two sides of the same coin: the Electroweak Force. By the mid-1970s, with the inclusion of the Strong Nuclear Force, the Standard Model appeared to be essentially complete. Also in the 1970s, the discovery of the Charm and Bottom quarks confirmed the fermion generation structure, completing three generations of quarks and leptons.

In 1983, the direct experimental validation of the Electroweak Theory and the massive nature of its force carriers earned Rubbia and van der Meer the Nobel Prize.[2] In 2012, CERN found the Higgs boson, confirming the mechanism that gives all fundamental particles their mass (but perhaps not neutrinos).[3]

The Standard Model is the result of unifying QED, the Weak Force, and the Strong Force into Quantum Field Theory. It describes two main types of particles: Fermions, the building blocks of matter, and bosons, which mediate the forces. Fermions are divided into six types of quarks and six types of leptons. They consist of three generations, with the first being (up and down quarks, electrons, and neutrinos) comprising all stable matter. Bosons mediate the four fundamental forces, except for gravity, and include the photon (electromagnetism), gluons (strong force), and W and Z bosons (weak force).

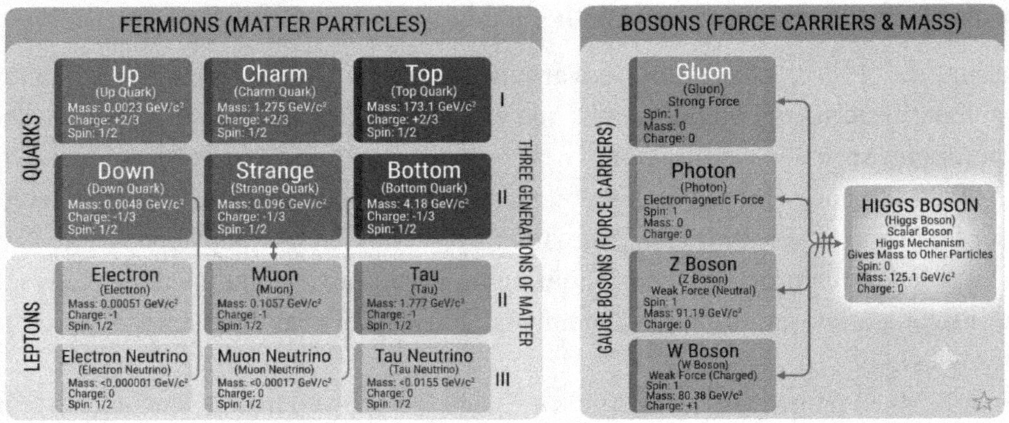

- ***Fermions*** consist of Quarks and Leptons. Quarks are grouped into three generations with six "flavors": up, down, charm, strange, top, and bottom. The up and down quarks, along with electrons, form the first generation and are the components of all ordinary matter. Protons are made of two up quarks and one down quark. Neutrons are made of one up quark and two down quarks.
- ***Leptons*** are fundamental particles that are not affected by the strong force. There are six types, also in three generations: the electron and electron neutrino, the muon and muon neutrino, and the tau and tau neutrino.
- ***Bosons*** (force-carrying particles) include the gauge bosons. They carry the fundamental forces. Photons mediate the electromagnetic force, Gluons mediate the strong force, and the W and Z Bosons mediate the weak force and are responsible for certain types of nuclear decay.
  - The ***Higgs*** is a unique boson that gives mass to other fundamental particles through the Higgs mechanism.

Might quarks themselves be composed of even smaller constituents? In 2026, experiments at the Large Hadron Collider investigated the internal structure of quarks at unprecedentedly small length scales.[4] Researchers probed distances approximately ten thousand times smaller than the diameter of a proton. The results provided no evidence for substructure within quarks. The team were able to establish a stringent upper limit: any potential subcomponents would only become apparent at energy scales exceeding 37 TeV.

Without going into too much detail, inside the Large Hadron Collider, violent collisions break protons apart into their constituent quarks and gluons. If a quark possessed internal subcomponents, scattering angles would deviate sharply from the prediction. The CMS data showed no statistical disagreement with the point-like quark model. Teams at CERN will push past 37 TeV with the LHC Run 3 and the move toward the High-Luminosity LHC upgrade. This will allow measurements of the scattering angle with lengths much smaller than 10^{-20} meters.

The Four Fundamental Forces

The four fundamental forces of nature serve as the structural framework for everything we observe in the universe, from the internal architecture of an atom to the rotation of entire galaxies. While these forces vary wildly in strength and range, they are governed by the exchange of messenger particles known as gauge bosons. (We don't know if gravity's gauge boson – the graviton – exists, so we cannot yet say that they are *all* governed by messenger particles)

Electromagnetism	Weak Nuclear	Strong Nuclear	Gravity
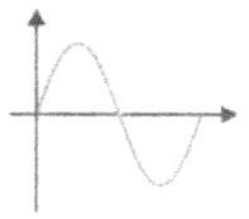	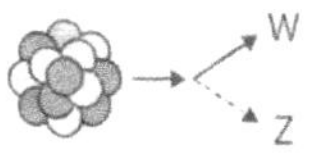	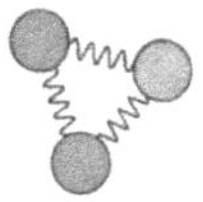	
Messenger: Photon (Massless) **Range:** Infinite **Cosmic Role:** Binds electrons to nuclei; governs light and chemical bonds.	**Messenger:** W & Z Bosons (Massive) **Range:** 1/1,000th of a proton **Cosmic Role:** Causes radioactive decay; allows particles to change identity.	**Messenger:** Gluon **Range:** Tiniest subatomic scale **Cosmic Role:** 100x stronger than EM. Binds quarks via Color Confinement (like stretching a rubber band until it snaps).	**Messenger:** Graviton (Undiscovered) **Range:** Infinite **Cosmic Role:** The great outlier. 10^36 times weaker than EM, yet dictates large-scale cosmic architecture.

The Electromagnetic Force

The Electromagnetic Force is the one most familiar to us in our daily lives, as it's responsible for the stability of atoms, the behavior of light, and the resistance we feel when we touch solid objects. It operates between charged particles and is mediated by virtual photon exchange.

In 1865, James Clerk Maxwell unified electricity and magnetism into a single set of equations – the electromagnetic field. Maxwell combined separate empirical laws discovered by other researchers into a single mathematical framework that predicted the existence of electromagnetic waves traveling at the speed of light.

Maxwell's laws can be expressed using four partial differential equations. They describe how electric and magnetic fields are generated and interact.

Gauss's Law for Electricity: This equation describes how electric charges produce electric fields. It states that the electric flux out of any closed surface is proportional to the total enclosed charge.

$$\nabla \cdot \mathbf{E} = \frac{\rho}{\varepsilon_0}$$

- E is the electric field vector.
- ρ is the charge density
- ε_0 is the vacuum permittivity.

Gauss's Law for Magnetism: This law states that magnetic monopoles do not exist. Magnetic field lines always form continuous closed loops, meaning the net magnetic flux through any closed surface is always zero.

$$\nabla \cdot \mathbf{B} = 0$$

Where B is the magnetic field vector.

Faraday's Law of Induction: This equation dictates that a time-varying magnetic field induces a spatially varying, curling electric field.

$$\nabla \times \mathbf{E} = -\frac{\partial \mathbf{B}}{\partial t}$$

Ampère's Law with Maxwell's Correction: Originally, Ampère's law stated that magnetic fields are generated by electric currents. Maxwell realized that this description was incomplete in the presence of changing circuits, such as a capacitor, and added a term for changing electric fields, known as the displacement current.

$$\nabla \times \mathbf{B} = \mu_0 \left(\mathbf{J} + \varepsilon_0 \frac{\partial \mathbf{E}}{\partial t}\right)$$

Where μ_0 is the vacuum permeability, J is the current density, and the second term within the parentheses is Maxwell's correction. Maxwell's addition of the term $\varepsilon_0 \frac{\partial \mathbf{E}}{\partial t}$ completed the mathematical symmetry between the fields. It established that a changing electric field generates a magnetic field, just as a changing magnetic field generates an electric field.

The equations, when solved for a vacuum where charge density ρ and current density J are zero, yield a wave equation for both fields. The calculated velocity of these self-propagating electromagnetic waves matched the measured speed of light exactly. This led Maxwell to conclude that light itself is an electromagnetic wave.

While Maxwell's equations provide a complete description of classical electromagnetism, physicists like Richard Feynman and Julian Schwinger reinterpreted it in the framework of quantum mechanics and special relativity in the

1940s. They developed Quantum Electrodynamics (QED), which is often said to be the most accurately tested theory in human history.

The Weak Nuclear Force

The Weak Nuclear Force operates at subatomic scales, roughly $\frac{1}{1000}$ the diameter of a proton. Unlike electromagnetism, which holds things together, the weak force mediates particle transformations and decays. It allows fundamental particles to change their identity, such as turning a neutron into a proton during beta decay.

When scientists observed radioactive beta decay, they saw a neutron decaying into a proton and an electron. They noticed that the emitted electrons possessed a continuous spectrum of kinetic energies. The conservation laws for energy and momentum dictate that the electron must exit with a single, fixed amount of kinetic energy. The energy should be divided equally between the proton and the electron every time. This puzzled researchers because it seemed to violate these two conservation laws.

If an electron had a discrete spectrum, it would only come out with precise, distinct packets of energy. Think of it like a staircase. You can stand on the first step or the second step, but you cannot float in the space between them. A continuous spectrum is like a ramp. An object on a ramp can exist at absolutely any height, smoothly moving from the bottom to the top without skipping any values. In their observations, the emitted electrons were not restricted to a few specific energy levels. Instead, after measuring millions of beta decays, they found electrons possessing 1.1 MeV, 1.2 MeV, 1.25 MeV, and every possible fractional value in between, up to a maximum limit. Basic mechanics dictates that when one stationary object breaks into exactly two pieces, the energy must split the exact same way every single time.

Imagine a firecracker at rest that splits into two fragments: a heavy rock (the proton) and a light BB pellet (the electron). Because momentum must balance out to zero, the two pieces have to fly apart in opposite directions. To satisfy both energy and momentum conservation, the light BB pellet must always carry away the exact same amount of kinetic energy in every single explosion.

Because scientists only detected the proton and the electron, they expected a single, sharp energy line (a discrete value). Seeing a smooth, continuous spread of energies suggested that energy was mysteriously vanishing from the universe during some of the decays.

In 1930, Wolfgang Pauli proposed that an undetected, neutral, light particle carried away the missing energy and momentum. Enrico Fermi named this particle the neutrino. In 1933, Fermi formulated the first formal theory of the weak interaction. He modeled beta decay as a direct contact interaction between four fermions occurring at a single point in space.

For many years, physicists had assumed that all fundamental forces respected parity, meaning physical processes would look identical if viewed in a mirror. In 1956, Tsung-Dao Lee and Chen-Ning Yang suggested that the weak force might violate this spatial symmetry.[5] Chien-Shiung Wu tested this hypothesis later that year. She cooled Cobalt-60 atoms to a few degrees above absolute zero and aligned their nuclear spins using a strong magnetic field. As the nuclei underwent beta decay, she measured the direction of the emitted electrons. The electrons emerged preferentially in the direction opposite to the nuclear spin, proving that the weak force violates parity symmetry. This asymmetric behavior is a unique property of the weak interaction.

Fermi's model worked well at low energies but predicted infinite-probability interactions at high energies. During the 1960s, Sheldon Glashow, Abdus Salam, and Steven Weinberg resolved this breakdown by unifying the weak force with electromagnetism into a single framework called the electroweak theory. Their model replaced the single-point contact interaction with the exchange of heavy mediator particles (gauge bosons). They predicted three massive vector bosons: the charged W^+ and W^- bosons and the neutral Z^0 boson. In 1973, experiments at CERN confirmed this theory by discovering neutral current interactions.[6] In 1983, they directly produced the W and Z bosons in high-energy collisions.[7]

The Strong Nuclear Force

As far as we know, the Strong Nuclear Force is the most powerful interaction in the universe, acting as the "glue" that binds quarks together to form protons and neutrons, which then bind to form atomic nuclei. It’s about 100 times stronger than electromagnetism but only occurs within the confines of the nucleus. Quark force carriers are called gluons. The three current quark bare masses sum to only $\sim 9\text{–}12\ \mathrm{MeV}/c^2$ (1% of the proton's 938 MeV/c^2 mass), with the remaining 99% generated dynamically via chiral symmetry breaking and QCD binding energy. This led to the development of Quantum Chromodynamics (QCD) in the 1970s, which introduced the idea of "color charge" (Red, Green, and Blue).

"Color confinement" is a unique property of the strong force, discovered through experiments at SLAC in the 1960s.[8] This property dictated that quarks could never be found in isolation. If two quarks are pulled apart, the energy increases like a stretching rubber band until it snaps, creating a new pair of quarks. This ensures that the building blocks of matter remain locked in stable, composite structures.

Physicists had figured out that electromagnetism alone couldn't explain how atomic nuclei remain stable. Because protons carry a positive electric charge, they repel each other with an immense force at short distances. For a heavy nucleus to hold together, an attractive force must exist that is strong enough to overcome the electromagnetic repulsion.

In 1935, physicist Hideki Yukawa proposed a possible model for this interaction. He realized that the nuclear force must operate by exchanging an undiscovered particle with a finite mass. Yukawa calculated that the short range of the nuclear force, which acts across distances of approximately 10^{-15} meters, required a mediator particle roughly 200 times heavier than an electron.

In 1947, a research team led by Cecil Powell discovered the predicted particle, the pi-meson or pion, in cosmic ray experiments. In the late 1960s, Deep Inelastic Scattering experiments at the Stanford Linear Accelerator Center revealed that protons and neutrons are not elementary particles but are made of smaller constituent particles called quarks. This discovery altered the understanding of the strong force.

The Yukawa force (or residual strong force) is considered a residual effect because it acts between "color-neutral" protons and neutrons, serving as a minor spill-over effect of the much more powerful fundamental strong force (QCD) that binds quarks together inside individual nucleons. Unlike photons in electromagnetism, gluons carry their own color charge, causing the force to strengthen as quarks move farther apart.

The Gravitational Force

The Gravitational Force is not like the other forces. It is unimaginably weak – approximately 10^{36} times weaker than the electromagnetic force – yet it dominates the large-scale structure of the cosmos. In 1915, Albert Einstein's General Relativity completely changed our understanding by showing that gravity is not a force in the traditional sense, but rather the curvature of spacetime caused by mass and energy. In 1919, Eddington's solar eclipse observation showed starlight bending around the Sun, verifying Einstein's prediction. In 2015, LIGO detectors provided direct proof by sensing gravitational waves from merging black holes.

Although a "graviton" has been theorized as the quantum messenger for gravity, it has never been detected. Reconciling gravity (general relativity) with quantum mechanics remains the primary challenge, often called the "Great Divide."

The Newtonian Gravitational Constant

Newton's law of universal gravitation describes the attractive force between two objects. It is expressed as:

$$F = G\frac{m_1 m_2}{r^2}$$

Where:

- F is the gravitational force acting between the two objects.
- G is the gravitational constant, approximately:
  $6.67 \times 10^{-11}\ m^3\ kg - 1\ s - 2$
- m_1 and m_2 are the masses of both objects in kilograms.
- r is the distance between the centers of the two masses in meters.

The Newtonian gravitational constant (*Big G*) is a fundamental number that determines the strength of the gravitational attraction between any two objects in the universe. It is the proportionality constant in Newton's Law, while *little g* represents the local acceleration due to Earth's gravity ($9.8m/s^2$). Big G is universal and stays the same whether on Earth, the Moon, or near a black hole. But finding its exact value has proven more difficult than envisaged.

In April 2026, a research team led by Stephan Schlamminger at the National Institute of Standards and Technology (NIST) published the results of a decade-long replication experiment.[9] The team used a sophisticated "scale", known as a torsion balance, which is hung from a very thin wire (which makes it very sensitive to change). It can measure the minute "twist" in the wire caused by the gravitational pull between a small and large metal cylinder. The experiment measured G at $6.67387 \times 10^{-11} m^3 \,.\, kg^{-1} \,.\, s{-}2$.

The result turns out to be a problem because it did not solve the mystery of Big G. The result is 0.0235% lower than that of a nearly identical experiment performed in France some years earlier. In the world of high-precision metrology, this is a massive discrepancy that indicates something is fundamentally wrong.

Gravity is about 10^{36} to 10^{40} times weaker than electromagnetism. This makes it extremely difficult to isolate from "noise" like the vibration of a passing truck, changes in air pressure, or even the static electricity on a scientist's clothing.

Because the force is so weak, small errors in measuring the mass of the objects or the exact distance between them can lead to wildly different results.

I recognize that certain fundamental constants cannot be derived from first principles, yet the gravitational constant (G) is so central that I sought to understand why its precise value remains unresolved. My research led me through a range of ideas, observations, and theories. Therefore, I have chosen to present a summary that appears reasonable based on current understanding.

In physics, when we "derive" a constant, we calculate it by combining other, more fundamental properties of nature. A good example is that the speed of light can be derived from electromagnetism using the vacuum permittivity and permeability. We cannot do this for G because we lack a complete underlying theory that connects gravity to the other three fundamental forces.

Gravity Has No Quantum Frame of Reference: The other three fundamental forces are described by the Standard Model of particle physics. Their strengths are tied to the properties of elementary particles, such as the elementary charge (e) of an electron. We can calculate how these forces behave because they are mediated by the exchange of quantum particles (like photons for electromagnetism).

Gravity is completely different. It's not a microscopic force pulling on particles; it is the geometric bending of smooth, continuous spacetime caused by mass and energy. Because G defines the stiffness of this spacetime fabric, there is no subatomic particle property to link it to.

The Extreme Weakness of Gravity: When we compare the gravitational attraction between two protons to their electrostatic repulsion, gravity is weaker by a factor of roughly 10^{36}. Because gravity is so incredibly faint at the atomic scale where the other forces rule, we cannot find a common baseline to compare them. To derive G, we need a "Theory of Everything" – a quantum theory of gravity – that unifies spacetime geometry with quantum mechanics.

The Planck Scale Problem: If you try to combine G with the other fundamental constants – the speed of light (c) and Planck's constant (h) – you can create a fundamental unit of mass, known as the Planck mass (M_P). The Planck mass is about 22 micrograms. In the quantum world, this is a colossal weight – equivalent to a small speck of dust, but billions of billions of times heavier than a proton.

The problem is that the energy required to probe a particle of this mass is far beyond anything we can produce. Without experimental data from this extreme energy scale, theorists have no way to verify if their mathematical frameworks (like string theory or loop quantum gravity) correctly predict the value of G.

Until a unified theory is mathematically proven and experimentally tested, G remains an arbitrary number that we simply measure from the universe around us.

The Grand Unified Theory (GUT)

A Grand Unified Theory (GUT), or Theory of Everything (TOE), conceptualizes all four fundamental forces as manifestations of a single, highly symmetric underlying framework. Electromagnetism and the weak force are already unified into the

Electroweak interaction. The next challenge is to incorporate the strong force. Most GUT models propose that at extremely high energies, comparable to those during the earliest moments of the Big Bang, these three forces converge into a single force governed by an expanded symmetry group. Integrating gravity remains the most formidable challenge.

The Higgs Field and the Higgs Boson

A Higgs boson enters a church. The priest says, "I'm sorry, we don't allow Higgs bosons in here." The Higgs boson replies: "But without me, you can't have Mass!"

During the development of the Standard Model, a central challenge was to explain why certain particles possess mass while others do not. The original equations implied that all fundamental particles should be massless, a prediction that was inconsistent with experimental evidence.

If the Higgs field were zero, the universe's structure would be fundamentally altered: electrons and quarks would be massless, the weak nuclear force would be as strong as electromagnetism, and force-carrying particles would no longer behave as they do now. The Higgs field is essential for giving matter's building blocks enough mass to form structures and for maintaining the separation of fundamental forces into distinct interactions. Without a non-zero Higgs field, quarks and electrons would travel at the speed of light, preventing the formation of atoms or complex structures.

Peter Higgs, François Englert, and Robert Brout were pioneering physicists who independently proposed the Brout-Englert-Higgs (BEH) mechanism in 1964. This theory explains how elementary subatomic particles acquire mass through their interaction with an invisible, all-pervading quantum field (the Higgs field), leading directly to the prediction of the Higgs boson.

An analogy compares this process to moving through molasses versus a vacuum. Moving through molasses is harder due to resistance. Similarly, particles interacting strongly with the Higgs field experience greater resistance and thus have more mass. Particles that do not interact with the Higgs field, like photons, are like moving through a vacuum: they face no resistance and remain massless.

Like other fundamental forces, the Higgs field has its own particle – the Higgs boson. It represents an excitation or ripple within the Higgs field, like a wave propagating through the "syrupy" field. When particles interact with the Higgs field, they may disturb it, creating ripples observed as Higgs bosons.

The 2012 discovery at CERN was a great achievement because it confirmed the existence of the Higgs field and led to the 2013 Nobel Prize in Physics to Peter Higgs and François Englert.

The Higgs boson does not directly give mass to particles – observing the Higgs boson is like detecting "molasses" by its ripples. Particles interacting with the Higgs field include leptons and W and Z bosons; the interaction strength determines their masses. Massless photons and gluons do not interact with the Higgs field.

It should be noted that quarks get their intrinsic rest mass from their Yukawa coupling to the Higgs field via electroweak symmetry breaking. In the Standard Model, their bare or "current" mass is generated when they interact with the non-zero vacuum expectation value of the Higgs field:

- An up quark has an intrinsic mass of only about $2.2\ \mathrm{MeV/c^2}$.
- A down quark has an intrinsic mass of about $4.7\ \mathrm{MeV/c^2}$.

Together, the three valence quarks in a proton (two up, one down) account for roughly $9.1\ \mathrm{MeV/c^2}$. Yet a proton has a measured rest mass of $938.3\ \mathrm{MeV/c^2}$. The three quarks provide less than 1% of the proton's mass. The remaining 99% of the mass of protons, neutrons, and everyday matter is generated dynamically by Quantum Chromodynamics (QCD):

- ***Massless Force Carriers:*** Individual gluons themselves are strictly massless gauge bosons ($\mathrm{m} = 0$).
- ***Field Energy & Confinement:*** Because gluons carry color charge, they interact with each other and form tight flux tubes between quarks. The tremendous potential energy stored in the gluon field, combined with the relativistic kinetic energy of quarks moving near light speed inside the confinement volume, manifests as invariant rest mass for the composite particle via Einstein's relation: $m = \frac{E}{c^2}$
- ***Chiral Symmetry Breaking:*** The QCD vacuum is populated by gluon fluctuations and quark–antiquark condensates ($\langle \bar{q}q \rangle \neq 0$). When light current quarks propagate through this vacuum, non-perturbative gluon interactions dress them into effective "constituent quarks" with an apparent dynamical mass of roughly $300\ \mathrm{to}\ 330\ \mathrm{MeV/c^2}$ each inside the hadron.

In Summary:

- ***Bare Quark Mass:*** Given entirely by the Higgs field.
- ***Gluon Rest Mass:*** Exactly zero.
- ***Nucleon / Hadron Mass:*** Generated almost entirely by the energy of the gluon fields, relativistic quark motion, and dynamical chiral symmetry breaking. Gluons are responsible for nearly all the mass of hadrons (like protons and neutrons).

The Higgs field is "on" everywhere and, unlike other fundamental fields, its lowest energy state – the vacuum of empty space – is not zero. It has a non-zero value permeating the entire universe. This unique feature is the source of all the Standard

Model's special effects. Every other known field has zero value in its lowest-energy state. When a light turns on, the field gets excited (emitting photons), and when it turns off, the field returns to zero.

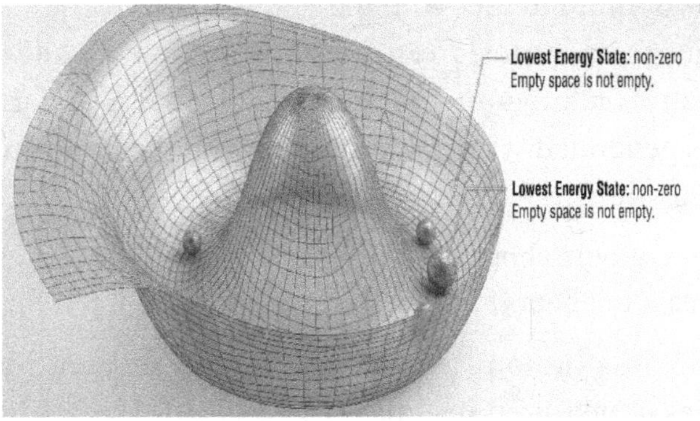

The Higgs field is different. Its energy potential resembles a "sombrero" or "Mexican hat." The center of the hat (zero value) is an unstable, high-energy point.

The hat's brim (non-zero value) is the lowest energy state.

Because the universe settles into its lowest-energy state, the Higgs field "rolled off" the center and settled at the brim, acquiring a non-zero value everywhere – a process called spontaneous symmetry breaking. Gauge Theory typically requires force carriers to be massless. The "Mexican Hat" potential is the only way to "cheat" this rule, allowing W and Z bosons to gain mass without "breaking" the underlying gauge architecture.

The Higgs boson is the only fundamental particle in the Standard Model with zero spin. Every other particle has some spin (like electrons) or is a force carrier (like photons), which makes them unique. A spin-0 particle means it possesses no intrinsic angular momentum, has perfect spherical symmetry, and its quantum state remains unchanged under spatial rotations. It behaves as a boson that doesn't follow the Pauli exclusion principle.

If the universe is the "syrup," the Higgs boson is a "glob" created by sloshing it, which then immediately dissolves back into the field. Even the Higgs boson gets its mass from interacting with the background Higgs field, like other particles. The Higgs boson's mass puzzles theoretical physicists. Quantum mechanics predicts it should be enormous – driven up to the Planck mass by quantum interactions. However, the measured mass is shockingly small. It's like expecting a boulder and finding a pebble.

For the Higgs to be so light, the known laws of physics suggest that its "bare" mass must be fine-tuned to an absurd precision – canceling out the quantum effects to ~1 part in 1×10^{34}. This is called the Hierarchy Problem – the mystery of why gravity is incredibly weak compared to the other fundamental forces, as reflected in the fact that the Higgs boson is so light. This extreme fine-tuning seems unnatural to physicists, and suggests our understanding of the Higgs is still incomplete.

There are several theoretical alternatives to the Higgs mechanism, known as Higgsless models, which propose that mass arises from strong interactions or from composite particles rather than an elementary Higgs field. Higgsless Models use strong dynamics (as in superconductivity) to break electroweak symmetry, making particles massive without a fundamental Higgs field, relying instead on composite particles. The Technicolor Theory suggests that new strong forces bind new particles, analogous to how quarks bind to form hadrons, creating composite particles that effectively break symmetry.

Composite and Little Higgs Models suggest that the Higgs itself isn't fundamental but is a composite particle whose properties arise from underlying physics. Some theories introduce multiple Higgs fields to explain mass, as in extensions of the Standard Model such as Supersymmetry (SUSY). Finally, some theories propose that mass arises from internal interactions within a particle rather than an external field, suggesting that those particles have substructure.

Massless Particles

Massless particles, like photons and gluons, do not couple to the Higgs field. The underlying quantum field theories describing these forces require the force-carrying particles (gauge bosons) to be massless to preserve a fundamental principle called "*gauge symmetry.*" According to special relativity, massless particles must always travel at the speed of light in any reference frame and can never be brought to a "stop" as they have no rest frame. They derive all their energy from their momentum.

In physics, mass is a form of "internal resistance" (inertia) to changes in motion. To accelerate an object with mass, energy must be applied. The closer it gets to the speed of light, the more energy it takes. To reach c, a particle with even a tiny amount of mass would require infinite energy.

Photons and gravitons (if they exist) travel at the speed of light, not by choice, but because the laws of physics leave them no other option. In a vacuum, (c) is not just a "speed limit"; it is the fundamental speed of causality. Why is the speed of light (in a vacuum) 299,792,458 m/s and not twice, or ten times as fast? The speed is determined by two physical "stiffness" constants of the vacuum itself. James Clerk Maxwell discovered these in the 1860s and proved that light is an electromagnetic wave and that its speed is the result of these two constants working together:

$$c = \frac{1}{\sqrt{\mu_0 \varepsilon_0}}$$

Where μ_0 is the permeability, and ϵ_0 is the permittivity. Permittivity is the property of every material which measures the opposition offered against the formation of an electric field. In electrostatics, permeability is the measure of a material's ability to support magnetic lines of force or a magnetic field within. (Since Maxwell was so influential to modern physics, you might want to take a look at his book, *A Dynamical Theory of the Electromagnetic Field.*)

Essentially, the "speed" of light is the speed at which electric and magnetic fields can "vibrate" into one another across empty space. If the vacuum were "stiffer" or more "resistant," light would travel more slowly. Because photons travel at the speed of light, they experience zero proper time. From the perspective of a photon, the moment it is emitted from a star and the moment it hits your eye happens at the same instant, regardless of how many light-years it traveled.

To look further at why particles (such as photons) can be massless, I'm using data from an article on the *Profound Physics* website, with permission from the authors Ville Hirvonen and Abir Ghosh. This is an excellent physics resource, and I highly recommend it: *https://profoundphysics.com/*.

Earlier, we saw that Special Relativity tells us that photons must always travel at the speed of light in any reference frame and can never be brought to a "stop", as they have no rest frame. Quantum Electrodynamics also predicts that photons cannot have mass due to gauge symmetry. It may seem obvious, but what is mass in this context? Mass is defined as the energy a particle has at rest (*rest energy*). This is a fundamental prediction of Einstein's special relativity, described by the energy-momentum relation:

$$E^2 = (pc)^2 + (m_0c^2)^2$$

The equation above states that a particle's total energy (E) equals its momentum (p) plus its rest mass (m_0). Scientists often use a shorthand called "natural units" where the speed of light (c) is set equal to 1. The Higgs boson's mass puzzles theoretical physicists. to: $E^2 = p^2 + m^2$. This, however, is only valid for a particle with no potential energy. If a particle is not moving, its momentum (p) is zero. The equation then becomes $E^2 = (m_0c^2)^2$, which simplifies to the famous $E = mc^2$. This represents the rest energy of the particle. For a particle with no mass, like a photon, the mass term drops out and we get $E^2 = (pc)^2$, which simplifies to $E = pc$. This shows that massless particles still carry energy purely through their momentum.

It also shows that mass is simply another form of energy, namely, the energy a particle has at rest. A photon *always* travels at the speed of light and cannot be brought to rest (technically, it cannot be brought to rest in a *vacuum)*. Since a photon doesn't have rest energy, which defines mass, photons are massless.

How exactly did special relativity predict this? A fundamental result of special relativity is the formula for relativistic total energy:

$$\mathrm{E} = \gamma m_0 c^2 = \frac{m_0 c^2}{\sqrt{1-\frac{v^2}{c^2}}}$$

Where γ is the Lorentz factor and m_0 is the rest mass. The Lorentz factor is a fundamental, dimensionless quantity in special relativity that represents how much measurements of time, length, and relativistic mass change for an object moving at a significant fraction of the speed of light.

From the above equation, we can solve for the mass and see what happens in the case of a photon (i.e., when the velocity v is the speed of light *c*):

$$m_0 = \frac{E}{c^2}\sqrt{1 - \frac{v^2}{c^2}}$$

If we now set the velocity equal to the speed of light, this makes the mass equal to zero. Move from the first equation on the left, see how it comes out in the middle equation, and then the final equation verifies that the mass (m_0) is zero:

$$\mathrm{m}_0 = \frac{E}{c^2}\sqrt{1 - \frac{v^2}{c^2}} = \frac{E}{c^2}\sqrt{1-1} \text{ thus } m_0 = \frac{E}{c^2} \times 0 = 0$$

This prediction applies to all particles that move at the speed of light, such as gravitons (if they exist) and gluons (although it's difficult to experimentally verify this, since gluons are never found by themselves). The three flavors of neutrinos do have very small masses but might not acquire it via the standard Higgs mechanism.

Quantum Electrodynamics also requires that photons don't have mass due to gauge symmetries. A gauge symmetry means that it is possible to have different mathematical expressions of a given field without actually changing any of the properties or dynamics of that field. Gauge symmetries and gauge fields describe the interactions between different particles (i.e., the fundamental forces of nature). The quanta (excitations) of these gauge fields are called gauge bosons and are the "exchange particles" for the different fundamental forces. (See the *Background Materials* for details on gauge theory and gauge symmetries)

The defining property here is that a photon is a gauge boson. Adding a mass term for the photon would break gauge symmetry, which is what the Standard Model is based on. Basically, gauge symmetry forbids the photon from acquiring mass. If we assume that the photon could get mass from the Higgs mechanism, we can construct alternate theories, which lead to some intriguing consequences. For example, photons with mass imply that electromagnetic waves should have three polarization states, but this is not seen in any experimentally verified phenomenon.

The simplest way to measure a photon's mass would be to measure the electric fields produced by charged particles. If the photon had a mass, it would change the way electric fields behave (since the photon is the "exchange" particle of electric forces) – the equation for an electric field produced by a charge Q (essentially Coulomb's law) would have some additional terms in it. The first part is the ordinary electric field, and the second part contains additional contributions from the photon's mass m_λ.

$$E(r) = \frac{kQ}{r^2}\left(1 + \frac{m_\lambda cr}{\hbar}\right) e^{-\frac{m_\lambda cr}{\hbar}}$$

Notice that if the mass of a photon is zero, this reduces back to the ordinary and correct electric field equation. These electric fields can be observed, and by measuring these so-called "damping factors" produced by the photon's mass, it is possible to work out an upper limit for the mass of a photon. These experiments provided upper limits of 10^{-42}kg to 8×10^{-51}kg for the photon, depending on the type of experiment. Using observations of planetary magnetic fields, data from the Pioneer 10 spacecraft, as it flew by Jupiter, set an upper limit of 7×10^{-52}kg.

Solar wind measurement experiments (plasma from the Sun that carries a magnetic field) show that if the photon had mass, this plasma would behave differently: a current would be produced, and the plasma would accelerate. If a photon has mass, it's at least twenty orders of magnitude lighter than an electron – the difference between the weight of a skyscraper and a single grain of dust.

If a photon carried even a tiny fractional electric charge, it would be deflected by magnetic fields as it traveled across the cosmos. Physicists have looked for this property by observing light from distant stars and the Cosmic Microwave Background (CMB). If photons had charge, the light reaching us would show evidence of dispersion or deflection that hasn't been observed.

A neutral particle with spin-1 (like a photon) can theoretically possess a magnetic dipole moment through higher-order quantum "loop" effects – which we won't go into here. While the predicted value for a standard photon is zero, some extensions of physics – such as those involving extra dimensions or specific types of dark matter – predict a tiny, non-zero magnetic moment.

Detecting this would require observing photons interacting with intense magnetic fields in ways that slightly alter their polarization or path, a phenomenon referred to as "vacuum birefringence" – quantum-mechanical phenomenon in which a strong magnetic field changes the refractive index of empty space.

The vacuum is actually a "quantum foam" with virtual particles – specifically virtual electrons and positrons – that pop in and out of existence. When an incredibly strong magnetic field is applied to a vacuum, it polarizes these virtual particles. They align along the field lines, effectively turning the "empty" space into a medium similar to a birefringent crystal, such as calcite.

As light passes through this magnetized vacuum, it splits into two different paths depending on its polarization relative to the magnetic field. The vacuum develops two distinct refractive indices. One index applies to light polarized parallel to the magnetic field, and another applies to light polarized perpendicular to it.

In classical physics, two beams of light pass through each other without interacting. Quantum Electrodynamics predicts that at extremely high energies, photons can actually "bounce" off one another by exchanging virtual particles. Such interactions could reveal new, undiscovered particles that act as "intermediaries," suggesting that the vacuum is even more active than current quantum theory predicts.

The Fundamental Constants (Fine-Tuning)

In the Standard Model of Particle Physics (and the Standard Model of Cosmology), there is a set of about 31 fundamental constants. Think of these as the "control dials" of the universe. What makes these numbers unique is that they cannot be derived or predicted by any mathematical theory. Physicists cannot explain why they have their specific values; they simply have to measure them in a laboratory and plug the raw numbers into their equations to make the math work.

This leads us to the Fine-Tuning problem – the hypothesis that, because "life as we know it" could not exist if the constants of nature had been even slightly different, the universe must be tuned specifically for life.

These physical constants, parameters, and initial conditions of the universe at, or around, the Big Bang played a key role in allowing life to emerge and evolve. Here are some prime examples of the constants:

1. ***Planck's Constant*** (h): The exact "size" of a quantum. It determines the minimum scale at which energy is quantized, setting the boundary between the classical world and the blurry quantum realm.
2. ***The Gravitational Constant*** (G): Determines the baseline strength of gravity and the extent to which mass curves spacetime.

3. ***Electron Mass:*** If the electron were slightly heavier or lighter, atoms would be drastically different sizes, and the chemical bonds that form DNA would be impossible.
4. ***The Fine-Structure Constant*** ($\alpha \approx \frac{1}{137}$): This number governs the strength of the electromagnetic interaction. It dictates exactly how tightly electrons are bound to atomic nuclei.
5. ***The Strong Force Coupling Constant*** (α_s): This determines how tightly gluons bind quarks together in the nucleus.

When scientists run simulations of the universe and adjust the "dials" of these constants, the results can be catastrophic. Changing almost any of these numbers by even a fraction of a percent, the universe becomes completely uninhabitable. Here are two of the most famous examples of this extreme fine-tuning:

1. ***The Carbon Resonance (The Strong Force):*** In the 1950s, Fred Hoyle realized that for life to exist, stars must be able to fuse helium into carbon. Fusion requires an incredibly precise "resonance" energy level in the carbon nucleus. If the strong nuclear force were even 2% stronger or 2% weaker, this resonance would not exist. The stars would either burn out instantly or they would never produce any element heavier than lithium. There would be no carbon, no oxygen, no planets, and no biology.
2. ***The Cosmological Constant:*** Quantum mechanics suggests that the vacuum of space should be boiling with energy, meaning the Cosmological Constant should be enormous. Measurements show it is unimaginably small, but not exactly zero. It is fine-tuned to 1 part in 10^{120}. If it were a little larger, the universe would have expanded so violently after the Big Bang that gases would never have coalesced to form galaxies or stars. If it were slightly smaller, the universe would have collapsed back in on itself in a "Big Crunch".

The margins for error are so astonishingly small that it gives rise to the Anthropic Principle, which asks why the universe is perfectly set up for observers like us to exist. Physicists generally fall into three camps to explain this oddity:

1. ***A Deeper Law (The Grand Unified Theory):*** Some physicists believe that once we unite quantum mechanics and gravity, we will discover that these 26 numbers are not "free" at all. A deeper mathematical logic will show that the universe must be exactly this way, and no other combination of numbers is mathematically possible.
2. ***The Multiverse:*** If our universe is just one of an infinite number of universes, each with randomly spun "dials," then almost all of them are dead, dark, and empty. We shouldn't be surprised that our dials are set perfectly for life, because we could only ever evolve in a universe that won the cosmic lottery.
3. ***The Teleological/Architect Argument:*** Without proof of a multiverse or deeper underlying equations, some argue that fine-tuning is exactly what it appears to be: a

setup. Whether interpreted as a divine creator, a cosmic architect, or an incredibly advanced simulation, the extreme precision of the fundamental constants is viewed as the signature of an intelligent designer who tuned the parameters to enable the emergence of life.

Our choices are to accept these exact values as a cosmic accident, search for an undiscovered law of nature that mandates them, or consider the possibility that our universe belongs to a multiverse where every numerical combination occurs.

String Theory: A Universe of Cosmic Threads

"It is said that papers in string theory are published faster than the speed of light. This, however, is not problematic since no information is being transmitted."

String theory is an incredibly ambitious idea, proposing a new picture of reality. It suggests that the fundamental constituents of the universe are not point-like particles, but relatively tiny, one-dimensional vibrating "strings" of energy. This seemingly simple idea offers a potential path to a unified "theory of everything".

By vibrating a violin string in different ways, you can produce a variety of distinct musical notes. String theory posits that elementary particles, such as electrons, photons, and quarks, are analogous to these notes. Each particle type corresponds to a unique vibrational mode of a fundamental string. The properties of each particle, like its mass and charge, are determined by the specific way its underlying string is vibrating. This elegant concept replaces the "particle zoo" of the Standard Model of Particle Physics with a single, underlying entity: the **string**.

The Kaluza-Klein theory is a 1919-1921 proposal by Theodor Kaluza and Oskar Klein that unifies gravity and electromagnetism by introducing a fifth spatial dimension, compactified in Einstein's general relativity. It suggests electromagnetic fields arise from the geometry of this hidden dimension, forming a foundation for modern higher-dimensional physics and string theory. The theory postulates a 5-dimensional spacetime (4 spatial dimensions and 1 temporal dimension). This extra dimension is "compactified" – curled into an unimaginably small, "rolled-up" circle, which explains why it is not directly observable. By extending Einstein's Field Equations to five dimensions, the spacetime curvature splits into a four-dimensional metric tensor (gravity) and a vector field that behaves exactly like the electromagnetic potential (Maxwell's Equations).

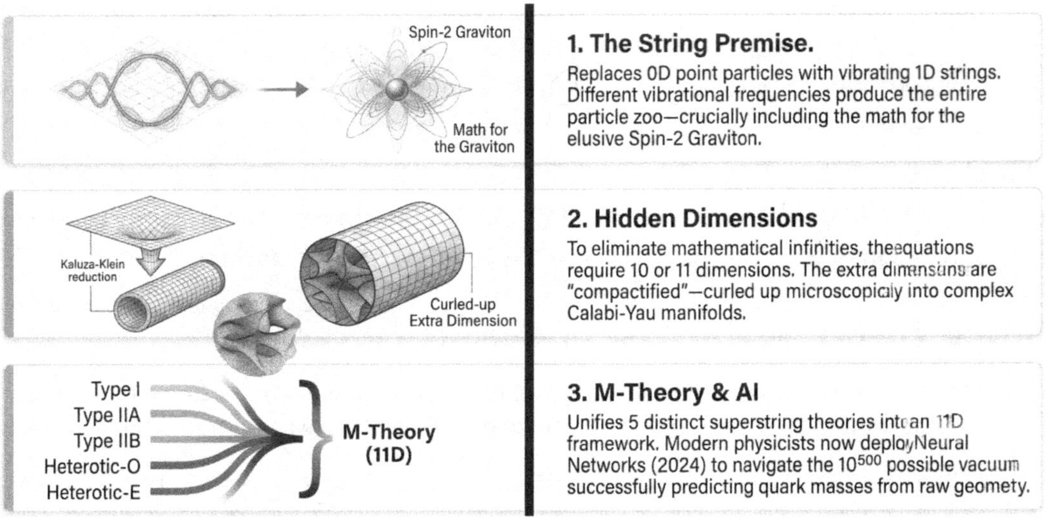

To explain why the fifth dimension isn't seen, Kaluza proposed that all fields are independent of the fifth coordinate, effectively "curling" the dimension. The five dimension vacuum Einstein equations, when projected into four dimensions, yield both gravity and Maxwell's Equations, a result often called the "Kaluza miracle". While it successfully unified gravity and electromagnetism, it could not explain quantum mechanics, weak/strong forces, or the mass/charge ratio of particles.

String theory offers a potential solution to the great divide. One of the vibrational modes of a closed string corresponds precisely to the properties of the hypothetical "graviton," the particle believed to mediate the force of gravity. This makes string theory the first quantum theory to incorporate gravity. By treating particles as extended strings rather than infinitesimal points, string theory smooths out the mathematical infinities that arise when trying to unite gravity with the quantum world. Physicists call this “compactification”. While 10 or 11 dimensions are common, some models, like bosonic string theory, require 26 dimensions.

For the mathematics of string theory to be consistent, the universe must have a total of 10 or 11 dimensions. Five consistent theories operate in 10 dimensions, utilizing supersymmetry to connect matter particles (fermions) and force particles (bosons).

The 11-dimensional theory proposes that when string interactions become strong, a new dimension opens, allowing strings to evolve into 2-dimensional membrane-like objects. Extra dimensions are required to eliminate mathematical inconsistencies (anomalies) in the equations, ensuring physical probabilities make sense. Here, the 11 coordinates are needed to pinpoint an event, just as we use coordinates to describe position and time in our daily lives. In our everyday world, we perceive four dimensions: Three dimensions of space: length, width, and height (up-down, left-right, forward-backward) – and one dimension of time – past to future.

To meet a friend at a specific location, you must provide four pieces of information: the street, the cross street, the floor number, and the time of the meeting.

In the late 20th century, physicists working on string theory discovered that their mathematical equations produced nonsense results, such as negative probabilities, when restricted to our four familiar dimensions. The math only balanced and became consistent when they calculated the equations in 11 dimensions. M-theory uses this 11-dimensional framework.

Because we do not see these extra seven dimensions of space, physicists propose that they are tightly curled up, or "compactified," at a scale far too small for human senses or scientific instruments to detect.

An analogy is a garden hose viewed from a distance. To an observer standing far away, the hose looks like a one-dimensional line; a point on the hose can move only forward or backward. However, if an ant walks on the hose's surface, it discovers a second, circular dimension. The ant can walk the length of the hose, or it can loop around the circumference.

It's possible that the extra seven dimensions are curled into complex, microscopic shapes at every point in our visible space. These tiny shapes, known as Calabi-Yau manifolds, dictate how fundamental particles vibrate and interact.

In standard quantum field theory, fundamental particles are treated as zero-dimensional points with no spatial extent. When calculating the gravitational force between two point particles, the calculation relies on the distance r between them. As these points approach each other and interact, the distance drops to zero ($r = 0$). Because the gravitational potential varies as $1/r$, dividing by zero yields an infinite energy density and gravitational force. When you try to calculate quantum corrections for gravity by summing all possible interaction paths, these short-distance infinities multiply uncontrollably, making the theory mathematically non-renormalizable.non-renormalizable.

String theory replaces these zero-dimensional points with one-dimensional objects that possess a finite minimum length, denoted as ℓ_s, which is near the Planck length of approximately 10^{-35} meters. This finite extension changes the geometry of particle interactions. When point particles interact, they form a sharp, distinct vertex in spacetime where the fields collide at a single point. When strings interact, they sweep a continuous two-dimensional surface called a worldsheet. The joining or splitting of two strings on a worldsheet does not possess a single, fixed coordinate in spacetime. Instead, the exact location of the interaction shifts depending on the observer's relativistic frame of reference.

Because the interaction is smeared over a non-zero spatial area defined by ℓ_s, forces at a distance of absolute zero are never calculated. The smooth geometry of the worldsheet replaces the $\frac{1}{0}$ vertex of point-particle mechanics, which *eliminates the infinities* from the calculation and yields finite quantum gravitational results.

While we move freely in the 3 large dimensions, the other dimensions are wound into a complex shape at every point in space. The geometry of these hidden dimensions is crucial because it determines the vibrational patterns of the strings and the physical laws and fundamental constants of our universe.

String theory is a complex subject, so here are some suggested books that will provide a lot more detail:

- *The Little Book of String Theory* by Steven Gubser.
- *A First Course in String Theory* by Barton Zwiebach.
- *String Theory and M-Theory: A Modern Introduction* by Becker, Becker, and Schwarz.

M-Theory

M-theory attempts to unify all five consistent versions of string theory into a single mathematical structure (and a particle description called Supergravity (SUGRA).

In the 1990s, physicists realized that there were five consistent but distinct versions of superstring theory. If it was to be a "theory of everything," why were there five of them? This led to the "second superstring revolution" and the proposal of M-theory by physicist Edward Witten.[10] M-theory, which operates in 11 dimensions, is a framework that tries to unify the five different superstring theories. These five theories are now seen as different approximations of a more fundamental underlying theory, much as different perspectives can describe the same object.

The energies required to probe the incredibly small scale of strings are far beyond the reach of our current particle accelerators. As a result, there is currently no direct experimental evidence to confirm or deny string theory.

String theory doesn't predict a unique universe. Instead, it seems to allow for a vast "panorama" of possible universes, perhaps as many as there are, each with its own set of physical laws. This makes it difficult to explain why our universe has the specific properties that it does. The complete mathematical structure of string theory, particularly M-theory, is not yet completely understood. It remains a work in progress, with many unanswered questions.

String theory (and M-theory) is still an active area of research, with many physicists exploring its complex mathematics, even though some aspects are no longer directly related to strings. Despite criticism, the theory has led to many significant breakthroughs in mathematics.

It is noteworthy that Artificial Intelligence is increasingly employed to address string theory problems by rapidly analyzing, classifying, and predicting the properties of potential universe-defining geometries. AI accelerates complex calculations, such as determining particle masses and vacuum stability, transforming months of computation into seconds, and assisting in navigating the vast number of Calabi-Yau manifold configurations.

Supersymmetry and Supergravity

Supersymmetry is a theoretical framework in particle physics that proposes a fundamental symmetry between bosons and fermions. Fermions are particles that obey the Pauli Exclusion Principle, which means that no two identical fermions can occupy the same quantum state simultaneously. Examples include electrons, protons, neutrons, quarks, and neutrinos. Bosons do not obey the Pauli Exclusion Principle, meaning that multiple identical bosons can occupy the same quantum state. Examples include photons, gluons, W and Z bosons, and the Higgs boson.

Supersymmetry attempts to address a problem with the Higgs boson – why is it much lighter than expected? Quantum corrections tend to drive the Higgs mass to the Planck scale, requiring a vast amount of fine-tuning to keep it at its observed value. Supersymmetry provides a mechanism to cancel these unwanted quantum corrections.

The strengths of the electromagnetic, weak, and strong forces change with energy. Extrapolating their strengths to very high energies, they almost, but not quite, converge at a single point. Supersymmetry predicts that, with the inclusion of superpartners, these forces can be unified at the scale of a Grand Unified Theory, suggesting a deeper underlying symmetry.

Supersymmetry postulates that for every fundamental particle in the Standard Model of Particle Physics, there exists a corresponding superpartner that differs in spin by one-half unit. For every fermion, there is a corresponding bosonic superpartner. For every boson, there is a corresponding fermion superpartner.

The Lightest Supersymmetric Particle (LSP) is a prime candidate for the non-baryonic dark matter that makes up a large portion of the universe's mass. While the simplest versions of the Standard Model predict massless neutrinos, experiments have shown that they do have mass, and supersymmetric extensions can naturally accommodate neutrino masses.

Many supersymmetric models incorporate a symmetry called R-parity. If R-parity is conserved, it implies that superpartners are always produced in pairs, and the lightest superpartner is stable. This stability could make the LSP a good candidate for dark matter.

Sfermions (Superpartners of Fermions): These are spin-0 bosons, named by adding an **s-** prefix to the particle name.

- ***Selectron:*** Partner of the electron.
- ***Smuon:*** Partner of the muon.
- ***Stau:*** Partner of the tau.
- ***Sneutrino:*** Partner of the neutrino.
- ***Squark:*** Partner of quarks (e.g., stop, sbottom, sup, sdown, strange, scharm).

Bosinos (Superpartners of Bosons): These are spin-$\frac{1}{2}$ fermions, named by adding an **-ino** suffix to the particle name.

- ***Gluino:*** Partner of the gluon.
- ***Photino:*** Partner of the photon.
- ***Wino:*** Partner of the W boson.
- ***Zino:*** Partner of the Z boson.
- ***Higgsino:*** Partner of the Higgs boson.

Other Significant SUSY Particles

- ***Gravitino:*** Partner of the graviton (hypothetical gravity particle).
- ***Neutralino:*** A combination of the photino, zino, and neutral higgsino. The lightest neutralino is a major dark matter candidate.
- ***Chargino:*** A combination of charged winos and charged higgsinos.

Supersymmetry increases the likelihood that the three fundamental forces (electromagnetic, weak, and strong) will unify into a single force at very high energies, and, for cosmology, it may explain the early universe and the formation of large-scale structures, primarily through its dark matter candidate.

Experiments at the Large Hadron Collider (LHC) have been searching for superpartners. So far, no new particles that fit the predictions of the simplest supersymmetric models have been found within the explored energy ranges. This lack of direct evidence has led to some debate within the physics community about the viability and exact form of supersymmetry. It could mean that Superpartners are much heavier than initially expected and beyond the reach of current experiments, or that they may not be the correct extension of the Standard Model.

Despite the lack of direct detection, research into supersymmetry continues. Physicists are exploring more complex supersymmetric models and investigating indirect methods to detect their effects. More powerful colliders may be needed to probe higher energy scales where superpartners could be found.

Some suggested reading:

- *"Supersymmetry and Supergravity"* by Julius Wess and Jonathan Bagger.
- *"Modern Supersymmetry: Dynamics and Duality"* by John Terning.
- *"A Supersymmetry Primer"* by Stephen P. Martin.

Supergravity (SUGRA)

Supergravity is a theoretical framework that attempts to unify the two pillars of modern physics: general relativity and supersymmetry. Essentially, supergravity is a field theory where the force of gravity is a necessary consequence of "local" supersymmetry. In a standard supersymmetric model, every matter particle (fermion) has a corresponding force-carrying partner (boson), but this is usually a global symmetry that applies the same way across all of space.

When this symmetry is made local – meaning the transformation between matter and force particles can vary from one point to another – the mathematics dictates that a gravitational field must exist to maintain the system's consistency. In this sense, gravity is not merely an added feature of the theory but a logical requirement of its internal structure.

While general relativity uses the graviton, a spin-2 boson, to mediate gravitational interactions, supergravity requires the graviton to have a "shadow" partner. This partner is the gravitino, a spin-$\frac{3}{2}$ fermion. The interaction between these two particles allows gravity to be integrated into the particle-heavy world of quantum field theory. The inclusion of the gravitino helps resolve the "hierarchy problem," which asks why gravity is so much weaker than the other fundamental forces. In supergravity, these superpartners often cancel the massive energy fluctuations that make quantum gravity calculations difficult.

One of the most important aspects of supergravity is its behavior in higher dimensions. Researchers discovered that the theory attains its most stable and mathematically elegant form when formulated in 11 dimensions (10 spatial and 1 temporal). This 11-dimensional supergravity is a landmark in theoretical physics because it is highly constrained; there are very few ways to alter it without breaking its internal consistency. While it was eventually realized that 11D supergravity is actually a "low-energy limit" of a more comprehensive framework called M-theory, it remains the primary tool for seeking to understand how all fundamental forces might be unified into a single, cohesive blueprint of the universe.

Summary

Chapter One introduces the fundamental framework of particle physics, often referred to as the "theory of almost everything". It outlines the 17 fundamental building blocks of nature, dividing them into matter particles and force-carrying particles.

The chapter details three of the four fundamental forces, while identifying gravity as the major outlier. It also comprehensively covers the Higgs field and the Higgs boson mechanism, explaining how particles acquire mass. In addition, the chapter explores speculative concepts aimed at expanding the Standard Model, such as String Theory, M-theory, Supersymmetry (SUSY), and Supergravity (SUGRA).

Chapter Two: Quantum Mechanics

Ψ

"Schrödinger's Cat walks into a bar.... and doesn't."

Early 1900s: The birth of Quantum Theory

In 1900, Max Planck introduced the idea of energy quantization to explain blackbody radiation. In 1887, Heinrich Hertz observed the photoelectric effect, and Philipp Lenard conducted detailed follow-up experiments. In 1905, working from these earlier results, Albert Einstein used Planck's idea to propose that light is composed of "quanta" (later called photons), thereby explaining the photoelectric effect. In 1911, Ernest Rutherford's gold-foil experiment demonstrated the existence of a dense, positively charged atomic nucleus. In 1913, Niels Bohr developed a quantum model of the atom in which electrons orbit the nucleus at specific, quantized energy levels.

1920s: The New Quantum Mechanics

In 1923, Arthur Compton provided further evidence for the particle nature of light by showing that X-rays scattered from electrons as if they were billiard balls. In 1924, Louis de Broglie proposed that matter, such as electrons, also exhibits wave-like properties. In 1925, Werner Heisenberg developed matrix mechanics, the first formulation of modern quantum mechanics. In 1926, Erwin Schrödinger developed wave mechanics, and Max Born provided the probability interpretation of the wave function. In 1927, Heisenberg formulated the uncertainty principle.

Late 1920s and beyond: Quantum Mechanics continues to expand.

In 1928, Paul Dirac created a relativistic equation for electrons, unifying quantum mechanics and special relativity. In 1931, James Chadwick discovered the neutron – crucial for understanding the nucleus. In 1935, Schrödinger devised his famous "cat" paradox thought experiment to highlight the absurdity of the standard (Copenhagen) interpretation regarding quantum superposition. In 1956, the Wu experiment demonstrated that parity is not conserved in weak interactions, meaning the universe has a "handedness". In 1957, Hugh Everett formulated the many-worlds interpretation of quantum mechanics. In 1962, experiments confirmed the existence of the electron and muon neutrinos.

Late 20th and 21st centuries: Quantum applications and technologies

In the 1980s, Richard Feynman proposed the concept of quantum computers. In 1994, Peter Shor's algorithm highlighted the potential to break current encryption. In the early 2000s, the first commercial Quantum Key Distribution (QKD) systems were developed for secure communication. In 2009, the first solid-state quantum processor using superconducting qubits was developed. In the 2010s, growth in trapped-ion research, larger qubit registers, and error-correction research occurred.

To set the scene for how quantum mechanics got started, we need to look at the ultraviolet catastrophe that forced Planck to rethink the classical view of energy, and into how Einstein built on this idea to develop the theory that light consists of particles that later became known as photons.

For a little background, German physicist Wilhelm Wien formulated the laws that bear his name. In 1893, he formulated his Displacement Law, and then in 1896, derived a Distribution Law. Wien's Displacement Law is a fundamental principle stating that the hotter an object is, the shorter the wavelength of its maximum radiant energy. As temperatures rise, an object's emitted color shifts across the spectrum from red to orange, yellow, and eventually blue.

Wien's law failed at long wavelengths and low frequencies, predicting zero energy when actual objects still emitted radiation. Relying on classical physics, it couldn't account for light's atomic nature, leading to Planck's theory of quanta.

Blackbody Radiation and Planck's Hypothesis

The concept of blackbody radiation explores how an idealized object absorbs and emits energy. In 1859, Gustav Kirchhoff introduced the concept of a perfectly "black body" during his investigations into thermal radiation. He defined it as an idealized material object that absorbs every ray of light falling upon it, regardless of wavelength or direction. In its original formulation, this body reflects and transmits no light, making it a perfect absorber of all incident electromagnetic radiation.

To understand the "catastrophe", imagine a hollow box heated to a high temperature and perfectly coated "black" on the inside. In this idealized scenario, the black coating ensures that any radiation being generated within the box is perfectly absorbed and re-emitted by the walls, allowing the energy inside to reach thermal equilibrium – a "blackbody". Any light entering a tiny hole in the box is trapped. When the box is heated, the radiation escaping through the hole depends only on its temperature and is unaffected by surface reflections.

Experimental measurements by physicists such as Lummer and Pringsheim revealed a clear, predictable pattern. The light intensity rises smoothly, reaches a distinct, temperature-dependent color peak, and then gently drops to zero at the shortest wavelengths. Classical physics, governed by the Rayleigh-Jeans Law, treated the electromagnetic radiation within this box as a series of standing waves.

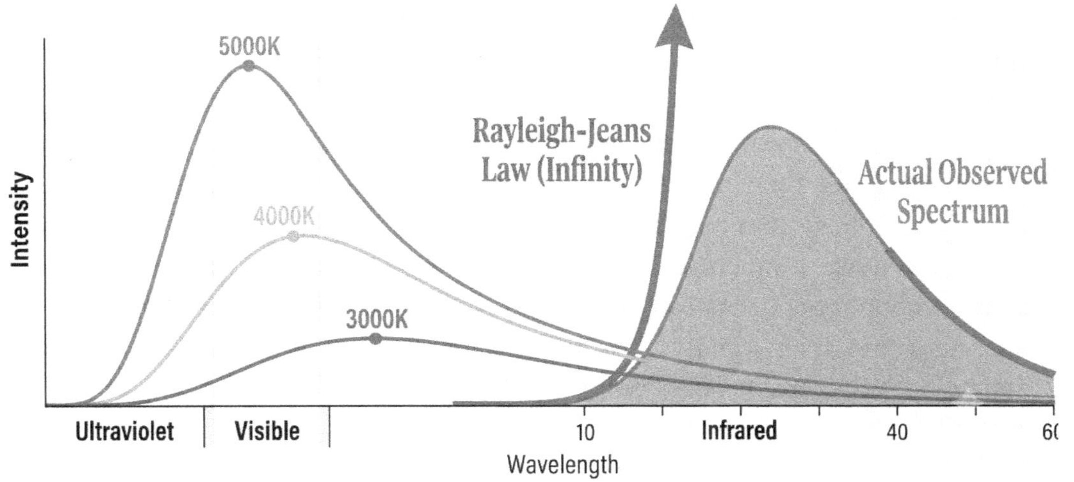

According to the principle of equipartition of energy, every possible vibration or "mode" of these waves should contain an equal amount of average thermal energy.

For the curious, the equipartition theorem states that in a physical system in thermal equilibrium, total energy is shared equally among all independent degrees of freedom. Each dynamic variable (quadratic degree of freedom) has an average energy of $\frac{1}{2}k_BT$ per molecule, where k_B is the Boltzmann constant and T is absolute temperature.

The problem occurs because shorter wavelengths are mathematically much more numerous than longer ones. As the wavelength decreases, more and more high-frequency modes can fit inside the box, suggesting a bias toward the high-energy end of the spectrum. If each of these infinite modes received an equal share of energy, the classical model predicted that the energy density would grow without bound as the frequency increased. As it moved into the ultraviolet and higher frequencies, the heat energy would, in theory, rise to infinity.

This "Ultraviolet Catastrophe" suggested that even a moderately hot object should instantly radiate an infinite amount of energy. This result contradicted physical reality and implied that the universe could not exist in a stable state.

Max Planck, obsessed with the Second Law of Thermodynamics, resolved this paradox in 1900 by proposing that energy is not continuous, but is instead divided into discrete, finite "quanta." In physics, *quantization* means that a specific property or value can only exist in separate, distinct amounts rather than a smooth, continuous range. Think of walking up stairs: you can stand on step one or step two, but you cannot float in the empty space between the steps.

Planck theorized that the energy of these packets is directly proportional to their frequency. In the context of the heated sphere, this meant that to excite a high-frequency (short-wavelength) wave, the system would need to "pay" a very high energy cost in a single packet.

At any given temperature, the box only has a limited amount of thermal energy to distribute. While there are infinite high-frequency modes available in theory, the probability of the system possessing enough energy to create the massive "quanta" required for those frequencies is extremely low. The energy emission does not increase without bound; instead, it peaks at a specific frequency determined by the temperature and then rapidly falls to zero in the ultraviolet and beyond. This quantization of energy successfully solved the difference between theory and observation, helping found the field of quantum mechanics.

The energy of each quantum is proportional to its frequency (v): $E = hv$, and this fits perfectly with the experimental data. The exchange rate between energy and frequency is governed by a fantastically small number: 6.63×10^{-34} Joule seconds. Planck's law, which defines the spectral energy density E_λ, represents the energy per unit volume radiated by a blackbody cavity at thermal equilibrium within a specific wavelength interval. The mathematical formula for the energy density per unit wavelength is:

$$E_\lambda = \frac{8\pi hc}{\lambda^5} \frac{1}{e^{\frac{hc}{\lambda k_B T}} - 1}$$

- λ is the wavelength of the radiation.
- T is the absolute temperature of the cavity walls in Kelvin.
- h is Planck's constant, which equals 6.626×10^{-34} J $\cdot$ s.
- c is the speed of light in a vacuum, which equals 2.998×10^8 m/s.
- k_B is the Boltzmann constant, which equals $1.381 \times 10^{-23} J \cdot K^{-1}$.

If you heat a blackbody cavity to a temperature of 1500K, the peak of the energy distribution occurs at a wavelength of 1.93×10^{-6} meters ($1.93\ \mu$m). Substituting these values into the formula allows you to calculate the exact radiant energy density present inside the cavity for any chosen wavelength interval.

The total energy density within the cavity can be found by integrating over all possible frequencies or wavelengths, a mathematical process that directly yields the Stefan-Boltzmann Law. The Stefan-Boltzmann Law is a fundamental physical principle stating that the total thermal radiation power emitted by an object is proportional to the fourth power of its absolute temperature. It explains how hot objects glow and radiate heat.

Observing the actual spectral energy distribution of radiation emitted by a heated object, Planck was forced to hypothesize that energy is not continuous but is exchanged in discrete packets, or quanta.

I'm going to take the liberty of adding some additional history that I believe is both relevant and interesting. It's unlikely that Planck really thought of his solution in terms of a new *quantum* theory as, in his landmark 1900 address to the German Physical Society, Planck didn't focus on the word "quanta". He referred to dividing energy into "*Energieelemente*" (energy elements). He used the term quantum more as a standard descriptive noun for an amount or parcel (e.g., *Elementarquantum* of electricity or matter).

I think it's fair to say that it was Einstein who deserves credit for kicking off quantum theory. He took the concept of "quanta" and gave it modern physical life, explicitly proposing localized "quanta of light" (*lichtquanta*, later termed photons). Planck did adopt the terminology more formally when naming (h) the "quantum of action" (*elementares Wirkungsquantum*).

Planck did care intensely about the resolution of blackbody radiation, but not in the way I assumed. He didn't view his *energy elements* as an exciting look into a new quantum reality. To him, it was an unwelcome mathematical contrivance forced upon him by experimental data. In a famous 1931 retrospective letter to American physicist Robert Wood, Planck candidly explained his frame of mind in the autumn of 1900:

"Briefly summarized, what I did can be described as an act of desperation... A theoretical explanation had to be found at any cost, no matter what it might be."

Measurements taken at the Physikalisch-Technische Reichsanstalt (PTR) in Berlin by Heinrich Rubens and Ferdinand Kurlbaum had conclusively broken Wilhelm Wien's radiation formula in the far-infrared. Planck had staked significant professional pride on Wien's law. When experimentalists showed him the new data, Planck immediately constructed an empirical interpolation formula. It matched the PTR curve perfectly, but it was an equation without a theoretical foundation.

To derive the formula from first principles, Planck turned to Ludwig Boltzmann's statistical methods. In 1877, Boltzmann had introduced discrete mathematical bins of energy simply to make combinatorial probability calculations possible, with the explicit plan to let the bin size shrink to zero ($\Delta\epsilon \rightarrow 0$) via calculus at the end.

Planck did the same thing with his cavity resonators: he divided the total energy E into discrete elements ϵ. But to make the math yield his empirical radiation curve, he found he could not take the continuous limit to zero. The element had to stay finite and proportional to frequency: $\epsilon = h\nu$.

Planck's true allegiance was to classical mechanics, thermodynamics, and James Clerk Maxwell's continuous electrodynamics. When Einstein suggested that light itself traveled and existed in real, localized particle-like bundles (the photoelectric effect), Planck was among the strongest opponents.

Even when Planck and others recommended Einstein for membership in the prestigious Prussian Academy of Sciences in 1913, Planck famously wrote an apology on Einstein's behalf:

"That he may sometimes have missed the target in his speculations, as, for example, in his hypothesis of light-quanta, cannot really be held too much against him, for it is not possible to introduce really new ideas.... without sometimes taking a risk."

Between 1911 and 1914, Planck formulated his so-called "second theory" to rescue as much classical physics as possible. He conceded that atomic resonators might *emit* energy discontinuously in quanta, but insisted that they still *absorbed* continuous electromagnetic waves smoothly and classically.

Planck cared deeply about the implications – so much so that he viewed the collapse of classical continuity as a tragedy for physics. He wanted his blackbody derivation to work, but spent much of his later career hoping someone would find a way to reconcile his constant h with smooth, continuous classical field equations.

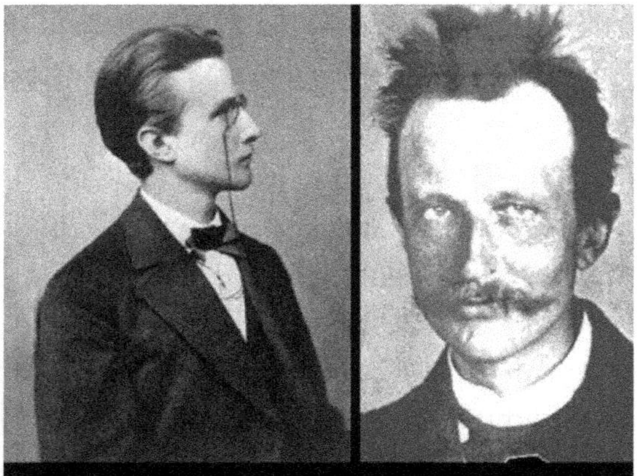

Max Planck in 1878 Max Planck in 1901

Some have suggested that Planck's discovery might have driven him to neglect his grooming. The truth is, Planck's weathered, exhausted look was the result of unrelenting personal tragedies, severe wartime hardships, and the psychological toll of living through two world wars in Germany, which systematically stripped away nearly everyone he loved.

The Photoelectric Effect

The photoelectric effect is a phenomenon where a material, most often a metal, emits electrons when exposed to light. At the turn of the 20th century, it posed a mystery that threatened to dismantle classical physics. In 1905, Einstein's explanation not only solved the mystery but also fundamentally changed our understanding of the universe by introducing the concept that light behaves as a particle.

In 1887, German physicist Heinrich Hertz was conducting experiments with radio waves. Hertz noticed that electric sparks jumped more readily between two metal electrodes when they were illuminated by ultraviolet light. By illuminating the electrodes with ultraviolet light, Hertz was effectively "pre-loading" the gap with free electrons. The electricity did not have to work as hard to jump the gap. This is why the sparks jumped more readily.

In 1902, Lenard took this research further by studying the specific properties of the electrons ("cathode rays") being emitted from the metal. Lenard's experiments revealed a direct conflict with the classical wave theory of light. According to classical physics, light was a continuous wave, and its energy was determined by its intensity (its brightness). Physicists assumed that shining a brighter light on a piece of metal would transfer more energy to the electrons, causing them to fly off with greater speed. However, Lenard found that the intensity of the light only changed the number of electrons emitted, not their maximum energy.

Even more confusing, the energy of the emitted electrons depended entirely on its frequency. A very bright red light could shine on a piece of metal all day and not eject a single electron, but a very dim beam of higher-frequency ultraviolet light would eject electrons instantly. Classical wave theory could not explain why a massive red-light wave failed where a tiny ultraviolet ripple succeeded.

In 1905, during his "miracle year," Albert Einstein provided the solution. He borrowed a controversial idea from Max Planck; that thermal energy is emitted in discrete, quantized units called "quanta." Einstein took this idea a big step further: he proposed that light itself is not a continuous wave, but rather a stream of discrete physical packets of energy. These packets of light would later be named "photons."

Einstein theorized that the energy of each photon is directly proportional to its frequency. High-frequency light (ultraviolet) consists of high-energy photons, while low-frequency light (red) consists of low-energy photons. He explained the photoelectric effect as a strict one-to-one interaction between a single photon and a single electron. Every metal has a specific "work function", which is the minimum amount of energy required to rip an electron away from the atomic lattice. If a photon of red light hits an electron, it lacks the energy to pay the "entry fee," and the electron stays put. Making the red light brighter just sends more low-energy photons, none of which can dislodge an electron.

However, if a photon of ultraviolet light strikes an electron, it carries more energy than is required to overcome the work function – to escape the metal, an electron must gain enough energy to overcome an electrostatic pull. Free electrons move throughout the metal, but they are bound to the material as a whole by an attractive force from the positive ions in the lattice. The electron absorbs the photon entirely and uses the required energy to break free. Any leftover energy from the photon is converted directly into the electron's kinetic energy (its speed). So a higher frequency of light results in a faster-moving electron.

Einstein's explanation was elegant, mathematically precise, and revolutionary. It proved that light has a dual nature, behaving as both a wave and a particle depending on how it's measured. Although not initially popular, his breakthrough, based on overwhelming evidence that light was a wave, laid the foundation for quantum mechanics and earned Albert Einstein the 1925 Nobel Prize in Physics.

The Schrödinger and Heisenberg Equations

When you look at the development of quantum theory, you see a sharp transition from classical mechanics to mathematical abstractions. Understanding this history requires examining a failed theory and a mathematical choice made by practical physicists. To see how we arrived at our current understanding, we need to look first at the BKS theory.

In 1924, Niels Bohr, Hendrik Kramers, and John Slater introduced the BKS theory. Bohr resisted Albert Einstein's 1905 proposal of light quanta, which you now know as photons. Bohr wanted to preserve the classical view that electromagnetic radiation travels as continuous waves.

The BKS theory proposed that atoms do not emit discrete particles of light. Instead, the theory argued that atoms contain virtual oscillators that generate a continuous electromagnetic field. This field continuously communicates with other atoms, determining the probability that an electron will jump between energy levels. Because the radiation field is continuous, whereas the atomic transitions are sudden, the theory abandoned strict conservation of energy and momentum in individual atomic events. It was proposed that energy and momentum are conserved only statistically, as averages over many interactions.

In 1925, Walther Bothe and Hans Geiger experimentally disproved the BKS theory. They built an automated coincidence counter to observe Compton scattering, a process where X-rays collide with electrons. If the BKS theory were correct, the emission of the scattered radiation and the recoil electron would happen independently, showing no timing correlation in individual events. Bothe and Geiger found that electrons and the scattered radiation emerged simultaneously. Arthur Compton and Alfred Simon confirmed these results using a cloud chamber. These experiments proved that energy and momentum are conserved in every individual quantum interaction, forcing Bohr to abandon the BKS theory and accept the photon concept.

In 1925, Werner Heisenberg, Max Born, and Pascual Jordan developed matrix mechanics. Scientists liked the conceptual description of matrix mechanics because it strictly relied on physical observables. It focused entirely on the frequencies and intensities of light that you can measure from atomic spectra. It discarded immeasurable concepts, such as the exact physical orbit of an electron around a nucleus. It provided a logically rigorous quantum description that avoided the internal contradictions of older models.

However, physicists preferred to use Erwin Schrödinger's 1926 wave mechanics equation for practical calculations. Matrix mechanics required complex linear algebra and the manipulation of infinite-dimensional matrices. In 1926, most physicists lacked training in these mathematical tools. Schrödinger's equation is a partial differential equation. This was a mathematical framework that physicists had used for decades to study acoustics, fluid dynamics, and classical optics. Solving a wave equation allowed physicists to use familiar boundary conditions and integration methods, making calculations for specific atomic systems much easier than computing matrix arrays.

Heisenberg's Matrix Mechanics

Werner Heisenberg formulated matrix mechanics in 1925 to describe atomic phenomena using only measurable quantities. He rejected classical concepts that cannot be observed, such as the continuous orbit of an electron. Instead, he focused entirely on the frequencies and intensities of light emitted or absorbed during atomic transitions.

Classical mechanics describes physical properties such as position x and momentum p as ordinary numbers that vary continuously over time. Heisenberg replaced these classical variables with arrays of numbers organized into infinite-dimensional matrices. The rows and columns of these matrices correspond to the specific stationary energy states of an atom. The individual elements within the matrix represent the physical transitions between those states. Because the values are tied to discrete transitions rather than continuous paths, the math naturally accounts for quantum jumps.

Because matrix mechanics represents physical quantities as matrices, the order of multiplication alters the result. Multiplying a position matrix by a momentum matrix yields a different outcome than multiplying the momentum matrix by the position matrix. Max Born and Pascual Jordan formalized this non-commutative property into the canonical commutation relation:

$$[\hat{x}, \hat{p}] = \hat{x}\hat{p} - \hat{p}\hat{x} = i\hbar\mathbf{I}$$

In this equation, x represents the position matrix, p represents the momentum matrix, i is the imaginary unit, and I is the identity matrix. The symbol $\hbar$ represents the reduced Planck constant, which has a specific value of $1.054\ \times 10^{-34}\ \mathrm{J}\cdot\mathrm{s}$.

This equation shows that the difference between xp and px is never zero. The non-zero product of these matrices forms the exact mathematical foundation for the uncertainty principle. In matrix mechanics, the state of the quantum system remains fixed in time, while the matrices representing physical observables change. The time evolution of any time-independent observable matrix A depends on its commutator with the Hamiltonian matrix H, which represents the total energy of the system. The Heisenberg equation of motion determines this change over time:

$$\frac{dA}{dt} = \frac{i}{\hbar}[H, A]$$

If the matrix A contains an explicit dependence on time independent of the quantum transitions, you add a partial derivative to the equation:

$$\frac{dA}{dt} = \frac{i}{\hbar}[H, A] + \frac{\partial A}{\partial t}$$

You can see how matrix mechanics works by examining the quantum harmonic oscillator, a system modeling a vibrating particle. The Hamiltonian matrix for an oscillator with mass m and angular frequency ω is:

$$H = \frac{p^2}{2m} + \frac{1}{2}m\omega^2 x^2$$

To solve this system, you do not solve a differential wave equation. Instead, you use algebraic methods to find matrices forx and p that satisfy the canonical commutation relation while making the energy matrix H diagonal. The resulting diagonal terms provide the quantized energy levels of the system:

$$E_n = \left(n + \frac{1}{2}\right)\hbar\omega$$

In this outcome, n represents a series of integers: 0, 1, 2, 3, and so on. The lowest possible energy state, where $n = 0$, yields an energy of $\frac{1}{2}\hbar\omega$. This configuration proves that a quantum system retains residual vibrational energy even at absolute zero, a direct consequence of the matrix commutation rules.

Schrödinger's Wave Mechanics

The Schrödinger wave equation is the fundamental equation of non-relativistic quantum mechanics, serving as the quantum-mechanical counterpart to Newton's second law of motion ($F = ma$). Developed by Erwin Schrödinger, it describes how the wave function (Psi: Ψ) of a quantum system – such as an electron or atom – changes over time and space, allowing the calculation of allowed energy levels and particle probability distributions. It's the foundation of wave mechanics.

Schrödinger's journey began with Louis de Broglie's hypothesis that all matter, not just light, possesses wave-like properties. If an electron is a wave, Schrödinger reasoned, there must be a "wave equation" that governs how that wave moves and interacts. In 1925, Schrödinger sought a mathematical framework for the electron in a hydrogen atom. He initially attempted to derive a relativistic version of the equation (now known as the Klein-Gordon Equation). However, because he had not yet accounted for "electron spin," his relativistic calculations did not match the known experimental data for hydrogen.

By returning to a nonrelativistic approach and treating the electron as a standing wave confined by the nuclear electric field, he successfully derived the equation that bears his name. It perfectly predicted the energy levels of the hydrogen atom, providing the first rigorous mathematical proof for the "quantum" nature of the subatomic world.

$$i\hbar\frac{\partial\Psi}{\partial t} = -\frac{\hbar^2}{2m}\frac{\partial^2\Psi}{\partial x^2} + V\Psi$$

The Schrödinger equation tells you how the wave function evolves over time. It does not predict a specific location for a particle; instead, it predicts the "shape" of the probability wave, indicating where we are most likely to find the particle if we measure it. The Schrödinger equation is *the* fundamental equation of quantum mechanics. It does for the quantum world what Isaac Newton's laws of motion do for the everyday world: it predicts how a system changes over time.

I have devoted considerable effort to understanding the equation. This is not a trivial exercise, and I acknowledge the length of the explanation that follows. My intention is to provide sufficient context to illustrate why it was such a significant breakthrough and to highlight the complexity faced by theoretical physicists.

A complete understanding of the equation is not required; readers may choose to proceed directly to subsequent sections if they wish to bypass the detailed explanation.

Let's start with the simplest type of wave – a sine wave. A classic sine wave represents a simple, repeating ripple. You can describe it mathematically using basic numbers that tell you its height (amplitude) and how close the crests are to each other (wavelength). Mathematically, a basic sine wave across distance is written as: $y = A\sin(x)$. In geometry, the sine function (sin) is tied to angles. A full rotation around a circle is 360 degrees, which equals 2π radians in standard scientific measurement. To include the specific wavelength (λ) in the formula so you can calculate the wave's height at any distance (x), physicists use this standard equation:

$$y = A\sin\left(\frac{2\pi}{\lambda}x\right)$$

The fraction $\frac{2\pi}{\lambda}$ acts as a scaling factor. It ensures that every time the distance x travels forward by one full wavelength (λ), the angle inside the sine function advances by exactly 2π radians (one full circle), causing the wave pattern to repeat perfectly.

In the quantum world, particles like electrons behave like waves. The Schrödinger equation uses a mathematical expression called the wave function, usually written as the Greek letter Psi (Ψ). The wave function is essentially a more sophisticated version of a sine wave. However, instead of tracking physical heights like a wave in the ocean, the wave function contains all the knowable information about a quantum particle. When you square the value of the wave function (Ψ^2), the result gives you the probability of finding the particle at a specific location. Where the wave is tall, the particle is highly likely to be found; where the wave is flat, the probability is near zero.

Time enters the equation because quantum waves are not static; they move, spread out, and change shape as time passes. There are actually two versions of the Schrödinger equation to handle this:

1. The Time-Independent Schrödinger Equation looks at "standing waves" that do not change their overall shape over time, like a plucked guitar string. It is used to find the allowed, stable energy levels of a system, such as an electron bound inside an atom.
2. The Time-Dependent Schrödinger Equation includes time as a variable. It acts like a movie camera, calculating exactly how the wave function evolves from one second to the next based on the forces acting upon it.

In ordinary algebra, variables represent fixed numbers. Quantum mechanics uses operators – a mathematical instruction that tells you to do something to the wave function, such as taking a derivative (measuring how fast the wave's shape changes).

Think of an operator like the "square root" symbol ($\sqrt{\ }$) or a kitchen appliance. A blender is an operator: you put an ingredient in, the blender performs an action, and you get a modified ingredient out. In the Schrödinger equation, specific operators extract physical properties from the wave function. For example, the total energy operator (called the Hamiltonian – see the *Background Materials*) looks at the shape of the wave function and calculates the particle's total energy.

Here again is the standard, time-dependent Schrödinger equation for a single particle moving in one direction:

$$i\hbar\frac{\partial\Psi}{\partial t} = -\frac{\hbar^2}{2m}\frac{\partial^2\Psi}{\partial x^2} + V\Psi$$

- ***i* (The Imaginary Unit)**: The square root of -1 ($\sqrt{-1}$). It sounds abstract, but multiplying by i rotates a wave mathematically. It allows the wave function to have two components (real and imaginary parts) that oscillate out of phase, which keeps the quantum wave moving forward smoothly like ripples on a pond. Without i, the equation would describe a spreading "blur" rather than a persistent wave.
- ***ħ* (Reduced Planck's Constant)**: The standard Planck's constant (h) is a tiny number ($6.626 \times 10^{-34} J \cdot s$) that defines the scale of the quantum world. Quantum equations constantly divide h by 2π, so physicists combine them into a single shorthand symbol: $\hbar = \frac{h}{2\pi}$. It dictates the size of the smallest possible "packets" of energy and sets the limit on how precisely we can know certain pairs of information (like position and momentum).
- ***Ψ* (The Wave Function)**: This represents the quantum state of the particle, which depends on its position (x) and time (t). The square of its magnitude ($|\Psi|^2$) gives the probability of finding the particle at a specific location.
- The derivatives $\frac{\partial}{\partial t}$ and $\frac{\partial^2}{\partial x^2}$ are the instructions to measure change. The term $\frac{\partial\Psi}{\partial t}$ measures how fast the wave function changes over time. The term $\frac{\partial^2\Psi}{\partial x^2}$ measures the curvature or sharpness of the wave across space.
- ***m* (Mass)**: This is the physical mass of the particle, such as the mass of an electron (9.109×10^{-31} kilograms).
- ***V* (Potential Energy)**: This represents the environment or external forces acting on the particle, like an electric field pushing on an electron or a physical barrier blocking its path.

When writing out the explicit wave equations that plug into the Schrödinger equation, physicists often use angular frequency (ω, the Greek letter omega) instead of standard frequency (f). Standard frequency measures cycles per second. Angular frequency measures radians per second. Because a full circle contains 2π radians, the relationship is $\omega = 2\pi f$. By using ω and $\hbar$ instead of f and h, physicists eliminate cluttered fractions containing 2π from their core equations, making the math much cleaner to read and solve.

In the Schrödinger equation, the derivatives are the mathematical engines that calculate change. While a standard number gives you a fixed value, a derivative tells you how quickly that value is rising, falling, or bending. Instead of providing a static value, it calculates an instantaneous rate of change.

In algebra, you find the average rate of change by calculating the slope between two separate points on a graph. You divide the change in the vertical position by the change in the horizontal position. This works well for a straight line, but physical systems constantly accelerate, decelerate, and curve.

To find the change at one exact moment rather than an average over time, calculus moves those two points closer together. As the distance between the points approaches zero, the line connecting them becomes a tangent line. The slope of this tangent line is the derivative. It represents the exact speed and direction of change at that specific instant.

On the right side of the Schrödinger equation, we see the spatial derivatives, often written as: $\frac{\partial}{\partial x}$ or $\frac{\partial^2}{\partial x^2}$. In quantum mechanics, spatial derivatives measure how the wave function bends and stretches across a region of space, which determines the kinetic energy of the particle. On the left side of the equation, you see this symbol: $\frac{\partial \Psi}{\partial t}$. This is a first derivative with respect to time. It measures the instantaneous speed of the wave function's transformation as time moves forward.

On the right side of the equation, we see an exponent and *psi* added: $\frac{\partial^2 \Psi}{\partial x^2}$. The small exponent 2 means this is a second derivative. While a first derivative measures the slope or steepness of a line, a second derivative measures the curvature of a line – how sharply it bends. In quantum mechanics, this spatial curvature relates directly to kinetic energy (the energy of motion).

If a wave function has sharp, tight peaks and valleys, it has a high second derivative. This high curvature translates to a high momentum and high kinetic energy. If the wave function is smooth, flat, and spread out, the curvature is low, meaning the particle has very little kinetic energy.

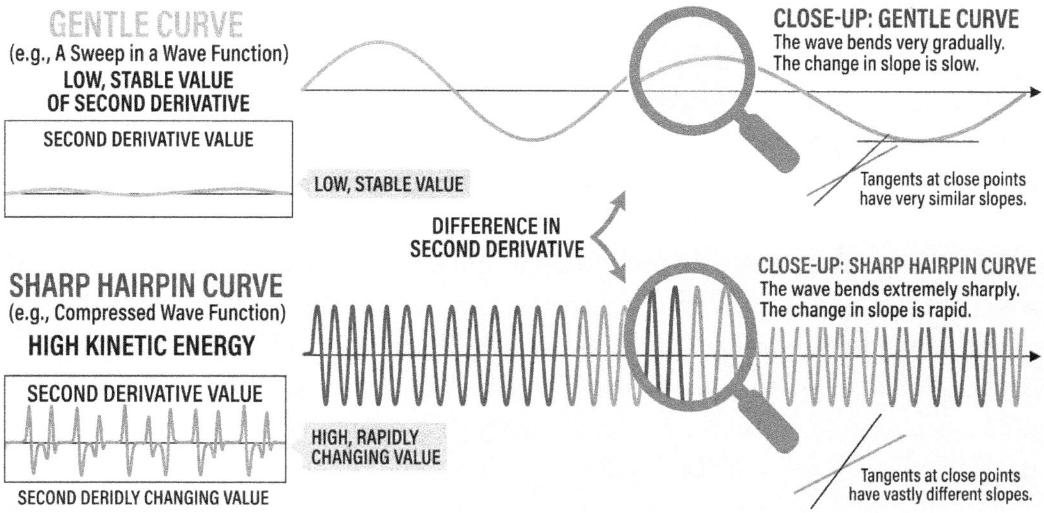

The Schrödinger equation essentially forces these two types of change to balance each other. It states that the way the wave function changes in time (the left side) must equal the curvature of the wave in space plus any outside forces acting on it (the right side).

A real-life example of this balancing act is a vibrating guitar string. When you pluck a guitar string, the tightness of the curves you create in the wire (the spatial curvature) determines how rapidly the string whips back and forth over time (the time change), which dictates the pitch of the note you hear. The Schrödinger equation applies a similar rule to electron waves, balancing their shape in space with their behavior over time.

Linearity means that the wave function and its derivatives appear only to the first power. The wave function does not interact with itself to create feedback loops. Linearity allows for quantum superposition. If state A is a valid solution and state B is a valid solution, then the linear combination of the two is also a valid solution. This allows quantum systems to exist in a combination of states until a measurement forces a transition to a specific state.

The Schrödinger equation is nonrelativistic because it uses the classical definition of kinetic energy ($E = p^2/2m$). It treats space and time as fundamentally different. In Einstein's relativity, space and time form a single, unified spacetime, and physical laws must treat them on equal footing. Because the Schrödinger equation treats time as a simple parameter and space coordinates as operators, it fails to describe particles moving near the speed of light. To fix this, Paul Dirac developed his equation, which reconciles quantum mechanics with special relativity for spin-1/2 particles.

The wave function is a complex-valued function, meaning it involves imaginary numbers. Probabilities, however, must be real, non-negative numbers. You cannot have an imaginary probability, nor can you have a negative chance of finding a particle. Multiplying the wave function Ψ by its complex conjugate Ψ^* yields the square of the magnitude, $|\Psi|^2$. This operation always results in a real, non-negative number, which satisfies the mathematical definition of a probability density.

Schrödinger developed his equation by treating particles like waves. In classical physics, a wave's intensity is proportional to the square of its amplitude. For a light wave, the square of the electric field amplitude dictates the intensity of the light. For a water wave, the square of the displacement dictates the energy density. Max Born extended this logic to quantum mechanics: if the intensity of a classical wave indicates energy concentration, the intensity of a quantum wave function indicates the probability density of finding the particle.

The square of the magnitude correctly accounts for wave interference. When two quantum wave paths overlap, they interfere constructively or destructively. If a particle can take path A or path B, the combined wave function is $\Psi_A + \Psi_B$. The total probability density is:

$$|\Psi_A + \Psi_B|^2 = |\Psi_A|^2 + |\Psi_B|^2 + \Psi_A^* \Psi_B + \Psi_B^* \Psi_A$$

The terms $\Psi_A^* \Psi_B + \Psi_B^* \Psi_A$ constitute the interference term. This term creates the alternating bands of high and low particle counts on a detector screen during a double-slit experiment. A simple linear absolute value would omit these cross-terms and fail to predict interference.

For a wave function to describe a real particle, the total probability of finding that particle somewhere in the universe must equal exactly 1. Mathematically, this requires normalization, where the integral of $|\Psi|^2$ over all space equals 1. Because the time-evolution operator preserves this specific quantity over time, using the square of the magnitude ensures that a particle does not spontaneously disappear or multiply as time passes. The square of the magnitude connects the complex-valued wave function to measurable physical probabilities.

To see the Schrödinger equation in action with real numbers, the best example is the *Quantum Dot*. These are tiny semiconductor crystals used in QLED televisions and medical imaging. A quantum dot acts as a "potential well" that traps an electron in a very small space.

The Schrödinger equation can be used to calculate the electron's energy, which determines the color of the light the dot will emit. We start by calculating the "confinement energy" (the minimum kinetic energy a particle must possess when it is restricted to a small, defined region of space) of an electron trapped in a cadmium selenide (CdSe) quantum dot that is 3 nanometers wide. For a particle trapped in a one-dimensional box (a standard simplification for a quantum dot), the Schrödinger equation yields this energy formula:

$$E_n = \frac{n^2 h^2}{8mL^2}$$

- n *(Energy Level):* We will calculate the ground state, so n = 1.
- h *(Planck's Constant):* $6.626 \times 10^{-34} J \cdot s$.
- L *(Width of the Dot):* 3 nanometers, which is 3×10^{-9} meters.
- m *(Effective Mass of an Electron):* Electrons in a crystal move differently than in a vacuum. In CdSe, the "effective mass" is about 13% of a normal electron's mass, which is 1.18×10^{-31} kg.

First we square the constants:

$$h^2 = (6.626 \times 10^{-34})^2 \approx 4.39 \times 10^{-67} \text{ and } L^2 = (3 \times 10^{-9})^2 = 9 \times 10^{-18}.$$

Then we multiply the denominator:

$$8 \times m \times L^2 = 8 \times (1.18 \times 10^{-31}) \times (9 \times 10^{-18}) \approx 8.50 \times 10^{-48}.$$

Finally we divide to find Energy (E):

$$E = \frac{4.39 \times 10^{-67}}{8.50 \times 10^{-48}} \text{ which becomes } E \approx 5.16 \times 10^{-20} \text{ Joules.}$$

In the world of physics, Joules are too large for single atoms, so we convert this to electron-volts (*eV*): E ≈ 0.32 eV. This 0.32 eV is the "extra" energy the electron gains simply because it is trapped in a tiny 3nm box.

When you combine this with the material's natural properties (its bandgap), you find the total energy of the light it emits. If you were to make the box smaller (say 2 nm), the L^2 in the denominator would get smaller, making the total energy higher. Higher-energy light shifts toward the blue end of the spectrum, while lower-energy light (from a larger 5nm dot) shifts toward the red. This is exactly how engineers "tune" the colors of a QLED screen – by using the Schrödinger equation to calculate the exact nanometer size needed for the perfect shade of green or red.

The Klein-Gordon Equation

The Klein-Gordon equation is the relativistic version of the Schrödinger equation. In 1925, before Schrödinger published his famous non-relativistic equation, he tried to create a relativistic one. When he applied it to the electron in a hydrogen atom, the results didn't match experimental data because he didn't yet know about "electron spin."

Shortly after, in 1926, Oskar Klein and Walter Gordon independently published the relativistic equation. While they were not the only ones to derive it, it became known as the Klein-Gordon equation. For several years, the equation was largely rejected for describing particles like electrons. It yielded "negative probabilities" and "negative energy states" that seemed physically impossible at the time. This led Paul Dirac to develop the Dirac Equation in 1928, which successfully described the electron and predicted antimatter.

In 1934, Wolfgang Pauli and Victor Weisskopf reinterpreted the Klein-Gordon equation, not as a description of a single particle's "probability wave," but as a description of a quantum field. In this context, the negative values made sense as representing particles and antiparticles (like pions or the Higgs boson). Feel free to skip the details below.

The most common form of the Klein-Gordon equation for a free particle is:

$$\frac{1}{c^2}\frac{\partial^2\psi}{\partial t^2} - \nabla^2\psi + \left(\frac{mc}{\hbar}\right)^2\psi = 0$$

Where:

- ψ is the Wave Function
- c is the speed of light.
- $\hbar$ (h-bar) is the reduced Planck constant.
- m is the mass of the particle.
- ∇^2 (the Laplacian) represents the 2nd spatial derivatives $(\frac{\partial^2}{\partial x^2} + \frac{\partial^2}{\partial y^2} + \frac{\partial^2}{\partial z^2})$.

In particle physics, this is frequently written in a more compact notation using natural units (where $c = 1$ and $\hbar = 1$) and the d'Alembertian operator ($\Box$): $(\Box + m^2)\psi = 0)$. Here, the d'Alembertian symbol $\Box$ represents $\frac{\partial^2}{\partial t^2} - \nabla^2$, which packages space and time derivatives together on equal footing, just as special relativity requires.

The Klein-Gordon equation is the fundamental description for spin-0 particles. These particles don't "spin" in a traditional quantum sense and are "scalars."

1. ***Massive Scalar Particles:*** They accurately describe particles such as the pion (a meson involved in the strong nuclear force) and the Higgs boson.
2. ***Massless Particles:*** If you set the mass (m) to zero, the equation becomes the standard Wave Equation, describing the propagation of light or sound waves.

3. ***Relativistic Energy-Momentum:*** The equation is essentially a quantum version of Einstein's famous energy-momentum relation: $E^2 = (pc)^2 + (mc^2)^2$. It ensures that as a particle's energy increases, its behavior remains consistent with the speed-of-light limit.

While it failed as a "single-particle" equation for electrons, the Klein-Gordon Equation remains a cornerstone of Quantum Field Theory. It proved that in a relativistic universe, we cannot think of particles as "points" moving through space; we must think of them as excitations in a field that exists everywhere at once.

The Dirac Equation

Dirac set out to find a version that was "first-order" in time (like Schrödinger's) but also mathematically consistent with the four-dimensional geometry of spacetime. His solution was a radical departure from previous methods: instead of using simple numbers, his equation required matrices. The Dirac equation is typically written as:

$$(i\gamma^\mu \partial_\mu - m)\psi = 0$$

- ψ ***(Dirac Spinor):*** Unlike the Schrödinger wave function, ψ is a "spinor" – a vector with four components. This additional mathematical "room" is what allows the equation to describe both spin and antimatter.
- γ^μ ***(Dirac Matrices):*** These are four 4×4 matrices (gamma matrices) that ensure the equation obeys the rules of relativity across all dimensions of space and time.
- ∂_μ: This represents the derivative across spacetime.
- m: The rest mass of the particle.

Real-world Example: Calculation of the Fine Structure of Hydrogen.

The Schrödinger equation provides a good estimate of the energy levels in a hydrogen atom but ignores relativistic effects and electron spin. The Dirac equation accounts for both. In this example, the relativistic correction to the ground-state energy of an electron in a hydrogen atom is evaluated. This correction explains the tiny energy shifts that define the "fine structure" of the atom's light spectrum.

The formula for the energy levels of a hydrogen-like atom is:

$$E_{nj} = E_n \left[1 + \frac{\alpha^2}{n^2}\left(\frac{n}{j+\frac{1}{2}} - \frac{3}{4}\right)\right]$$

To find the correction, we use these specific values:

- E_n ***(Schrödinger Energy):*** For the ground state of hydrogen, this is exactly -13.605eV.

- ***n* (Principal Quantum Number):** For the ground state, n $= 1$.
- ***j* (Total Angular Momentum):** For the ground state (s orbital), the electron spin makes $j = \frac{1}{2}$.
- ***α* (The Fine-Structure Constant):** This is a fundamental dimensionless constant, approximately 0.007297 (or roughly $1/137.036$).

Since n $= 1$ and j $= 1/2$: $\left(\frac{1}{1/2+1/2} - \frac{3}{4}\right) = \left(\frac{1}{1} - 0.75\right) = 0.25$, we then square the Fine-Structure Constant: $\alpha^2 = (0.007297)^2 \approx 0.00005325$. Now multiply to find the Correction Factor: $\frac{\alpha^2}{n^2} \times 0.25 = \frac{0.00005325}{1} \times 0.25 \approx 0.00001331$, and finally apply the correction to the Energy (E_n): $= -13.605eV \times 0.00001331$, and we get: $\approx -0.000181eV$.

This calculation shows that the electron's energy is shifted by about $0.00018eV$ due to relativistic effects. While this seems tiny, it is a massive discovery in physics. It proves that the electron has "spin" and that space and time are linked even at the scale of an atom.

The Fifth Solvay Conference (1927)

I think it's important to cover some things that happened at the Solvay conferences – it's relevant to how quantum mechanics is interpreted. The Fifth Solvay International Conference on Electrons and Photons, held in Brussels in October 1927, is widely regarded as the most famous meeting in the history of physics.

Between 1925 and 1927, modern physics saw a massive explosion of progress. Heisenberg formulated matrix mechanics, Schrödinger published wave mechanics, Born introduced the statistical interpretation of the wave function, and Dirac unified quantum mechanics with special relativity.

Although the mathematics of quantum mechanics was working with astounding precision, the physics community was deeply divided over what the mathematics actually meant. The 1927 conference was explicitly called to tackle this crisis and answer a key question: *Does quantum mechanics provide a complete, objective description of physical reality, or is it simply a statistical tool?*

The conference split into two primary camps:

- ***The Copenhagen School:*** Led by Niels Bohr, Werner Heisenberg, Wolfgang Pauli, and Max Born, arguing that classical determinism was dead, physical quantities do not exist until measured, and the wave function represents the ultimate limit of physical knowledge.

- ***The Realists:*** Led by Albert Einstein, Erwin Schrödinger, and Louis de Broglie, contending that physical reality exists independently of observation and that quantum theory, while correct as far as it went, remained incomplete.

While formal papers were delivered in the main hall during the day, the real drama unfolded over breakfast and late evening discussions at the Hotel Métropole. Einstein continuously created clever thought experiments meant to expose internal conflicts in the Copenhagen view. Bohr spent his days analyzing Einstein's setups and successfully defending the standard interpretation.

By the end of the conference, Bohr and Heisenberg had persuaded most of the physics community that the Copenhagen Interpretation was sound, establishing it as the dominant paradigm for decades. However, it was not as clear-cut as that sounds.

Einstein's Hemisphere with Pinhole

Einstein presented a thought experiment that involved a single particle passing through a screen with a small pinhole. Beyond the pinhole sat a large photographic plate curved into a hemisphere centered on the hole.

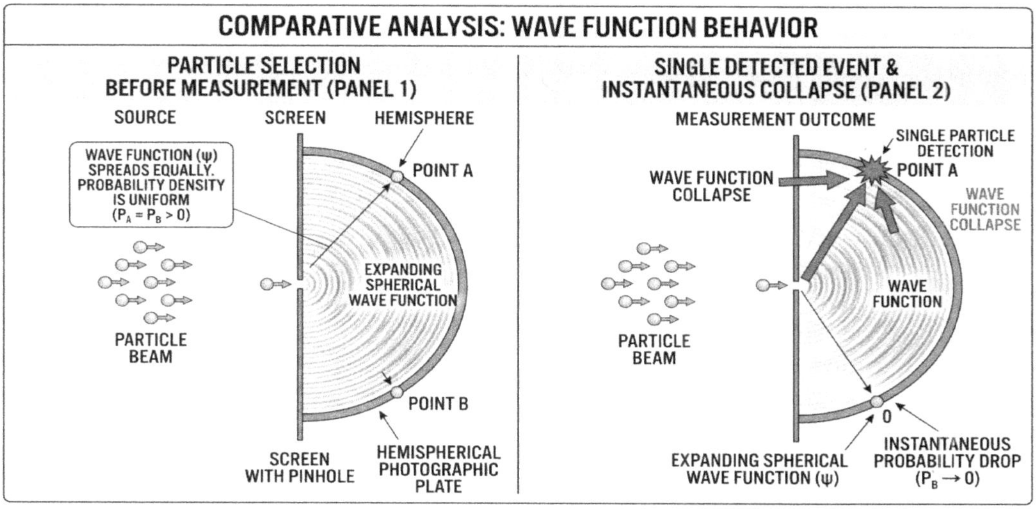

According to quantum mechanics, the particle's wave function (ψ) spreads out as an expanding spherical wave after passing through the pinhole. Just before hitting the film, the wave function distributes equal detection probability across the entire inner surface of the hemisphere. When the particle hits the film at point A, a single spot darkens. At that moment, the probability of detecting the particle at any other point on the hemisphere – such as point B, meters away – instantly drops to zero.

Einstein wasn't arguing that quantum mechanics allowed superluminal information transfer; he understood no controllable signal was sent faster than light. Nor was determinism his central objection. His concern was locality (that an action at point A cannot instantaneously alter physical reality at point B) and completeness (whether ψ offers a full description of a single system).

Einstein offered two interpretations of the experiment:

1.The wave function ψ does not describe an individual particle. It describes a statistical collection of many particles. The detection at point A reveals where that particular particle was located all along. Point B is unaffected because no real physical wave collapsed.

2, The wave function ψ fully describes a single particle. In this case, detecting the particle at point A requires a non-local mechanism across the entire hemisphere to prevent a second detection at point B.

Einstein rejected this action-at-a-distance and concluded that if the standard interpretation was correct, quantum mechanics depended on non-local effects, making the theory incomplete. However, Bohr and Heisenberg had persuaded the physics community that the Copenhagen Interpretation was the correct view.

Bohr did not resolve the non-locality problem. Instead, he redefined what quantum mechanics was supposed to describe. Bohr insisted the wavefunction did not represent an objective physical reality moving through space, but rather our mathematical knowledge of the entire experimental arrangement. Because no physical signal or energy traveled across the screen, Bohr claimed relativity was not violated – which was not what Einstein was claiming.

This response did not satisfy Einstein, who viewed it as dodging the physical reality of space and time. History proved Einstein right to press the issue. This hemisphere argument was the direct ancestor of the 1935 Einstein-Podolsky-Rosen (EPR) paradox and John Bell's 1964 theorem, which later seemed to confirm that quantum mechanics genuinely requires non-local connections.

The Sixth Solvay Conference (1930): Einstein's Photon Box

This time, Einstein targeted the energy-time uncertainty principle: $\Delta E\,\Delta t \geq \frac{\hbar}{2}$. He designed a thought experiment to prove that both energy and time could be measured simultaneously with exact precision.

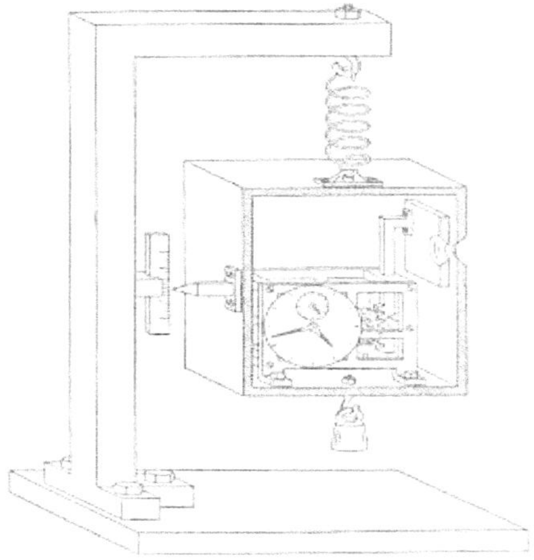

1. A box filled with radiation (photons) is suspended from a spring scale.
2. Inside the box, a clock controls a shutter covering a small hole.
3. The clock opens the shutter for a brief, precise duration Δt, allowing a single photon to escape, and records the exact time t of release.
4. Weighing the box before and after the photon leaves reveals the change in mass Δm.
5. Applying $\mathrm{E} = \mathrm{mc}^2$ calculates the exact energy $\mathrm{E} = \Delta m \cdot c^2$ of the escaped photon.

Einstein's Photon Box Thought Experiment. Courtesy of *Interpretations, Jim Baggott, Substack*

Einstein claimed that because t is read directly from the clock and E is calculated from the balance weight, both quantities could be measured without uncertainty, violating quantum theory.

Niels Bohr spent the night examining the setup and discovered a flaw using Einstein's own theory of General Relativity.

When the photon escapes, the box loses mass (Δm) and moves upward on the spring scale. According to General Relativity, a clock positioned higher in a gravitational field ticks faster because of gravitational time dilation. The upward displacement creates uncertainty in the box's position within the gravitational field, which translates directly into uncertainty in the clock's time reading (Δt). When Bohr calculated the exact gravitational time dilation effect, the uncertainty relation $\Delta E\, \Delta t \geq \frac{\hbar}{2}$ was restored.

Albert Einstein, Boris Podolsky, and Nathan Rosen published their argument in 1935 under the title "*Can Quantum-Mechanical Description of Physical Reality Be Considered Complete?*" They constructed a formal mathematical definition of physical reality and completeness to demonstrate that mainstream quantum mechanics contained a strict logical contradiction.

The Einstein-Podolsky-Rosen (EPR) Premises

The paper established two logical conditions:

1. If, without in any way disturbing a system, we can predict with 100% certainty the value of a physical quantity, then an element of physical reality corresponds to that quantity.

2. A physical theory is complete only if every element of physical reality has a direct counterpart in the theory.

To maintain locality, EPR assumed that a measurement performed on system A cannot instantaneously alter the physical reality of a spatially separated system B.

EPR framed their paper using two particles, 1 and 2, that interact briefly at time $t = 0$ and then travel far apart in opposite directions. Because of their initial interaction, the wave function describes conserved quantities between the pair:

1. Total momentum is zero: $p_1 + p_2 = 0$
2. Relative distance is fixed: $x_1 - x_2 = D$

An observer at location A chooses which property to measure on particle 1:

- ***Measuring Position:*** Measuring x_1 allows the observer to calculate $x_2 = x_1 - D$ with 100% certainty. Because particle 2 was untouched at location B, its position x_2 must be an element of physical reality.
- ***Measuring Momentum:*** Measuring p_1 allows the observer to calculate $p_2 = -p_1$ with 100% certainty. Because particle 2 was untouched, its momentum p_2 must also be an element of physical reality.

The Contradiction

In quantum mechanics, non-commutative operators like position and momentum cannot simultaneously have definite values: $[\hat{x}, \hat{p}] = i\hbar$.

EPR argued that because an observer can choose to predict either x_2 or p_2 without altering particle 2, both x_2 and p_2 must simultaneously exist as elements of physical reality. Since the standard wave function ψ cannot specify both quantities at once, EPR concluded that quantum mechanics does not tell the full story and is incomplete.

In 1951, David Bohm reformulated the EPR thought experiment in terms of discrete spin values rather than continuous position and momentum. This rephrasing made the physical principles sharper and provided the exact setup John Bell used in 1964 to show that EPR's assumption of local realism could be experimentally tested.

David Bohm's Spin Formulation (EPRB)

Bohm replaced position and momentum with a pair of spin-1/2 particles (such as electrons or photons) produced by a source in a total spin-singlet state (S=0):

$$|\psi\rangle = \frac{1}{\sqrt{2}} (|\uparrow\rangle_1 |\downarrow\rangle_2 - |\downarrow\rangle_1 |\uparrow\rangle_2)$$

This singlet state means that the total spin of the system along any measurement axis is zero. The two particles travel in opposite directions toward two distant detectors, operated by Alice and Bob.

- ***The Measurement:*** If Alice measures the spin of particle 1 along the z-axis and finds it pointing "up" ($|\uparrow\rangle$), she knows with 100% certainty that Bob's particle will be found pointing "down" ($|\downarrow\rangle$) if measured along the same z-axis.
- ***The EPR Conclusion:*** Because Alice can choose *any* arbitrary angle θ for her measurement axis and immediately know the outcome for Bob's particle without interacting with it, Bob's particle must possess predetermined spin values for all possible measurement directions simultaneously.

Since standard quantum mechanics forbids a spin-$\frac{1}{2}$ particle from having simultaneous definite spin values along non-parallel axes ($[_x, \hat{S}_z] \neq 0$), EPR and Bohm concluded that the wave function description was incomplete and that additional "local hidden variables" determined the outcomes ahead of time.

John Bell's Theorem (1964)

For nearly three decades, local hidden variable theories were considered a matter of philosophy rather than physics. In 1964, John Stewart Bell proved mathematically that local hidden variable theories make fundamentally different statistical predictions from quantum mechanics.

Bell defined *Local Realism* with two assumptions:

1. Particles possess objective physical properties (hidden variables λ) before measurement.
2. The measurement outcome at Alice's detector depends only on its setting a and the state λ, completely independent of Bob's detector setting b or outcome.

Deriving the Inequality

Bell considered correlations E(a, b) between measurements made along different detector angles a and b. Under any local hidden variable theory governed by a probability distribution $\rho(\lambda)$, the correlation is written as:

$$E(a, b) = \int A(a, \lambda) B(b, \lambda) \rho(\lambda) d\lambda$$

where

$$A(a, \lambda) = \pm 1 \text{ and } B(b, \lambda) = \pm 1.$$

By comparing measurements made across three separate detector angles (a, b, and c), Bell showed that any local realistic theory strictly obeys a mathematical bound known as *Bell's Inequality* (or in its expanded CHSH form):

$$|E(a,b) - E(a,c)| \leq 1 + E(b,c)$$

The Quantum Violation

When calculating the same correlation functions using standard quantum mechanics for a spin-singlet state, the correlation depends strictly on the angle between the detectors:

$$\mathrm{E}_{QM}(a,b) = -\cos(\theta_{ab})$$

If the detector angles are chosen such that $\theta_{ab} = 45°$, $\theta_{bc} = 45°$, and $\theta_{ac} = 90°$:

- $\mathrm{E}_{QM}(a,b) = -\cos(45°) \approx -0.707$
- $\mathrm{E}_{QM}(b,c) = -\cos(45°) \approx -0.707$
- $\mathrm{E}_{QM}(a,c) = -\cos(90°) = 0$

Plugging these values into Bell's inequality yields:

$$|-0.707 - 0| \leq 1 + (-0.707) \implies 0.707 \leq 0.293$$

This is a direct mathematical contradiction. Quantum mechanics predicts a level of correlation that no theory based on local hidden variables can ever reproduce. Subsequent experiments – beginning with Stuart Freedman and John Clauser (1972), Alain Aspect (1982), and culminating in loophole-free tests in 2015 – consistently violated Bell's inequalities and confirmed quantum predictions, demonstrating that nature cannot be both local and realistic.

This would seem to be the end of Einstein's belief about non-locality and that quantum mechanics is incomplete. However, Philip Mannheim (see Quantum Gravity Theories) agrees with the core logic of Einstein's 1927 Solvay objection. Mannheim maintains that Einstein's objection about local causality is correct – specifically that Copenhagen wavefunction collapse requires an instantaneous physical change across space that conflicts with relativity. I cover some of this later when discussing Einstein and Superdeterminism. Suffice to say I already had some sympathy for Einstein and Pilot-Wave Theory because of non-locality issues (yes, I know Pilot-Wave Theory may require non-locality, but then again).

Mannheim also addresses EPR, Bell's theorem, and quantum non-locality in his work on quantum theory foundations and local quantum field theory. His position on these topics breaks down as follows:

1. Rejection of Physical Non-Locality. Mannheim doesn't accept the conclusion that nature contains physical non-locality or action-at-a-distance. He argues that relativistic quantum field theory is local by construction. In quantum field theory, field operators at space-like separated points commute with one another:

$$[\Phi(x), \Phi(y)] = 0 \quad \text{for } (x-y)^2 < 0$$

This commutator makes certain that no measurement at position x can physically alter a measurement or state at position y outside its light cone.

2. Interpretation of Bell's Theorem. Mannheim argues that Bell's theorem does not prove physical non-locality. Instead, it demonstrates the failure of classical probability theory to model quantum systems. Bell's inequalities rely on the assumption that hidden variables behave as classical numbers that satisfy standard commuting probability axioms. Mannheim emphasizes that quantum observables are non-commuting operators rather than classical c-numbers. In his view, violating a Bell inequality proves that quantum mechanics cannot be mapped onto classical probability distributions, not that spatial non-locality exists in the physical fields.

To summarize, Mannheim argues that Bell's theorem doesn't prove spooky action at a distance (physical non-locality) in quantum mechanics. Instead, it just shows that quantum systems operate under alternative mathematical rules that standard, classical probability laws can't accurately describe.

3. Alignment with Einstein. Mannheim agrees with Einstein on the requirement for local, relativistic field equations. However, he differs on how to resolve the problem. Where Einstein sought a classical deterministic underlying theory to restore locality, Mannheim argues that standard quantum field theory is already fully local when formulated through local operator actions and local symmetries – such as local conformal invariance – without needing non-local wave collapse mechanisms.

Local conformal invariance is a mathematical symmetry where physical laws remain unchanged even if you stretch or shrink the scale of spacetime by a different factor at every individual point.

I admit Mannheim is very smart and has a wealth of experience – and he's not working alone. But after reviewing his data in some detail, it's a little beyond my level of understanding. I was following along, then it started sounding like trying to understand this:

> *Modern field theory derives equations from a Lagrangian density, converting it to a Hamiltonian that acts through Hermitian operators on a complex state space. When two observables commute, they share a common eigenvalue*

spectrum, while introducing an imaginary time coordinate converts wave functions into real probability distributions.

The Strange World of Quantum Mechanics

Quantum Mechanics is not intuitive, and the experimental realities make us question what we're seeing. But the mathematics works, and the theory is probably the most successful ever derived by mankind.

Wave-Particle Duality

Wave-Particle Duality is a cornerstone of quantum mechanics, challenging our classical intuition by suggesting that every fundamental entity – such as an electron or a photon – possesses the characteristics of both a discrete particle and a continuous wave.

Note: Mathematics has no single universal definition for duality; rather, it is a foundational framework that translates concepts, theorems, or mathematical structures into corresponding structures in a one-to-one fashion. This is not really a good definition for the idea of wave-particle duality. Admittedly, it's not intuitive. My answer is to stick with the modern view that uses Quantum Field Theory (QFT), because then the apparent duality dissolves completely:

Fields are the fundamental building blocks of the universe, not particles. There are no isolated, zero-dimensional point particles bouncing around in a void. There are only continuous, spacetime-filling quantum fields (e.g., the electron field, the electromagnetic field). These fields obey relativistic wave equations. Because the fields are quantized, their energy states can only change in discrete steps. What we call a "particle" is simply a localized, single-quantum excitation of a field. The wave-like nature is the natural dynamics of the continuous field propagating through space; the particle-like nature is the discrete jump rule governing how the field exchanges energy.

In our macroscopic experience, we tend to categorize objects as one or the other. A pebble is a particle with a definite location, while a ripple on a pond is a wave that spreads across space. However, at the subatomic scale, this distinction disappears. A quantum object travels through space like a wave, interfering with itself and exploring multiple paths simultaneously. Yet, it behaves as a localized particle the moment it interacts with a detector. The "true" nature of matter is neither a solid ball nor a simple vibration, but a more complex reality that changes depending on how we choose to observe it.

This duality is most famously demonstrated through the Double-Slit Experiment (covered in detail later). When a stream of electrons is fired at a barrier with two slits, one might expect them to pile up in two distinct bands behind the openings, much like paintballs would. Instead, the electrons act as waves, passing through both slits at once and creating an "interference pattern" on the far screen – a series of light and dark stripes where the peaks and troughs of the electron waves have either reinforced or canceled each other out. Even more baffling is that if scientists fire electrons one by one, the interference pattern still gradually builds up over time. This suggests that each electron is interfering with itself, behaving as a "probability wave" that exists in multiple places until it strikes the screen and "collapses" into a single, definite point.

The reconciliation of these two seemingly contradictory identities lies in the Wave Function, a mathematical description that maps the probability of finding a particle in a particular state or location. Before a measurement is made, the particle is in a state of superposition, meaning it exists as a "cloud" of possibilities described by its wavelength. The wavelength of a particle is inversely proportional to its momentum, a relationship discovered by Louis de Broglie. This explains why we do not see wave behavior in baseballs or cars: their mass is so great that their "de Broglie wavelength" is infinitesimally small and undetectable. In the quantum world, however, mass is small enough that the wave nature becomes the dominant mode of operation, forcing us to view "matter" as a series of rhythmic excitations in the underlying fields of the cosmos.

The Observer Effect (see also the "measurement problem") in quantum mechanics is perhaps the most misunderstood concept in modern physics. It refers to the phenomenon in which measuring or observing a quantum system can fundamentally change its state. In our daily lives, looking at a thermometer does not change the room's temperature. But in the subatomic world, "looking" requires a physical interaction – such as hitting an electron with a photon – that disturbs the delicate wave-like state of the particle.

When we attempt to determine which path a particle takes in the double-slit experiment by placing a detector at the slits, the very presence of that measurement device causes the electron's "wave of probability" to vanish. The electron stops behaving like a spread-out wave. It begins to behave like a classical, solid particle, passing through only one slit or the other, thereby destroying the interference pattern on the back screen. This transition from a cloud of possibilities to a single, definite reality is known as the Collapse of the Wave Function. Before the measurement, the particle exists in a state of superposition, mathematically described as a wave that occupies all possible locations at once. The moment information about the particle's "which-path" is extracted from the system, the universe essentially "forces" the particle to choose a single state. This "observer" need not be a conscious human being; any physical interaction that links the quantum particle to the macroscopic world – be it a camera, a sensor, or even a single stray molecule – is sufficient to cause this collapse (decoherence).

A quantum particle is the fundamental building block of the universe that follows the laws of quantum mechanics rather than the classical laws of physics. While we often call them "particles," as we have seen, they are better understood as excitations in a field that exhibit a dual nature: they act like particles in some experiments and like waves in others. I like the word *wavicle*, as popularized by Sir Arthur Eddington in 1948, or perhaps *partiwave*.

Heisenberg's Uncertainty Principle

Newton's First Law of Motion says "an object should have a definite position and momentum at all times". However, the uncertainty principle states that the position and momentum of a particle cannot be known with perfect accuracy at the same time. Formulated by Werner Heisenberg in 1927, it is a fundamental limit to what we can know about the subatomic world.

The principle states that the position (Δx) and momentum (Δp) of a particle cannot be known simultaneously with arbitrary precision. It is expressed as $\Delta x \Delta p \geq \frac{h}{4\pi}$ or $\Delta x \Delta p \geq \frac{\hbar}{2}$. Where $\hbar = \frac{h}{2\pi}$ is the Reduced Planck's constant. It also applies to energy and time, angular momentum and angular position, and others. This is not a limitation of our measuring tools or a lack of technological sophistication; rather, it is an intrinsic property of the universe itself. In the classical world, we assume that if we look closely enough, we can measure exactly where a car is and exactly how fast it is going.

However, because quantum entities behave like waves, they do not occupy a single point in space. To measure a particle's position more accurately, we must use a shorter-wavelength light, which carries more energy. This energetic "kick" from the light inevitably changes the particle's momentum, making it impossible to know its speed with the same level of certainty.

In quantum mechanics, a particle is a "wave packet" – a localized disturbance made of many different waves combined. The more you try to squeeze that packet to pin down its location, the more varied the wavelengths must be, which spreads out the possible values for its momentum. Mathematically, this relationship is expressed as an inequality where the product of the uncertainty in position and the uncertainty in momentum must be greater than or equal to a fixed value involving Planck's constant. This trade-off ensures that the more the universe "reveals" about one aspect of a particle, the more it "hides" about the other. Since it also applies to energy and time, it means that energy can be "borrowed" from the vacuum for extremely short periods. This allows virtual particles to pop in and out of existence (from the "quantum foam"), a phenomenon experimentally verified and essential to our understanding of the fundamental forces.

What people often get wrong is thinking that Heisenberg's uncertainty principle is caused by the physical act of measuring a particle. Textbooks and popular science often explain that if you want to find an electron's position, you have to bounce a photon (light) off it, and that collision kicks the electron, changing its momentum.

The uncertainty is not a technical limitation of our instruments, nor is it caused by the "disturbance" of shining a light or poking the particle. Quantum particles (like electrons) do not just have a hidden precise location and speed that we fail to measure; they fundamentally behave like waves. A wave cannot be localized at a single precise point while also having a single precise wavelength (which defines momentum).

The uncertainty principle stems from the mathematics of waves and Fourier transforms. Compressing a wave's spread in space inherently broadens the spread of its frequency components (momentum). This trade-off exists in classical wave mechanics (like sound and signal processing) long before you even talk about an observer or a measurement.

Modern physics distinguishes between *intrinsic uncertainty* (the fundamental spread of a quantum state) and *measurement disturbance*. Experiments have even shown that a measurement can sometimes disturb a system by less than Heisenberg's original formula predicted, but the foundational quantum fuzziness of the particle itself remains untouched.

Quantum Superposition

Quantum superposition is a fundamental principle of quantum mechanics that dictates that a physical system – such as an electron or a photon – exists in all its theoretically possible states simultaneously until it is measured. Unlike classical systems, where an object must be in one distinct state (either "on" or "off"), a quantum system can exist in a linear combination of these states.

In quantum mechanics, the state of a particle is described by the wave function, denoted by the Greek letter psi (Ψ). Superposition arises from the fact that the

Schrödinger equation is a linear differential equation. In mathematics, if two different functions are solutions to a linear equation, then any combination of those two functions is also a valid solution.

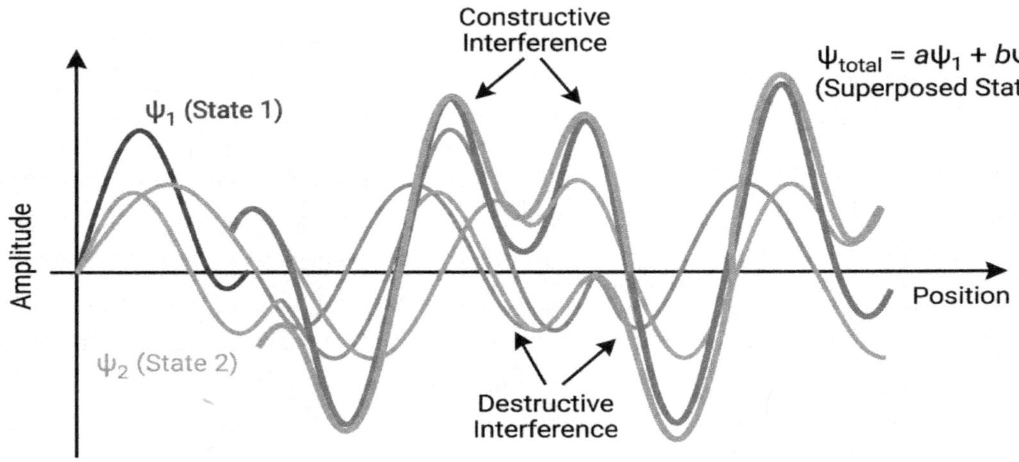

You could have a wave function that describes a particle being in one location, and another that describes it being somewhere else. Add them together, and we get a new totally valid solution where the particle is in both places at once.

So when we say that a particle could be in multiple places (or states) at the same time, that is fundamentally true because it's a linear equation we're working with and you can add different quantum states to each other to get the notion that the particle can be spinning clockwise and counterclockwise or the particle can be in this energy state and also this energy state all at once in these quantum superposition states.

If state $|A\rangle$ and state $|B\rangle$ are possible for a system, then the state: $|\Psi\rangle = \alpha|A\rangle + \beta|B\rangle$ is also a valid state. Here, α and β are complex numbers known as probability amplitudes. In the expression $|A\rangle$, the vertical bar denotes a "ket" or bra-ket, representing a specific state vector in a Hilbert space. Superposition is essentially the "tilting" of a vector so that it points partially toward one state and partially toward another. The most counterintuitive part of superposition is what happens when we try to look at the system:

- ***Wave Function Collapse:*** When a measurement is made, the system instantly "chooses" one of the basis states. This transition from a blur of possibilities to a single, definite reality is called the collapse of the wave function.
- ***The Born Rule:*** We cannot predict with 100% certainty which state the system will collapse into. Instead, we use the probability amplitudes (α and β). The Born Rule states that the probability of finding the system in state $|A\rangle$ is the

square of the magnitude of its amplitude ($|\alpha|^2$). To ensure the particle is found *somewhere*, the sum of these probabilities must always equal 1 ($|\alpha|^2 + |\beta|^2 = 1$), a process called normalization.

Quantum Entanglement

Quantum entanglement is a phenomenon in which particles interact such that the quantum state of each particle can't be described independently of the others, even if they are separated by large distances. Einstein famously dismissed this as "spooky action at a distance." Yet, it has since become one of the most rigorously tested and proven foundations of modern physics.

In classical physics, if you have two gloves and put them in separate boxes, the moment you open one box and see a "right-hand" glove, you know the other is "left-hand." This is just correlation – the gloves were always that way; you just didn't know it – quantum entanglement is different: Before measurement, the particles do not have a definite state (like "left" or "right"). They exist in a "blur" of all possibilities at once (superposition). Instantaneous correlation occurs the moment you measure one particle – its state is "fixed." Simultaneously, the state of its entangled partner is also fixed, regardless of whether it is in the next room or across the galaxy. Crucially, the particles aren't "hiding" their state. The outcome is truly determined only at the moment of measurement.

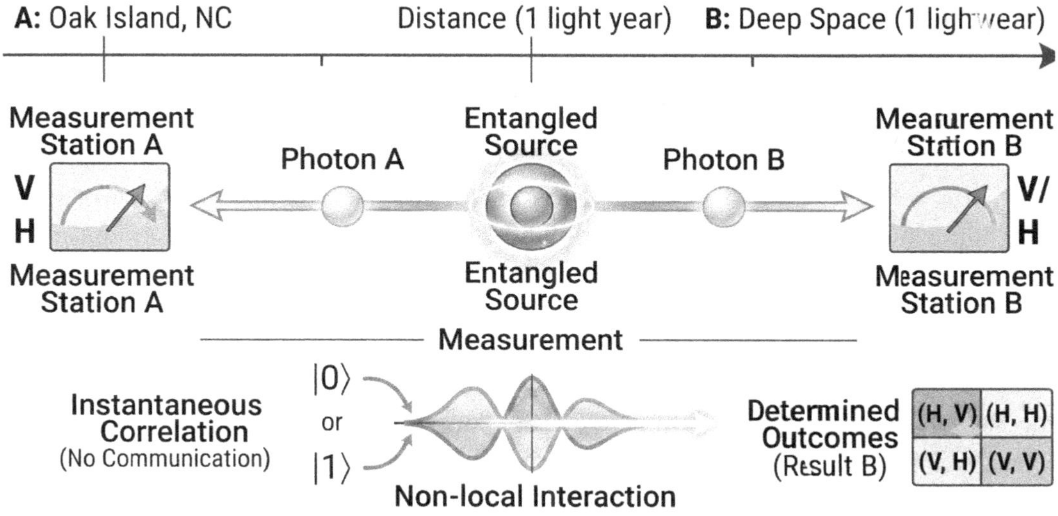

As we saw earlier, the discovery of entanglement led to one of the most famous intellectual battles in the history of theoretical physics. Albert Einstein was a proponent of local realism, which states that particles always have definite properties. Entanglement is just a sign that our theory is "incomplete" and missing "local hidden variables."Niels Bohr supported the Copenhagen Interpretation, which states that reality is probabilistic. Properties don't exist until they are measured. This nonintuitive connection is a fundamental part of nature. For decades, this was just a philosophical debate. In 1964, physicist John Bell proposed a mathematical test of it, known as Bell's Inequality. He proved that if Einstein were right (that there must be local hidden variables), there was a limit to how correlated the particles could be. If Bohr were right, that limit would be broken. Several key experiments would help determine the answer.

In 1972, Clauser and Freedman performed the first experimental "Bell Test," which favored quantum mechanics over Einstein's hidden-variable theory. In 1982, Alain Aspect closed a major "loophole" by changing the measurement settings while the particles were in flight, making sure they couldn't "communicate" via light-speed signals. In 2015, Hanson et al performed the first "loophole-free" Bell test, confirming entanglement with absolute statistical certainty. In 2022, the Nobel Prize in Physics was awarded to Alain Aspect, John Clauser, and Anton Zeilinger for their definitive proofs of entanglement.

Here are two interesting experiments that demonstrate Entanglement:

Using 100,000 Video Gamers to Outsmart Physics: One persistent objection to entanglement tests is the "freedom-of-choice" loophole. Skeptics argued that physical random number generators used in laboratories might share some unseen, pre-existing physical connection with the particles being measured. In November 2016, the Institute of Photonic Sciences organized the Big Bell Test. Researchers enlisted roughly 100,000 video gamers across the globe to mash buttons on their computers and phones, generating over 90 million unpredictable zeroes and ones in real time. These human choices were beamed to 12 laboratories worldwide (including facilities in Vienna, Zurich, and Munich) to set the measurement angles for entangled atoms and photons. Quantum mechanics passed every single test.

Using 12-Billion-Year-Old Quasars as Detectors: To push the freedom-of-choice test even further back in time, a team led by Anton Zeilinger set up an experiment on the Canary Island of La Palma in 2018. They directed two large astronomical telescopes toward distant quasars located 8 billion and 12 billion light-years away on opposite sides of the sky. The incoming color fluctuations of this ancient light determined the polarization angles used to measure entangled photon pairs in a mobile laboratory between the telescopes. For any classical, non-quantum mechanism to explain the correlated results, the setup would have had to be preprogrammed before the quasars emitted their light – over 12 billion years ago, when the universe was in its infancy.

Quantum Tunneling

Quantum Tunneling is a phenomenon that defies our classical intuition, allowing a subatomic particle to pass through a barrier it should not be able to cross. In everyday experience, if you throw a ball at a solid brick wall, it will always bounce back because it lacks the energy to penetrate the brick. However, because quantum particles like electrons and protons possess a wave-like nature, their position is defined by a "cloud" of probability rather than a single, fixed point.

When a particle encounters an energy barrier – such as an electric field or a chemical bond – its wave function does not simply stop at the edge of the barrier. Instead, a small portion of that wave "leaks" through to the other side. This means there is a non-zero mathematical probability that the particle will instantly vanish from one side of the wall and reappear on the other.

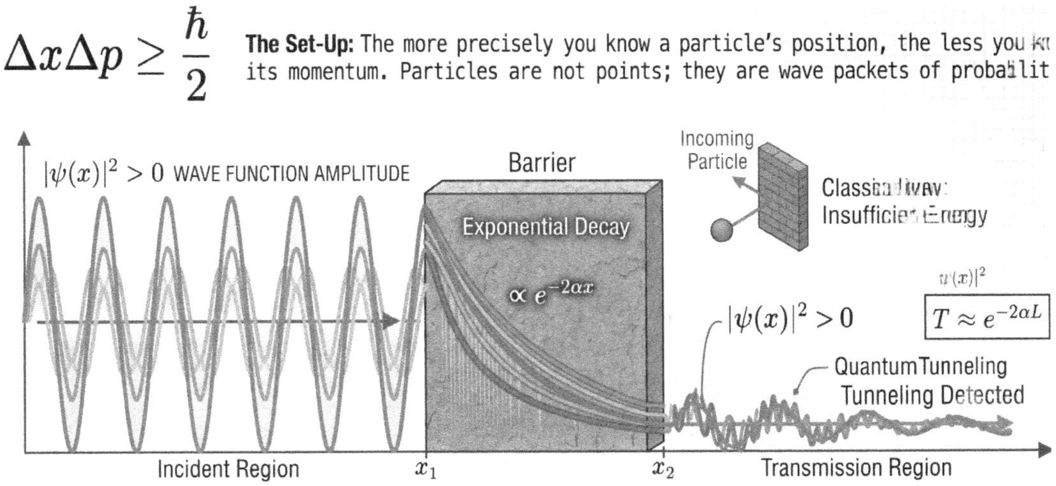

The probability of this event depends heavily on the barrier's thickness and "height" (energy level). If the barrier is thin and the particle's energy is close to the barrier's peak, the likelihood of tunneling increases. This process is essentially the universe's way of "cheating" the laws of classical mechanics by exploiting the Uncertainty Principle. Because we cannot know both the exact position and the exact energy of a particle at the same moment, the particle can "borrow" enough energy to get through the gap, provided it returns that energy in an incredibly short amount of time. It is as if the particle finds a secret passage through the mountain instead of having to climb over the summit.

Quantum tunneling is the mechanism that allows the Sun to shine. It produces energy through nuclear fusion, in which protons overcome their mutual repulsion (the Coulomb barrier) to fuse into helium. Classically, the Sun's temperature is too low for this, but quantum tunneling allows protons to "tunnel" through this barrier, providing the sustained heat and light necessary for life on Earth.

In the world of computer chips, quantum tunneling is both a major challenge and a valuable tool. The most famous use of quantum tunneling is in NAND Flash memory (the storage in a smartphone and solid-state drives). To save a bit of information, the chip needs to move electrons onto a gate that is surrounded by an insulator. The chip applies a specific voltage that makes the insulating barrier thin enough (in a quantum sense) for electrons to tunnel through it and get trapped on the other side. Because the electrons "tunneled" in and are now blocked by the insulator, they stay there even when the power is turned off. This is how files stay saved for years without a battery.

Vacuum Tunneling and the Collapse of the Universe

The universe started in the Big Bang, but it's possible a tiny quantum event could end it – and we wouldn't even realize it was happening. Talk about "spooky action at a distance"! Here one moment, gone the next.

In quantum field theory, a vacuum is not empty space. It represents the lowest possible energy state of a quantum field. The stability of the universe depends entirely on whether the fields that govern physical laws have settled into their absolute lowest energy configuration.

If a quantum field, such as the Higgs field, is sitting in a local minimum rather than the global minimum, the universe exists in a state called a false vacuum. This configuration is metastable (i.e., in apparent equilibrium). It can persist for billions of years, giving the appearance of permanent stability. However, quantum mechanics allows for a process known as quantum tunneling. A quantum field can spontaneously transition through the energy barrier separating the false vacuum from the true vacuum, which is the absolute lowest energy state.

If this tunneling occurs at any single point in space, it creates a tiny bubble of true vacuum. If this bubble exceeds a critical threshold size, it becomes unstable and expands outward in all directions. The boundary of the bubble accelerates until it propagates at nearly the speed of light. As the bubble expands, it overwrites the existing laws of physics. The properties of particles inside the true vacuum bubble would change completely. Mass, charges, and nuclear forces would alter or cease to exist in their current form. Because the bubble moves at the speed of light, you would have no advance warning. The boundary would pass through the solar system instantly, destroying all atomic matter, molecules, and biological life without any perceptible delay. Cosmologists calculate that if the Standard Model parameters are completely accurate, the current universe is likely metastable, though the expected lifetime of this false vacuum extends for trillions of years.

In March 2026, researchers at Tsinghua University in Beijing successfully simulated the dynamics of false vacuum decay using a tabletop quantum simulator.[11] To create a controllable analog of a quantum field, the experimental team used an array of Rydberg atoms arranged in a ring configuration. Think of them as a ring of tightly packed, oversized spinning tops that push against one another until a single laser tap forces them out of alignment. In technical terms, a Rydberg atom is excited by a laser so that its valence electrons occupy exceptionally high principal quantum numbers. This state expands the physical size of the atom, making it roughly 100,000 times larger than a normal ground-state atom. The exaggerated size amplifies the van der Waals interactions between neighboring atoms, allowing precise control over their behavior.

The researchers placed an even number of these mutually repulsive Rydberg atoms into a precise ring. Initially, the atoms settled into a highly stable, alternating spin alignment pattern, which served as the analog for the true vacuum ground state. The team then used a targeted laser system to inject energy and manipulate the system's symmetry, forcing the entire ring into an aligned spin configuration. This newly engineered state functioned as the experimental false vacuum.

By adjusting the strength of the symmetry-breaking laser field, the scientists observed the system transition out of the unstable configuration. The ring decayed back toward the preferred lower-energy state through the formation of localized quantum bubbles. The experiment confirmed a primary mathematical prediction of quantum field theory: the false-vacuum decay rate decreased exponentially with the inverse of the symmetry-breaking field. This tabletop setup provides physicists with a highly addressable platform to test complex mathematical models where quantum mechanics and relativity intersect.

As long as research remains focused on simulating the dynamics of false vacuum decay, I have no concerns. Even if such a transition were to occur in reality, it is likely that we would remain unaware.

The Quantum Zeno Effect

"I'm reading a book about Zeno's Paradox. I'm not quite halfway through it."

The Quantum Zeno Effect is one of the most striking examples of how measurement does not merely record reality but actively shapes it. Named after the ancient Greek philosopher Zeno of Elea – who famously argued that motion was impossible because an object must first reach the halfway point of any distance – this effect proves that, in the quantum realm, a "watched pot" truly does not boil. You may have heard of the tortoise or the arrow analogies as Zeno's paradox.

Normally, a quantum system, such as an unstable atom, will naturally evolve, eventually decaying or transitioning to a new state. However, if that system is observed often enough, the transition is repeatedly interrupted. Each measurement forces the particle to "snap back" into its original state, effectively freezing it in time and preventing it from ever changing.

The underlying cause of this freezing is the Collapse of the Wave Function that occurs during observation. When a quantum system is left alone, its wave function spreads, indicating a growing probability that the particle has moved or decayed. If you check on the particle almost immediately after it starts this evolution, the probability that it has changed is mathematically minuscule. By observing it, you "collapse" that tiny probability back to zero, forcing the particle to start its journey all over again from the beginning. If the measurements are performed rapidly enough, the system never gains the "momentum" required to transition. The universe is prevented from moving forward because it is being asked too frequently to define exactly where it is.

In the field of Quantum Computing, the Quantum Zeno Effect could be used to protect delicate "qubits" from environmental disruptions. By constantly "watching" the qubits, engineers might be able to freeze them in their programmed states, preventing the errors that often occur when quantum information leaks away. This suggests a deep link between information and existence. If the evolution of the universe depends on the intervals between interactions, then time itself might be seen as a sequence of "snapshots" rather than a continuous, unbreakable flow.

The Quantum Anti-Zeno Effect

"Watching the pot actually makes it boil faster."

The Quantum Anti-Zeno Effect is the paradoxical counterpart to the Quantum Zeno Effect, demonstrating that under certain conditions, frequent observation can actually accelerate the transition or decay of a quantum system. While the Zeno Effect occurs when we "catch" the system early in its evolution, forcing it back to its starting point, the Anti-Zeno Effect occurs when measurements are timed to align with the system's natural fluctuations.

Instead of freezing the particle, these observations provide a series of "prods" that help it transition out of its initial state more quickly. It is as if, by checking the pot at exactly the right intervals, we are effectively turning up the heat on the stove, forcing the water to boil faster than it ever would have on its own. This phenomenon occurs because every quantum system is coupled to an environment – a "bath" of other particles and fields. When we perform a measurement, we aren't just looking at the particle; we are temporarily changing how the particle interacts with that environment. If the "disturbances" caused by our measurements occur at a frequency that matches the energetic "noise" of the surroundings, they create a resonance that pushes the particle out of its stable state. This highlights a critical truth in quantum physics: the rate of change is not just a property of the object itself, but a product of its engagement with the rest of the universe. In this context, the "observer" acts as a catalyst, using measurement to inject or redistribute energy within the system.

The discovery of the Quantum Anti-Zeno Effect has shifted our understanding of how information and time interact at the smallest scales. It suggests that time is not a smooth, independent river flowing at a single speed, but a variable rate that can be manipulated through interaction and information exchange. This leads us directly to the Thermal Time Hypothesis (TTH), which proposes that time itself is not a fundamental part of the universe's makeup. Instead, it is an emergent phenomenon – a "statistical illusion" that happens because we, as observers, are unable to see the full, infinite complexity of quantum fields. Because we can only perceive a blurred, "thermal" average of all the quantum activity around us, we experience that blurring as the flow of time and the increase of entropy.

In this view, the universe at its most basic level is "timeless." Equations like the Wheeler-DeWitt equation describe a world in which everything exists as a giant, static map of possibilities. Time only "appears" when a macro-observer – like a human or a measuring device – interacts with these fields. Much as the Zeno and Anti-Zeno effects show that our measurements can freeze or speed up processes, the Thermal Time Hypothesis suggests that our very existence as finite beings creates the "clock" we live by. We don't experience time as "passing"; we experience it as we gather and process information from a quantum soup that is far too complex for us to fully comprehend.

Quantum Mechanics Interpretations

There are several interpretations of quantum mechanics, including the Copenhagen interpretation, the Many-Worlds interpretation, the de Broglie-Bohm theory, and objective-collapse theories. These differing frameworks aim to explain basic quantum concepts, including wave-particle duality, superposition, uncertainty, and entanglement. They offer different philosophical explanations for the underlying reality and the role of observation in it. We will also consider the Hidden Variables and superdeterministic theories.

Before getting into these theories, let's consider the *measurement problem* in quantum mechanics. It is the unresolved question of how a quantum system that exists in a superposition of multiple states collapses into a single, definite state upon measurement. This is a conflict between the deterministic evolution of a quantum system described by the Schrödinger equation and the seemingly random, nondeterministic collapse that occurs during a measurement.

An electron's spin could be both "up" and "down" simultaneously until it is measured. The Schrödinger equation describes how this superposition evolves continuously and deterministically. However, when a measurement is performed, the system is always found in a single, definite state. This "collapse" of the wave function appears to be a discontinuous, random event that is not described by the Schrödinger equation.

The problem is compounded by the ambiguity of what constitutes a "measurement." It's unclear at what point the collapse happens – when a detector interacts with the system, or only when a conscious observer looks at the detector?

Einstein's Hidden Variables Interpretation

Albert Einstein was deeply unsettled by the random nature of quantum mechanics. While he acknowledged the theory's mathematical brilliance and predictive power, he fundamentally rejected the idea that the universe operated on pure chance at its lowest levels. This sentiment culminated in his famous assertion that "God does not play dice." Einstein used the word "God" as a poetic metaphor for the fundamental laws of nature and the rational order of the universe. Einstein said: "I believe in Spinoza's God, who reveals himself in the lawful harmony of all that exists, but not in a God who concerns himself with the fate and the doings of mankind."

To Einstein, the apparent randomness of a quantum system–where a particle exists in a blur of probabilities until it is measured – was not a fundamental feature of reality, but a flaw in the theory. He argued that quantum mechanics was simply an incomplete picture. He proposed the existence of "hidden variables," which are underlying (local) physical properties or mechanisms that we cannot yet see or measure. In his view, if we could somehow detect these hidden variables, we would find that the subatomic world is just as deterministic and predictably structured as the classical physics that governs planetary motion.

To prove his point, Einstein, along with Boris Podolsky and Nathan Rosen, published the EPR paradox in 1935. They focused on quantum entanglement, in which two particles interact and become inextricably linked. Quantum theory suggested that measuring one particle instantly determined the state of the other, regardless of the distance between them. Most historians and physicists think that Einstein dismissed entanglement as "spooky action at a distance". He is believed to have said this as it seemed to require communication faster than the speed of light, which would violate his own theory of relativity. Instead, he argued that the particles *must* contain local hidden variables – essentially hidden instructions created at the moment of interaction – that dictate their future states without any need for a faster-than-light signal.

For decades, this debate remained purely philosophical because there was no obvious way to test it. This changed in 1964, when the physicist John Bell formulated Bell's Theorem, which established a strict, testable boundary known as Bell's Inequality. He proved mathematically that if Einstein's local hidden variables existed, the statistical correlation between measurements of entangled particles could never exceed a certain absolute limit. However, the pure probabilistic math of quantum mechanics predicted a stronger correlation that would violate this limit.

Since the 1970s, physicists have conducted increasingly rigorous experiments to test Bell's Inequality, culminating in the 2022 Nobel Prize in Physics awarded to Alain Aspect, John Clauser, and Anton Zeilinger. Every single definitive experiment has shown that Bell's Inequality is violated.

Most physicists would now argue that Einstein has been definitively proven wrong regarding his specific belief in "local" hidden variables. The universe does not pre-program entangled particles with hidden, independent instructions. We would do well to be very careful not to entirely rule out any ideas that come from Einstein! There remains a possibility that Einstein's broader desire for a deterministic universe is correct.

The first possibility is "non-local" hidden variables, as in the De Broglie-Bohm pilot-wave theory. This framework is entirely deterministic and relies on hidden variables to guide particles along precise paths. However, it only works mathematically if the universe allows for instantaneous, non-local connections across vast distances. It successfully preserves determinism, but it fully embraces the very "spooky action at a distance" that Einstein sought to avoid in the first place. I still think the pilot-wave theory may yet be resurrected as our understanding of quantum physics improves.

The second, highly controversial possibility is "superdeterminism." This loophole suggests that free will, or random choice, does not exist anywhere in the universe. If the history of the cosmos was predetermined at the Big Bang, then the statistical tests of Bell's Theorem are inherently flawed. Let's see why that is...

Superdeterminism

In a superdeterministic universe, particles and measurement devices share a hidden historical variable that dates back to the dawn of time. This technically makes the universe deterministic as Einstein wished, but most modern physicists reject it.

The scientific method is invalidated by Superdeterminism because it eliminates the possibility of independent, isolated experiments. It also means that quantum mechanics is not a fundamental theory – it can be derived from an underlying deterministic theory. Superdeterminism represents one of the most radical and controversial interpretations of quantum mechanics.

To understand it, we first have to look at the fundamental problem it attempts to resolve: Bell's Theorem and quantum entanglement. When Bell proved that measuring one entangled particle instantly affects the other, his mathematical proof relied on a massive, unquestioned assumption called Statistical Independence. This is the assumption that the experimenter possesses the free will to choose which measurement to make (e.g., measuring the particle's spin on the X-axis versus the Y-axis), and that this choice is completely independent of the hidden variables of the particles themselves.

Superdeterminism simply rejects the assumption of Statistical Independence.

It proposes that the universe is entirely predetermined. From the exact moment of the Big Bang, every single interaction, every trajectory of every particle, and every state of energy was fixed. In a superdeterministic universe, the particles being measured and the scientist's brain, which chooses how to measure them, share a hidden, historical variable. The scientist's "choice" was established 13.8 billion years ago to perfectly match the state the particle was always going to be in.

If superdeterminism is true, it fundamentally alters quantum mechanics, turning it back into a classical, deterministic theory. That would mean the end of "Spooky Action" – Einstein's ultimate dream is realized. There is no faster-than-light communication between entangled particles. They appear correlated not because they are mysteriously communicating across space, but because their states – and the exact measurements we subject them to – were strictly choreographed from the dawn of time.

The "blur" of the wave function and the mystery of the collapse disappear. Particles always have definite properties. We only use probabilities and wave functions because we lack the deep knowledge of the underlying, deterministic variables. The Measurement Problem is also solved.

Most physicists reject superdeterminism because it is seen as a "conspiracy theory" of nature. If the universe actively coordinates our measurement choices with the systems we are trying to measure, it essentially invalidates the scientific method. How can you perform an isolated, objective experiment if the system and testing tools are secretly aligned?

In this framework, consciousness and choice are merely the macroscopic outputs of a microscopic, deterministic chain of events. The feeling of making a decision – whether it is choosing to measure a particle's spin, deciding what to eat for breakfast, or writing a book about physics – is an illusion created by the complex, localized physics of your brain. You are a biological entity playing out a physical sequence that was set in motion at the Big Bang, with zero capacity to alter the outcome.

Looking at just the facts, a superdeterministic universe can feel intellectually and emotionally heavy. If the universe's master blueprint was completely finalized at the Big Bang, it is easy to slip into fatalism – the belief that our choices do not matter. However, philosophers and physicists have spent centuries developing frameworks to reconcile a deterministic reality with the human experience of freedom and the pursuit of happiness. Here are some ways to find peace within a predetermined universe:

The most widely accepted philosophical solution is Compatibilism, which argues that determinism and free will are not mutually exclusive; we simply need to upgrade our definition of free will. Compatibilism is the belief that free will and determinism are mutually compatible. It suggests that humans can still act freely and be held morally responsible even in a universe where every event – including human thought – is caused by preceding events.

In a libertarian view, free will means "under the exact same physical conditions, I could have chosen otherwise", but Superdeterminism destroys this. However, compatibilism redefines free will as "the ability to act according to your own internal motivations without external coercion."

Each individual is an extraordinarily complex biological system. Decision-making is not a random process, but rather the precise execution of neural architecture shaped by experiences, memories, and preferences. Individuals are not passive agents manipulated by external forces; rather, they represent the unfolding of deterministic processes. While I find this perspective intellectually compelling, I remain uncertain about the implications for the concept of free will as merely the manifestation of a deterministic universe.

Another idea is the Epistemic Horizon. I like this a little more. The Epistemic Horizon is the fundamental limit of what can be known, predicted, or understood by an agent (person, AI, or system) at any given time. Similar to a physical horizon, it acts as a boundary beyond which information is inaccessible, often due to physical constraints or the nature of quantum mechanics. Even if the universe's timeline is completely fixed, the computational capacity required to predict it is physically impossible to achieve from inside the universe itself. We are bound by our arrow of time and a strict limit on information.

Because the future is inaccessible, individuals must make decisions in real time. The subjective experience of deliberation, uncertainty, and eventual resolution reflects the physical processes that determine outcomes. This illusion of free will is sufficiently robust that, for all practical purposes, it is indistinguishable from genuine autonomy. The uncertainty of future events persists, as one must still experience each outcome sequentially.

While it could be argued that superdeterminism sounds like a philosophical dead end, some theoretical physicists argue that taking it seriously could lead to massive technological breakthroughs. If quantum mechanics is just an approximation of a deeper, deterministic layer of reality, then that layer has strict physical rules that we might be able to uncover and use.

Currently, Heisenberg's Uncertainty Principle dictates a hard limit on what we can know; the universe is fundamentally blurry. But if superdeterminism is true, the blur is just missing data. If we can isolate and identify these underlying "hidden variables," we could build predictive models that far surpass the statistical guessing game of current quantum mechanics.

Theoretical physicist Sabine Hossenfelder suggests that if superdeterminism is valid, it may be possible to identify microscopic systems in which standard quantum mechanics fails. By cooling large molecules to near absolute zero and searching for correlations that violate quantum predictions, researchers might observe evidence of underlying determinism. While such a discovery would be extraordinary, I am personally ambivalent about uncovering such deterministic details, as it would be akin to knowing the precise circumstances of one's own mortality.

Current quantum computers battle with "decoherence" – the fragile quantum states collapsing due to environmental noise. If we understood the deterministic variables dictating that noise, we could theoretically predict and perfectly cancel it out, leading to flawless, infinitely scalable quantum computation and unbreakable deterministic cryptography.

In short, superdeterminism suggests that the universe is not playing dice; the dice are loaded, and the table is rigged. But if we can figure out exactly how the table is rigged, we could theoretically rewrite the boundaries of physics. It can be a difficult pill to swallow – that lack of free will.

The Copenhagen Interpretation

The Copenhagen Interpretation is one of the oldest and most widely taught ways to understand the strange rules of quantum mechanics. It emerged between 1925 and 1927, primarily shaped by Niels Bohr and Werner Heisenberg, with contributions from Max Born. It emerged as physicists grappled with quantum phenomena, proposing that particles exist in a state of superposition (a probability distribution) until measured, at which point the wave function "collapses" into a definite state, fundamentally linking observation to reality and establishing quantum mechanics as a probabilistic, incomplete yet complete theory of measurement.

It's less a single, rigid set of rules and more a collection of related ideas that form a "*shut up and calculate*" approach. It tells us how to use mathematics to get the correct answers, without worrying too much about what's "really" happening beneath the surface.

In quantum mechanics, a particle isn't a tiny ball. Instead, it describes the wave function (ψ). The wave function represents *all possible states* of the particle at once. For example, the wave function for an electron could describe it as being in many different positions simultaneously. This state of existing in multiple possibilities simultaneously is called superposition. The particle doesn't have a definite position, momentum, or spin; it exists in a "blur" of all possible outcomes.

In 1926, German physicist and mathematician Max Born proposed his rule that particles behaving like waves were not physical objects, but "probability waves". For this statistical interpretation, he was awarded the 1954 Nobel Prize in Physics. The Born Rule is a fundamental principle of quantum mechanics stating that the probability of finding a particle at a given location (or in a given state) is given by the square of its wave function (or amplitude) at that point. Essentially, it's the crucial link connecting abstract quantum math (the wave function) to real-world experimental outcomes, telling us how likely different measurement results are, even though the Schrödinger equation describes how the wave function evolves. Thus, the wave function doesn't tell you where the particle *is*. It only tells you the probability of finding it in a specific place if you look. If the wave function is "bigger" in one area, it means you have a higher probability of finding the particle there. The particle's reality is purely statistical *until* it is measured.

The "Measurement Problem", as we discussed earlier, is the most famous – and controversial – part of the interpretation. Before a measurement, the electron is in a superposition (a wave of probabilities). During measurement, when an observer (or a measuring device) interacts with the particle to find its position, the wave function "collapses." Once measured, the superposition is gone. The particle "chooses" one single, definite state (e.g., it is *right here*).

The Copenhagen Interpretation creates a clear divide between the quantum world and the classical world (the world we live in). In the quantum world, systems are described by wave functions and superpositions, whereas in the classical world, observers and their measuring devices are "classical" and obey standard physics. They can only ever see one definite outcome, not a superposition.

The Heisenberg Cut is the imaginary boundary between the quantum system being studied and the classical observer performing the measurement. The "collapse" happens at this cut. Random events, with known limits, like rolling dice, average out to an exact, predictable reality. Say we agree to split 10,000 gold coins. We will determine each person's coin percentages by coin flip. With a single flip, one person gets all the coins. If we agree to flip each coin fifty times, we are (almost) guaranteed to get 5,000 coins each. Quantum Mechanics works like this on the gross scale of tables, chairs, rocks, and observers. It does not take many events for an increasingly stable reality to arise from a very random set. Thus, the "Heisenberg cut" comes into existence because it must?

The rationale that this cut "must" exist, given the sheer volume of events, is supported by *Decoherence Theory*. Quantum states are incredibly fragile. In a vacuum, a single particle can exist in multiple states at once. However, as soon as that particle interacts with its environment – bumping into an air molecule or being hit by a photon – it is "measured" by the world around it.

That reality happens "because it must" suggests that the classical world is an emergent property of the quantum world. Just as a single pixel doesn't look like a face, but a million pixels do, our stable reality is the inevitable result of an astronomical number of quantum events resolving into a collective average. It turns the "mystery" of the Heisenberg cut into a mathematical certainty. Once you have enough "flips," the individual's chaos has no choice but to surrender to the group's order. Well, it sounds reasonable.

Crucially, "measurement" or "observation" does not require a conscious human. The interaction of the quantum particle with any large enough classical object – like a detector, a screen, or even an air molecule – is enough to force it to "choose" a state. There's even a theory of gravity-induced wave function collapse (specifically the Diósi-Penrose model). It proposes that wave-function collapse is a physical process induced by gravity rather than by measurement or observation. It suggests that superpositions become unstable and collapse when the mass distribution creates conflicting spacetime geometries, with larger objects collapsing faster than small particles. However, the role of the conscious observer in wave function collapse remains (for some) an accepted theory and is used to support a variety of claims. See the "consciousness-causes-collapse" hypothesis later.

As an analogy for the Copenhagen interpretation, imagine a special coin spinning in a closed, dark box. As long as the box is closed, the coin is in a superposition of both "heads" and "tails" at the same time. It's not one or the other; it's a 50/50 probability wave. The measurement happens when you open the box to look. The instant you open it (and light photons bounce off it), the coin "collapses" into a single, definite state. It is now either heads or tails. The Copenhagen interpretation says that it is meaningless to ask what state the coin was in before you opened the box. It didn't have a definite state. The only reality we can talk about is the one we measure, which, in this case, was "heads."

And then there's Einstein, whose objections are often summed up by his famous quote, "*God does not play dice*," reflecting his distaste for the theory's inherent randomness. However, Einstein's deeper objection was philosophical. He believed in an objective, local reality – that objects have definite properties (like position) whether we are looking at them or not. The Copenhagen interpretation, by contrast, suggests that properties are not "real" until they are measured.

Einstein argued that quantum mechanics was "incomplete" and that there must be "hidden variables" we had not yet discovered, which would restore determinism and objective reality to the universe. As we saw earlier, Bell's Inequality and subsequent experiments seem to have proven Einstein wrong – but see the explanation earlier on Superdeterminism (be careful saying Einstein is wrong!).

The main concerns are that the Copenhagen interpretation feels incomplete, logically inconsistent (by having two different sets of rules), and philosophically unsatisfying, as it fails to describe the nature of reality independent of an observer.

Schrödinger's Cat Thought Experiment

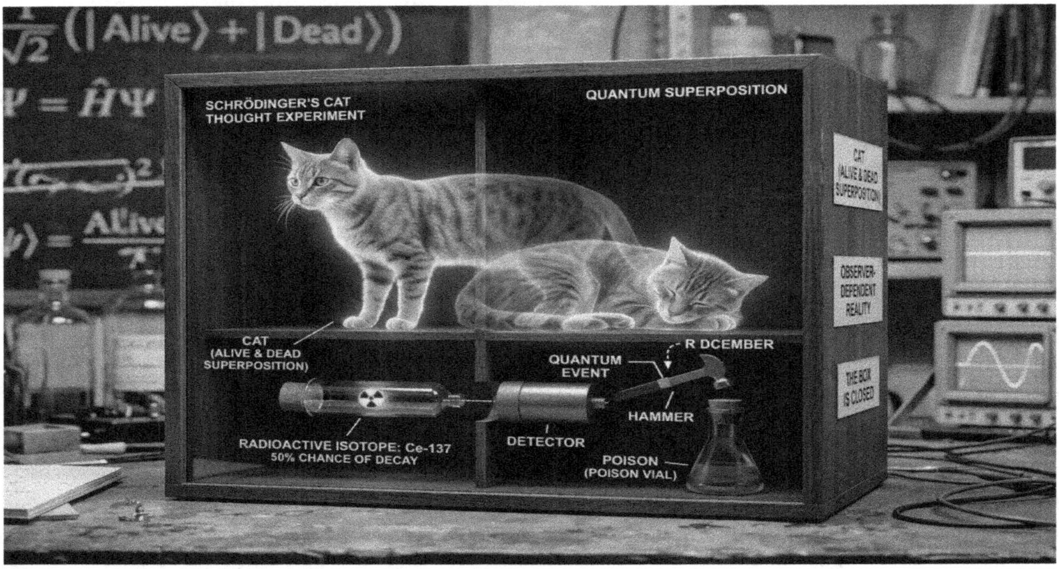

The Schrödinger's Cat thought experiment is a famous paradox proposed by Erwin Schrödinger in 1935. It was designed not to describe reality but to highlight the absurd consequences of applying the rules of quantum mechanics to macroscopic objects (such as a cat). The experiment challenges the interpretation of quantum superposition and the observer's role in quantum mechanics. Schrödinger imagined sealing a cat inside a steel box with a contraption that has a 50% chance of killing the cat within one hour.

The probability for the system is: 50% that the atom decays (cat is dead) and 50% that the atom does not decay (cat is alive). According to the Copenhagen Interpretation, before any measurement is made, the cat exists in a state of superposition – it is simultaneously both alive and dead. It is only when an observer looks that the wave function collapses and commits to one definite state.

The thought experiment forces us to ask: At what point does the quantum world end and the classical world begin? When the box is opened, the act of observation forces the quantum state to collapse. The observer measures the result, and the cat instantly resolves into one definite classical state: either alive or dead.

Schrödinger argued that it is illogical to assume a cat can be both alive and dead simultaneously. The simple application of superposition to macroscopic systems must be flawed or incomplete. The thought experiment highlights the complex and unresolved "measurement problem" in quantum mechanics.

Schrödinger's thought experiment was specifically designed as a critique to expose what he viewed as the absurd implications of the Copenhagen Interpretation.

The Consciousness-Causes-Collapse Interpretation

The Von Neumann–Wigner interpretation is a controversial hypothesis in quantum mechanics that asserts that a conscious observer is required to collapse the quantum wave function. Often summarized as "consciousness causes collapse," it attempts to solve the quantum measurement problem.

In 1932, John von Neumann's foundational book established the rigorous mathematical framework of quantum mechanics. He established that the wave function can logically collapse at any point in the chain from a measuring instrument to subjective perception. In 1961, Eugene Wigner, a Nobel Prize-winning physicist, took von Neumann's work a step further. In his article "*Remarks on the Mind-Body Question*,"[12] Wigner formally argued that a physical measurement can't be complete until a conscious mind perceives it. That is, consciousness is fundamental, and quantum mechanics requires a conscious observer to collapse the wave function. He rejected materialism, asserting that the mind cannot be explained by physics alone, but rather, must include the mind.

As we have seen, in quantum mechanics, a subatomic particle does not have definite properties until it is observed. Instead, it exists in a "superposition" of all possible states, described mathematically by a wave function. The Schrödinger equation perfectly describes how this wave of probabilities flows and evolves over time. But the Schrödinger equation does not contain any mathematical mechanism for a "collapse." It doesn't explain how a blur of probabilities snaps into a single, definite reality the moment we measure it.

In the 1930s, the brilliant mathematician John von Neumann rigorously formalized the mathematics of quantum mechanics and exposed a logical paradox. If a quantum particle is in a superposition, and you use a machine to measure it, that machine is also made of atoms. Meaning the machine should also be governed by quantum rules. According to the math, the machine does not collapse the particle; instead, it becomes entangled with the particle, entering a superposition itself (e.g., the machine's dial points to both "Up" and "Down").

If a camera records the machine, it too enters the superposition. If a human eye looks into the camera, the retina and optic nerve enter superposition. This infinite sequence of entanglement is known as the "Von Neumann Chain." Decades later, physicist Eugene Wigner took von Neumann's mathematics to its ultimate logical conclusion. Wigner argued that the chain must break somewhere because we never experience a blur of superpositions; we experience definite, single outcomes.

Wigner pointed out that everything in the physical universe – machines, cameras, biological eyes, and even a human brain – is made of physical matter and must obey the Schrödinger equation. So no physical object can break the Von Neumann chain. Wigner concluded that the only thing capable of causing the wave function to collapse is the subjective, non-physical experience of a conscious mind.

According to this interpretation, the universe remains in a state of suspended, probabilistic animation until the exact moment when a conscious observer registers the information, the act of human awareness that forces the universe to "choose" a definite state. To illustrate this, Wigner proposed the "Wigner's Friend" thought experiment. Wigner places a friend inside an isolated laboratory to measure a quantum particle. To the friend inside the lab, the wave function collapses the moment they look at the result. However, to Wigner, who is completely isolated outside the lab, his friend, the particle, and the room remain in a massive quantum superposition until Wigner opens the door and asks his friend for the result.

While the Von Neumann–Wigner interpretation was highly influential in the mid-20th century and perfectly consistent with the mathematics of the time, today it is largely a minority view. The primary criticism is that it places humans at the center of the universe, leaning heavily into mysticism or dualism by treating consciousness as a magical entity separate from physical reality.

The Many-Worlds Interpretation

The Many-Worlds Interpretation (MWI) is a theory of quantum mechanics that proposes that all possible outcomes of a quantum measurement actually occur in separate, parallel universes. Instead of a quantum system's wave function "collapsing" into a single reality upon measurement, the MWI suggests that the universe splits, or branches, to accommodate every possible result.

This means that for every quantum decision, a new, separate universe is created. Unlike other interpretations, the MWI rejects the notion of wave function collapse. The universal wave function is considered objectively real and never disappears. When a quantum measurement is made, the universe splits into multiple branches, each representing a different outcome. Every possible outcome of a quantum experiment is physically realized in its own separate "world."

The MWI holds that the random nature of quantum mechanics is an illusion. While all outcomes occur, an individual observer experiences only one branch, making it seem as if only one result were certain. These "other worlds" exist simultaneously with our own, occupying the same space and time yet remaining separate and noninteracting.

Physicist Hugh Everett III introduced the "Many-Worlds" concept, initially called the "relative state" formulation of quantum mechanics, in his 1957 doctoral dissertation.[13] Everett's revolutionary work sought to solve the "measurement problem" in quantum physics without relying on the traditional Copenhagen interpretation, which holds that the act of observation forces a quantum system's wave function to "collapse" into a single definite state.

Everett proposed that the Schrödinger equation should be taken literally everywhere, meaning no wave functions ever collapse. Instead of a system collapsing into a single outcome, Everett's mathematical framework describes the observer and the object becoming entangled upon interaction. Each outcome exists relative to a specific state of the observer. It implies that every time a quantum event has multiple potential outcomes, the universe splits so that all possibilities occur simultaneously in separate, non-communicating branches ("many worlds").

It wasn't until the 1970s and 80s – specifically following the development of quantum decoherence – that Everett's ideas were revived, popularized by physicist Bryce DeWitt, and eventually accepted as one of the major interpretations of quantum mechanics.

Many physicists and philosophers find the MWI attractive and potentially correct. The main appeal is its mathematical elegance and simplicity. Its only rule is that everything in the universe (including us) follows the Schrödinger equation at all times. This is contrary to the Copenhagen interpretation, which requires two contradictory rules. The first is that unobserved quantum systems evolve smoothly in superpositions. The second is that when "measured," this wave magically and instantaneously collapses into a single, random outcome.

The MWI supporters argue that "wave function collapse" is an unnecessary, ad-hoc addition. They propose that the Schrödinger equation is the only rule governing the universe. The MWI offers a clean solution to the biggest problem in the Copenhagen interpretation: In the Copenhagen view, a "measurement" causes the collapse. But what counts? A human? A cat? A Geiger counter? This is undefined.

In Many-Worlds, there is no collapse and no measurement problem. When a quantum system in a superposition interacts with a larger system (like a detector, a cat, or you), the superposition doesn't collapse. Instead, the larger system simply gets entangled with the quantum state. The entire system (atom + detector + you) evolves deterministically into a new, larger superposition. The superposition state equals

(Atom Decayed + Detector Fired + You See "*Decayed*") + (Atom Not Decayed + Detector Silent + You See "*Not Decayed*").

You, the observer, have now split into two versions, each existing in a separate, non-interacting branch of reality. The Copenhagen interpretation requires an arbitrary "Heisenberg Cut" that separates the quantum world from the classical one (where observers live). MWI argues this is an artificial and arbitrary line. After all, observers and measuring devices are themselves made of quantum particles.

Personally, I feel this view of the “cut” being arbitrary is incorrect. Werner Heisenberg proposed a theoretical "cut" in which the quantum wave function describes everything below the line. In contrast, everything above it (including the measurement apparatus and the observer) is treated classically. Heisenberg argued that, *within limits*, one had the freedom to choose where to place this dividing line, as the result of a calculation would be the same. In other words, he meant that the dividing line between the quantum-governed microscopic system and the classical-governed measuring apparatus is not fixed by nature, but is a flexible, conceptual choice made by the physicist.

I remain unconvinced by this explanation. It may be more appropriate to assume that, if such a boundary exists, its nature is not yet fully understood, even in the context of decoherence, and that a valid scientific explanation may eventually be established.

The MWI is entirely consistent: everything is quantum. It treats observers, cats, and planets the same way it treats electrons. There is no special "classical" realm where the rules of quantum mechanics suddenly break. While quantum mechanics feels random, the MWI is fundamentally deterministic.

The universal wave function (the mathematical object describing all possible worlds) evolves smoothly and predictably according to the Schrödinger equation. There is no randomness or "dice-rolling." The illusion of probability is simply because an observer exists in only one of the many branches and cannot know which one they will end up in beforehand. From their perspective within a branch, the outcome appears random, but the overall "multiverse" contains every possible outcome and evolves with perfect predictability.

There are several objections to the Many-Worlds Interpretation of quantum mechanics, but we should be careful not to use our “common sense” – that’s always a risk in the quantum world. The most common and intuitive objection is that the MWI avoids the "wave function collapse" by positing that every quantum measurement creates an entirely new universe – an astronomical number of them splitting off every fraction of a second. Critics argue this is an incredible violation of

Occam's razor (the principle that the simplest explanation is usually the best). It seems extraordinarily "extravagant" and wasteful to invent an infinite number of unobservable worlds just to avoid the collapse of one wave function.

While Occam's razor appeals to my sensibilities, I'm not sure I can justify the objection that the MWI is not the most straightforward explanation. In quantum theory, what does "simple" really mean? Perhaps splitting off new universes *is* the easier way, and we just don't understand why that is the case.

The next objection is that you cannot make an observation or falsify the theory. The "other worlds" are completely inaccessible and unobservable to us. We can never contact, measure, or collect data from them. This leads many scientists to argue that the MWI is unfalsifiable. If you can't design an experiment to test for the existence of these other worlds, it's not truly a scientific theory. It becomes ascientific, a philosophical preference rather than physics.

The most serious technical objection is that the MWI is a deterministic theory: everything that *can* happen *does* happen (in some branch). So, where do probabilities come from? If you measure an atom's spin with a 70% chance of being "up" and a 30% chance of being "down," the MWI says the universe splits into two branches: one "up" and one "down." If both outcomes happen, why should you expect to find yourself in the "up" branch 70% of the time? Why isn't it a 50/50 toss-up between which "you" you become?

The MWI struggles to explain why we experience the Born Rule (which states that the probability of an outcome is the square of its amplitude). Proponents have sought to derive the Born Rule within the MWI (using concepts such as "self-locating uncertainty"), but many critics find these arguments unconvincing.

When a system splits, why does it split into the specific states we observe (like "alive cat" and "dead cat")? Why not some bizarre superposition of the two? This is known as the "preferred basis problem." Why does the universe split along a specific basis (e.g., position, or "alive/dead")? The MWI's answer to this is decoherence. It states that interaction with the environment rapidly forces the split to happen along stable, classical-like states (the "pointer basis"). However, critics argue that decoherence explains why a system appears classical, but it doesn't fully resolve *which* specific basis is the "real" one for the split. To some, it feels like a way of reintroducing the "measurement problem" in a more subtle form.

Quantum Immortality

Quantum immortality is a concept derived from the Many-Worlds interpretation, proposing that an individual's consciousness persists by transitioning into a parallel universe in which they survive a life-threatening event. While most versions of the individual would not survive, the conscious experience would continue in a universe where survival occurs. I find the notion of consciousness 'transitioning' into alternate universes challenging to accept. Notably, even in popular culture, such as the television series Quantum Leap, the protagonist remains within a single timeline.

The MWI is the foundation of this theory, which holds that the universe splits into parallel universes at every quantum event, with each possible outcome occurring in its own universe. Quantum immortality applies the MWI to an individual's consciousness. It proposes that a person's subjective experience of being alive continues through the branches of reality where they survive, rather than ceasing in a universe where they die. The theory suggests that when faced with a fatal event, a person's consciousness doesn't die but "moves on" to a parallel universe where they have survived. From that individual's perspective, they simply experience a fortunate survival, not a transition. According to this theory, if there's a non-zero probability of survival in each situation (no matter how small), consciousness would continue along the timeline in which it survives, leading to a form of subjective immortality. If the multiple universes are splitting off (I asked myself), why wouldn't consciousness split off too and continue in multiple timelines where "you" had survived?

The 'quantum suicide' thought experiment popularized by physicist Max Tegmark in 1997 proposes that quantum randomness (such as a particle's spin) leads to a fatal outcome, but consciousness always follows the branch on which survival occurs, creating an endless, improbable string of lucky escapes.[14]

The setup involves an observer standing before a "quantum gun" triggered by a subatomic particle. Each time the trigger is pulled, the outcome (the gun fires or doesn't fire) depends on a quantum measurement. According to the Many-Worlds Interpretation, the universe instantaneously splits into two realities every time the trigger is pulled. In one universe, the particle triggers the gun to fire, and the observer dies; in the other, it triggers the gun not to fire, and the observer survives.

Tegmark noted that because an individual cannot consciously experience their own death, the observer in the "alive" timeline is the only conscious experience they can ever have. Thus, from the observer's subjective perspective, the trigger would continually click without ever firing, surviving against all odds. To make the thought experiment mathematically and physically sound, Tegmark outlined three strict criteria:

1. The trigger must rely on a purely quantum (not classical) event so that it genuinely places the observer in a superposition of dead and alive states.

2. The gun must kill the subject (or cause unconsciousness) faster than the brain can process the outcome so that the observer doesn't experience dying.
3. It must be guaranteed to be 100% fatal, not merely cause injury.

While the thought experiment suggests a participant will subjectively survive indefinitely across parallel universes, Tegmark explicitly rejects the idea that this makes an individual functionally immortal in real life. He pointed out that real-world deaths (such as aging or fatal illnesses) do not meet his criteria. Dying in the real world is a gradual, progressive process rather than a sudden binary event. Because a dying person loses self-awareness and consciousness gradually, Tegmark argues that one will not feel subjectively immortal or snap into a parallel reality where they are miraculously saved from natural causes. And cosmologist Anthony Aguirre warns that it "would be foolish (and selfish) in the extreme to let this possibility guide one's actions in any life-and-death question".

Some other theories include the Subjective Continuity/Branch Selection theory, which argues that because an individual cannot observe their own death, they will always subjectively experience the continuation of their life in a universe in which they survived. Consciousness effectively filters out branches that are, at a minimum, in a "barely alive" state, leading to a sense of perpetual survival.

Quantum immortality is not an accepted scientific theory but a speculative philosophical and physics concept. Most physicists view it as a thought experiment rather than a literal possibility. Yes, it's a bizarre idea, but is it bizarre enough to be a quantum theory? For a different and amusing look at quantum theory and this idea of immortality, read *Quantum Physics for Hippies* by Dr. Lukas Neumeier and Dr. James Douglas. It's a delightful book and fun because it's written as a novel.

The Pilot-Wave (De Broglie-Bohm) Theory

In this theory of Bohmian mechanics, every electron always has a definite position, even if observers are ignorant of its exact value. An electron is pushed around by a guiding "pilot-wave" that influences its position. While each electron travels through one slit or the other, the pilot wave passes through both slits simultaneously. Interference in the pilot wave leads to the observed interference pattern. A measurement at the slits will collapse the pilot wave and reveal where the electron was all along.

The Pilot-Wave Theory is a fascinating and somewhat controversial interpretation of quantum mechanics. It offers a fundamentally different way of understanding what quantum mechanics is telling us about the world compared to the more mainstream Copenhagen interpretation.

Bohmian mechanics (often called the de Broglie–Bohm theory or pilot-wave theory) was created by David Bohm in 1952.[15] Louis de Broglie created Bohmian mechanics (often called the de Broglie–Bohm theory or pilot-wave theory) originally proposed The foundation for this interpretation in 1927, but he abandoned it. Bohm later rediscovered and expanded upon de Broglie's "pilot-wave" concept, providing a complete, deterministic alternative to the standard Copenhagen interpretation of quantum mechanics. The theory provides a consistent description of quantum phenomena by treating particles as real physical objects that have exact positions at all times, guided by a "pilot-wave" or quantum potential rather than behaving randomly.

The central idea is that quantum reality consists of two fundamentally real entities: Particles that are actual, point-like particles with definite positions and trajectories at all times. Think of them as being like classical particles in some ways. This contrasts with interpretations in which particles are merely probabilities (excitations in a field) or in which their positions are defined only upon measurement. The Pilot-Wave (function) is a real, physical wave (described by the wave function, ψ) that exists in configuration space (not ordinary three-dimensional space, but a higher-dimensional space representing the positions of all particles in the system). This wave guides the motion of the particles.

Pilot-Wave Theory is fundamentally deterministic. Given the initial positions of all particles and the initial pilot wave, the system's future evolution is entirely determined. There's no inherent randomness in the fundamental laws themselves. Quantum probabilities arise from our ignorance of the precise initial conditions (particle positions and initial wave function). It states that both particles and the pilot wave are objectively real entities existing in the world, independent of observation. There is no fundamental collapse of the wave function. The wave function constantly evolves according to the Schrödinger equation. What appears to be wave-function collapse in measurements is explained by the specific way the pilot wave and particles evolve during interactions, especially with measuring apparatuses. Because particles have definite positions and are guided by the pilot wave, they follow unique, definite trajectories in space and time. These trajectories are complex and non-classical due to the influence of the pilot wave.

Pilot-Wave Theory is inherently non-local. This means that the velocity of a particle at one location can instantaneously depend on the positions of other particles, even if they are very far away. This non-locality is a consequence of the pilot-wave's existence in configuration space, which is inherently entangled in multi-particle systems. This non-locality is consistent with Bell's Theorem and experimental violations of Bell's inequalities, which show that any realist theory that reproduces quantum mechanics must be non-local. Measurement is simply a physical interaction between systems. There's no special "measurement postulate" or a need for a conscious observer to trigger anything. The outcome of a measurement is determined by the initial conditions of the system and the measuring apparatus, and by the deterministic evolution of the system and the apparatus.

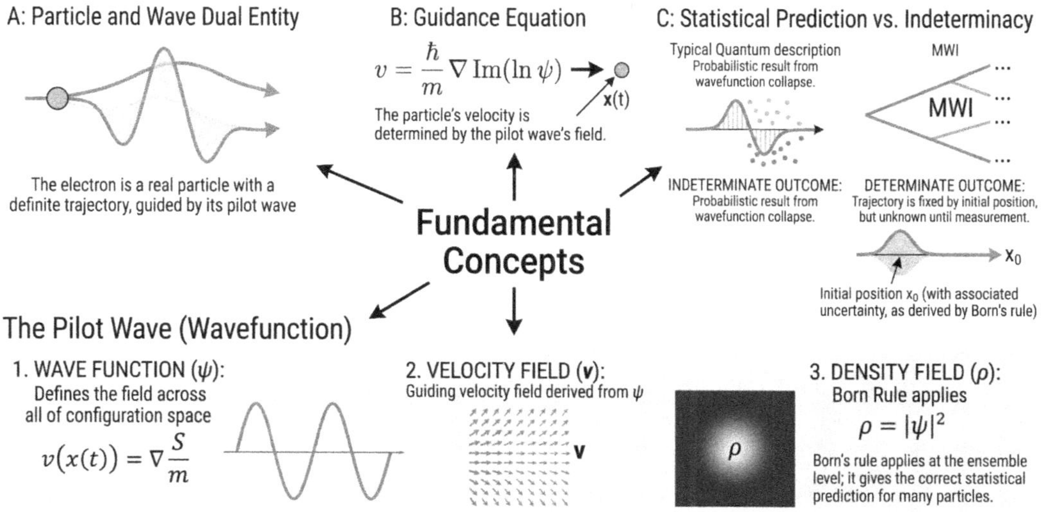

Lastly, Pilot-Wave Theory reproduces all the empirical predictions of standard quantum mechanics. It aims to describe all observed quantum phenomena, such as wave-particle duality, superposition, interference, diffraction, tunneling, and entanglement. Particle motion is governed by a guiding equation derived from the Schrödinger equation. Essentially, the velocity of each particle is determined by the phase of the pilot wave at the particle's location in configuration space. The Guiding Equation (or "equation of motion") is the central piece of Pilot-Wave Theory. It's the unique rule, added to the Schrödinger equation, that dictates how a real, physical particle moves.

For a single particle with mass m, the Guiding Equation is expressed as:

$$\mathbf{v} = \frac{d\mathbf{Q}}{dt} = \frac{\hbar}{m}\mathrm{Im}\left(\frac{\nabla\Psi}{\Psi}\right)$$

Where:

- v (or q) is the velocity of the particle at position Q.
- Ψ is the complex wave function (the "pilot-wave").

- $\nabla\Psi$ is the gradient of the wave function, showing how it changes across space.
- Im denotes the imaginary part of the term in the parentheses.
- $\hbar/m$ is a scaling constant based on the particle's mass and the reduced Planck constant.

In short, the equation defines a particle's velocity at a specific point in space. This velocity is determined by the particle's "pilot-wave" (the standard quantum wave function). The theory is governed by three primary mathematical components that relate the wave field to the particle's motion. The pilot wave itself evolves according to the standard non-relativistic Schrödinger equation. It explains why a particle can "feel" the presence of two slits even if it only passes through one – the pilot wave passes through both and exerts a force on the particle. Think of the wave function (ψ) as a complex, invisible "weather map" that fills space. This map has peaks, valleys, and "currents."

The guiding equation is the instruction that tells the particle, "At your *exact* location, look at the slope *right here* and move in this specific direction with this specific speed." The particle is like a passive "surfer" or a cork on the water. The pilot wave completely and deterministically guides its motion.

Unlike the Copenhagen interpretation, this theory says a particle *always* has a precise, real position, even when we aren't looking. If you knew the particle's exact starting position and the shape of its pilot wave, you could predict its *entire* future path with 100% certainty. The "randomness" of quantum mechanics comes from our ignorance of the particle's exact initial position, so the theory is deterministic.

It's also non-local, as the wave function can be spread over a large area. This means the particle's velocity *right here* can be instantly affected by something that changes the *shape* of the wave far away (like in an entanglement experiment) and, in this theory, it's a real, physical connection through the pilot wave.

Wave-particle duality is explained by the coexistence of both fundamental particles and a real pilot wave (the wave aspect). Particles always have definite trajectories, but these trajectories are influenced and guided by the wave, leading to wave-like behavior (interference, diffraction). Before measurement, a particle has a definite position (even if we don't know it), and the pilot wave might be in a superposition. During measurement, the interaction with the measuring apparatus causes the pilot wave to evolve, guiding the particle into one of the possible "eigenstate" regions in configuration space, corresponding to a definite measurement outcome. An eigenstate is a specific, stable quantum state of a system where a measurable property – such as energy, momentum, or spin – has a definite, fixed value. "Collapse" is just the wave evolving and guiding the particle into a specific region; there's no actual collapse of the wave function as a fundamental process.

Entanglement is explained by the fact that the pilot wave is a wave in configuration space. For entangled particles, the pilot wave is a single entity that spans the configuration space of both particles. The guiding equation for each particle depends on the positions of *all* particles in the entangled system. This leads to correlations in their behavior, even when separated by large distances, due to the non-local nature of the pilot wave's influence.

Advantages of this theory include providing a clear account of measurement without invoking wave function collapse or a special role for observers. It appeals to a classical intuition of a deterministic and realist description of nature. Many physicists find this philosophically attractive. It also provides a clear picture of what exists in quantum reality, comprising particles and pilot-waves. Lastly, it can reproduce all the empirical predictions of standard quantum mechanics.

Some physicists find the idea of "hidden variables" and the deterministic nature of Pilot-Wave Theory philosophically unappealing or unnecessary, and, in some interpretations of Pilot-Wave Theory, particle trajectories can exhibit surprising, sometimes counterintuitive "surreal" features that seem to run counter to classical intuition. Surreal "Non-Local" Trajectories (ESSW) were identified by Englert, Scully, Süssmann, and Walther in an analysis of a 1992 double-slit experiment.[16] In an entangled system, a particle's trajectory can be influenced by the measurement of another particle, even if that particle is far away. The "surreal" aspect occurs when a trajectory through one slit is guided by the wave branch passing through the other slit. This seemingly "non-local" influence challenges the classical idea that a particle reacts only to its immediate, local environment.

In 2025, Nature published a report on an experiment that tested Bohmian mechanics (*"Energy–speed relationship of quantum particles challenges Bohmian mechanics"*).[17] Its authors used photons in two coupled waveguides to create a region where Bohmian mechanics predicted that particles move with zero velocity. But the authors reported measured speeds of about one million meters per second.

In a follow-up to the 2025 Nature report, a 2026 comment in Nature ("*Bohmian mechanics remains unchallenged by tunnelling experiment*"), a group of physicists argued that the 2025 experiment does not rule out Bohmian mechanics.[18] The authors of the comment argue that this was not the correct particle velocity used in Bohmian mechanics, and that the result is then consistent both with standard quantum theory and with Bohmian mechanics.

Pilot-Wave Theory remains a somewhat minority view, but there's ongoing work to develop relativistic versions of the theory (which are challenging due to tensions with Lorentz invariance and non-locality), exploring applications in quantum field theory and quantum gravity, investigating the philosophical implications of determinism and realism in quantum mechanics, and developing computational methods for Bohmian mechanics, especially for complex systems.

The summary, at this point, is that Pilot-Wave theory remains a compelling alternative interpretation of quantum mechanics, offering a deterministic and realistic picture of quantum reality. It provides explanations of key quantum phenomena, such as wave-particle duality, superposition, measurement, and entanglement.

Further research: In pilot-wave theory, the universe contains a single, overarching universal wavefunction, denoted as $\Psi(Q)$. The variable Q represents the actual physical positions of every single particle in existence. Because you cannot easily calculate or manipulate a wavefunction that includes the coordinates of the entire universe, you must isolate smaller systems. This is achieved by dividing the universe into two parts: the specific subsystem coordinates (x) and the environment coordinates (y). To find the effective wavefunction of your isolated subsystem, you take the universal wavefunction and substitute the actual, exact physical positions of the environment particles (Y) at that specific moment. This operation yields the *conditional wavefunction*: $\psi(x,t) = \Psi(x, Y(t), t)$.

As the environment particles move along their trajectories, their changing positions ($\mathrm{Y(t)}$) alter the conditional wavefunction. This mechanism explains why a quantum subsystem appears to undergo wavefunction collapse during a measurement. The particles of the measurement apparatus change their physical positions, which alters the conditional wavefunction of your subsystem without requiring a wave-collapsing external observer.

Another common point of confusion involves how Bell's theorem affects pilot-wave theory. Bohmian mechanics is a hidden variable theory. The actual, precise positions of the particles serve as the hidden variables because standard quantum mechanics omits them from its formalism. Bell's theorem does not rule out all hidden variables. It specifically shows that no *local* hidden-variable theory can reproduce the statistical predictions of quantum mechanics.

Pilot-wave theory avoids the restrictions of Bell's theorem because it is explicitly *non-local*. The guiding equation states that the velocity of any single particle depends instantaneously on the positions of all other particles in the system, regardless of their distance. Because the theory violates locality, it satisfies Bell's theorem while maintaining definite particle positions. Quantum Mechanics is also non-local, so this is not necessarily a show-stopper.

One of the big problems is that the standard pilot-wave theory lacks Lorentz invariance. The guiding equation relies on a single, universal time parameter (t) to calculate how non-local interactions occur across space. This formulation requires an absolute frame of reference to define which events happen at the exact same time. This reliance on a preferred reference frame conflicts with special relativity, which states that no reference frame is unique. Physicists use three primary solutions to address the lack of Lorentz invariance in pilot-wave theory:

1. ***Hypersurface Foliations:*** Detlef Dürr, Sheldon Goldstein, and Nino Zanghì developed a version where the guiding equation no longer depends on flat time slices. Instead, it uses a frame-independent set of space-like hypersurfaces. The universal wavefunction itself dynamically determines the preferred foliation, thereby preserving the underlying physical laws' relativistic character.
2. ***Bohmian Quantum Field Theory:*** The pilot-wave model can be extended to quantum field theories. In these models, the hidden variables shift from particle positions to continuous field configurations or fluctuating particle numbers. This accommodates particle creation and annihilation while matching the statistical Lorentz invariance of standard quantum field theory.
3. ***Hidden Preferred Frames:*** Some formulations accept that a physical, preferred reference frame exists at the microscopic level. Because quantum equilibrium randomizes particle distributions according to the Born Rule ($|\Psi|^2$), you cannot use non-local connections to send signals faster than light. This statistical equilibrium hides the preferred frame from your laboratory instruments, rendering all observable predictions perfectly Lorentz invariant.

The Bad News: Physicists built Quantum Field Theory on the foundation of special relativity. It treats space and time symmetrically, ensuring that physical laws remain identical across all moving reference frames. Pilot-wave theory struggles with this symmetry because its non-local interactions require an instantaneous connection between particles. To calculate how a particle at one location immediately responds to another, the theory requires a hidden, absolute time frame. This reliance on absolute simultaneity violates the core structure of special relativity, which contains no preferred frame of reference.

In Quantum Field Theory, the number of particles in a system fluctuates constantly. For example, when an excited hydrogen atom drops to a lower energy level, it emits a single photon. In high-energy particle accelerators, an electron and a positron collide and annihilate each other, generating two gamma-ray photons.

QFT handles these changes naturally using mathematical operators that create and destroy particles. Louis de Broglie and David Bohm originally engineered the pilot-wave theory to track a fixed number of permanent particles moving along continuous paths. Attempting to track definitive trajectories when particles pop in or out of existence instantly makes the mathematics awkward and inefficient.

The fundamental reality in QFT consists of continuous fields, such as the electron field or the electromagnetic field, which spread across the entire universe. Particles are merely localized bundles of energy within these fields. The main appeal of pilot-wave theory in basic quantum mechanics is that it preserves the intuitive concept of concrete, localized particles. When physicists try to upgrade pilot-wave theory to match QFT, they often must abandon particle positions entirely. Instead, they must claim that the continuous field configurations themselves possess definite hidden values at every single point in space. This shift complicates the mathematics and destroys the original conceptual simplicity that made the theory appealing in the first place.

Pilot-wave theory does not alter the actual statistical predictions of quantum physics. To use a Bohmian version of a field theory, the full standard quantum wavefunction to find the probabilities of your experimental outcomes must first be calculated. After completing that step, additional, intensive mathematical calculations must be performed to simulate how the pilot wave guides the hidden field configurations or particle positions. Because the physical results remain identical to those of standard QFT, a major computational effort is expended without yielding any new testable predictions or experimental advantages.

My final summary is that Pilot-Wave Theory might actually be a better theory than standard Quantum Mechanics. But with the development of Quantum Field Theory, the majority of prior problems were solved – something that pilot-wave does not do. Attempts to "fix" the theory are making it overly complicated and less attractive. QFT was essential for the Standard Model of Particle Physics; it moved the goalposts, and any new theories must improve upon it. I don't think we can argue that the pilot-wave theory does that.

I appreciate the original formulation and its approach to addressing wave-particle duality and the measurement problem. It is possible that future developments may refine this theory or introduce a related framework. At present, it remains a compelling and thought-provoking perspective in the ongoing search for a deeper understanding of quantum phenomena.

Objective Collapse Theories

Also known as spontaneous-collapse models and dynamical-reduction models. Objective Collapse Theories propose that the wave function (a mathematical description of a quantum system) undergoes a spontaneous physical collapse, reducing a superposition of states to a single, definite state without requiring a conscious observer. These models aim to resolve the measurement problem by treating wave-function collapse as an inherent physical process, distinct from observer-dependent interpretations such as the Copenhagen interpretation.

They can be further categorized as:

Orchestrated Objective Reduction (Orch-OR): *A specific hypothesis stemming from the Diósi-Penrose model, which suggests that objective reductions occur within the microtubules of the human brain, forming the basis of consciousness.*

Continuous Spontaneous Localization (CSL): A leading theory stating that the wave function undergoes a gradual, continuous collapse rather than a sudden jump.

GRW Theory: Named after physicists Giancarlo Ghirardi, Alberto Rimini, and Tullio Weber, this was the first specific physical-collapse model. It suggests that particles experience tiny, random, spontaneous collapses into definite locations.

Diósi-Penrose Model: An objective collapse theory formulated by physicists Lajos Diósi and Roger Penrose, which proposes that wave function collapse is physically caused by gravitational interactions.

Unlike interpretations that suggest collapse is triggered by observation or measurement, these theories hold that collapse occurs naturally, driven by physical mechanisms that are not yet fully understood but that affect all quantum systems.

In these models, the wave function is treated as a real physical wave. Its collapse is a physical, objective process that happens irrespective of any external measurement device or conscious observer. A central feature is that the collapse does not depend on consciousness or observation. The system collapses on its own, resolving quantum ambiguities into classical reality.

These theories avoid the measurement problem by explaining how quantum superpositions of multiple possible states evolve into a single, definite outcome observed in the macroscopic world, thereby addressing the fundamental question of why we don't see large objects in multiple places at once. This is achieved by modifying the Schrödinger equation with additional nonlinear, stochastic terms that become significant for massive objects, thereby bridging quantum fuzziness and classical reality. Objective collapse theories, also known as dynamical reduction models, introduce new physical laws that dynamically reduce the superposition of states over time.

Developed by Roger Penrose and anesthesiologist Stuart Hameroff, the Orch-OR (Orchestrated Objective Reduction) theory posits that microtubules, the internal structures of neurons, act as quantum computers.[19] The "Orch" (orchestrated) part refers to the idea that the brain somehow organizes these quantum processes. The "OR" (objective reduction) refers to Penrose's theory that quantum superpositions spontaneously collapse due to gravitational effects, a process that he suggests is linked to consciousness. Each Orch-OR event is theorized to be a moment of consciousness. The sequence of these events, which occurs at about 40 per second, creates the stream of conscious experience. The theory suggests a link between the brain's quantum processes and the fundamental structure of the universe, specifically, spacetime geometry.

The Ghirardi–Rimini–Weber (GRW) theory is an objective collapse model that solves the measurement problem by modifying the Schrödinger equation with spontaneous, random, and nonlinear localization "hits".[20]

Under GRW, the wave function of any particle evolves normally most of the time. However, at random intervals, the wave function undergoes spontaneous localization, collapsing into a highly localized region of space.

The primary feature of GRW theory is its reconciliation of the quantum behavior of subatomic particles with the classical behavior of everyday objects via two fundamental constants of nature. The first constant is the collapse rate (λ). For a single, isolated microscopic particle like an electron, a spontaneous collapse occurs roughly once every 10^{16} seconds, which translates to about once every 30 million years. Because these events are incredibly rare at the atomic scale, quantum particles remain in wide superpositions for long periods, allowing you to observe interference patterns in laboratory experiments.

The second constant is the localization width (a). When a collapse happens, the wave function shrinks to a region of about 10^{-7} meters. This width is small on a human scale but large compared to the scale of an atom.

When you group particles together into a macroscopic object, the mathematics change entirely. A standard laboratory instrument or a cat contains roughly 10^{23} particles. While the chance of a single particle collapsing is minimal, the total collapse rate of the macroscopic system is the sum of all its constituent particles. For a large object, one of its particles will undergo a spontaneous localization every 10^{-7} seconds. Because all the particles are bound together in a rigid structure, the collapse of a single particle instantly drags the wave function of the entire macroscopic object into a definite location. This explains why large objects never appear in a spatial superposition.

Unlike standard interpretations of quantum mechanics, GRW is a distinct physical theory that makes unique, testable predictions. The random, rapid hitting of particles during spontaneous localization imparts a tiny amount of kinetic energy to the system. This energy injection causes a faint, constant warming of matter and the emission of low-energy radiation, known as spontaneous X-ray emission. Modern laboratories search for this faint background radiation to place strict experimental bounds on the exact parameters of the GRW constants.

Continuous Spontaneous Localization (CSL) is a theoretical framework in quantum physics that proposes a physical, objective mechanism for wave-function collapse. Formulated largely by physicist Philip Pearle as an extension of the discrete Ghirardi-Rimini-Weber (GRW) model, CSL modifies the standard Schrödinger equation by adding non-linear and stochastic (random) terms.19 These modifications cause quantum superpositions to spontaneously suppress themselves over time without needing an external observer or measurement apparatus.

Wave function collapse occurs smoothly and continuously in time rather than through instantaneous mathematical "jumps". Localization explicitly occurs in position space, meaning that quantum states are forced to collapse into definite physical locations. The effect scales with the system's mass density. For a single microscopic particle (like an electron), the spontaneous collapse rate is so incredibly weak that it behaves according to standard quantum mechanics. However, for a macroscopic object composed of roughly 10^{23} particles, the combined effect causes almost instantaneous collapse into a classical, localized state. Unlike earlier models, CSL preserves the permutation symmetry properties required for systems of identical particles.

Objective collapse theories modify the Schrödinger equation by adding a mathematical term that causes the wave function to collapse spontaneously. Critics argue this is "ad hoc" – meaning it is a "brute force" fix explicitly designed to match our observations of the macro-world, rather than being derived from a more fundamental physical principle.

Most objective collapse models imply a tiny, constant "jitter" in particle motion. Over vast spans of time, this would lead to a gradual increase in the universe's energy, violating the Law of Conservation. While proponents argue that this increase is currently too slight to detect, it remains a theoretical hurdle.

These theories usually assume that collapse happens in position space (particles "choose" a location in physical reality. In this framework, you define the state of a particle or a system by its spatial coordinates – typically (x, y, z). Critics argue that there is no inherent reason in quantum math why position should be more special than momentum or any other property. Yet objective collapse forces this choice, which explains why objects appear solid to us.

Transactional Quantum Physics Interpretation (TI)

Transactional Quantum Physics, more accurately known as the Transactional Interpretation (TI) of Quantum Mechanics, is an interpretation of standard quantum mechanics that offers an alternative way to understand these concepts, particularly measurement, wave function collapse, and the nature of time in quantum processes. John Cramer developed it in the 1980s as an alternative to the Copenhagen interpretation and other interpretations.[21]

It's essential to distinguish it from theories that modify quantum mechanics or attempt to unify it with general relativity. TI accepts the standard mathematical formalism of quantum mechanics. TI proposes that a quantum event, such as the absorption or emission of a photon, or any other quantum interaction, is not a one-way process but a transaction or "handshake" between an emitter and an absorber.

This transaction involves two types of waves: The Offer Wave (Retarded Wave) is similar to the standard forward-in-time wave function in quantum mechanics. An emitter (like an excited atom) sends out an "offer wave" into the future, propagating in the usual forward direction of time. This offer wave is a potential for a quantum transition.

The Confirmation Wave (Advanced Wave) is an absorber (such as a detector or another atom) that responds to the offer wave and sends back a "confirmation wave" into the past. This confirmation wave is a response to the offer and signifies the absorber's willingness to participate in the transaction. The actual quantum event, referred to as the "transaction," occurs when the offer wave and the confirmation wave constructively interfere in spacetime. This interference is the physical manifestation of the quantum event. It's the "handshake" in which energy, momentum, and other quantum numbers are exchanged between the emitter and the absorber.

TI introduces a fundamental time symmetry into quantum mechanics that is often obscured in other interpretations. It treats past and future more symmetrically, at least at the fundamental interaction level. Advanced waves, which travel backward in time, are considered real physical waves rather than mere mathematical constructs.

In TI, wave function collapse is not a mysterious, instantaneous process triggered by observation or measurement. Instead, it's seen as the completion of the transaction. Before the transaction, there are only potential offers and confirmations. Once the transaction occurs, the potential is actualized, and the system transitions to a definite state. Collapse is not "put in by hand"; it is part of the transaction process itself.

Unlike some interpretations, TI does not require a conscious observer or a classical measuring apparatus to cause "collapse." Any absorber that can respond to an offer can initiate a transaction. Measurement is just a particular type of transaction. An absorber is any particle or object capable of interacting with the offer wave – e.g., an atom, a photo-detector, or dust in the environment.

TI offers a way to understand quantum entanglement and "spooky action at a distance" without resorting to faster-than-light signaling or influences. In entangled systems, transactions are seen as non-local in spacetime, occurring across space and time. The transaction itself can be considered instantaneous or non-local, even if the individual offer and confirmation waves propagate at or below the speed of light. It emphasizes that it's the transaction that is non-local, not necessarily the underlying waves, in a way that violates relativity.

Once a transaction occurs, it is a definite, deterministic event. However, the choice of *which* transaction to perform from a set of possible offers and confirmations is probabilistic, governed by quantum-mechanical probability rules.

TI offers a more physically intuitive representation of quantum events, particularly measurement and wave function collapse, than other interpretations. It provides a way to understand measurement without invoking mysterious collapse postulates or assigning a special role to consciousness. It aligns better with the time-symmetry of many fundamental laws of physics – except for thermodynamics and entropy. It provides a framework for understanding entanglement that doesn't necessarily imply faster-than-light communication.

There are some issues with the theory. The concept of advanced waves traveling backward in time is counterintuitive and potentially problematic for causality. Cramer argues that these advanced waves are not used to send signals into the past and do not violate macroscopic causality because they are part of the transaction mechanism and are not directly controllable or observable in isolation.

TI is not a mainstream interpretation. While it has its proponents and has been explored in various contexts, it has not achieved the widespread acceptance of interpretations such as Copenhagen, Many-Worlds, or Bohmian Mechanics. Many physicists are comfortable with standard quantum mechanics and do not feel the need for a radically different interpretation (remember "*Shut up and calculate*"?)

While TI employs the standard quantum-mechanical formalism, certain aspects, particularly the detailed dynamics of transaction formation, may be less rigorously defined than the established mathematical framework of quantum mechanics itself.

As with many interpretations of quantum mechanics, it's challenging to devise experiments that would definitively distinguish TI from other interpretations, since they often make the same empirical predictions.

Transactional Quantum Physics is not, in itself, a theory of quantum gravity. It's an *interpretation* of quantum mechanics itself. However, interpretations of quantum mechanics can inform or inspire approaches to quantum gravity by suggesting different ways of thinking about spacetime, causality, and quantum processes at a fundamental level. We cover this in Chapter Four.

I find Transactional Quantum Physics (TI) a fascinating, less mainstream interpretation of quantum mechanics. It offers a unique perspective on quantum events, as transactions involve offer waves (moving forward in time) and confirmation waves (moving backward in time). Interestingly, most fundamental laws are time-reversible, so maybe waves moving backwards in time are not so improbable?

The theory addresses some conceptual challenges in standard quantum mechanics, particularly regarding measurement and non-locality, but also faces criticisms, particularly regarding advanced waves and causality. It is not a quantum gravity theory, but its principles might influence some speculative ideas at the intersection of quantum mechanics and gravity for researchers interested in this interpretation.

Interestingly, Carlo Rovelli is a proponent of Relational Quantum Mechanics (RQM) and holds that reality is purely relational, in which properties exist only in relation to other systems, not objectively "in themselves". Both aim to resolve quantum paradoxes by moving beyond classical notions of reality and observers. Rovelli's view is that any physical system acts as an observer, making properties relative and resolving issues such as wave function collapse, a mechanism distinct from the TI's quantum handshake.

Quantum Bayesianism

QBism, or Quantum Bayesianism, is a relatively modern interpretation of quantum mechanics that fundamentally redefines what the wave function represents. Developed in the early 2000s by physicists like Christopher Fuchs, Rüdiger Schack, and Carlton Caves, it shifts the focus of quantum theory from the "external world" to the internal expectations of the "agent" (the observer).[22]

QBism is an "agent-centric" interpretation that treats quantum mechanics not as a description of nature itself, but as a manual for making better bets on the outcomes of experiments. The name is derived from Bayesian probability, a branch of statistics where probability is not a fixed frequency (like how many times a coin lands on heads over a thousand trials), but a subjective degree of belief – which is an odd choice of words. In Bayesianism, if you say there is a 70% chance of rain, you are describing your current state of information. If you see clouds, you "update" that belief. QBism applies this logic to the quantum world.

In most interpretations, the wave function (Ψ) is seen as something real–either a physical wave (like a water wave) or a fundamental field of probability.

In QBism, the wave function is epistemic, meaning it exists only in the agent's mind. It is a mathematical tool that an individual uses to quantify their personal expectations about the consequences of their actions. Because it is a "personal" tool, two different observers could technically assign two different wave functions to the same system based on their unique information, and neither would be "wrong."

In standard quantum mechanics, "measurement" is a mysterious process that causes a wave function to "collapse." In QBism, measurement is much simpler:

- It is an action an agent takes on the world.
- The result is a personal experience for that agent.
- The "collapse" is simply the agent updating their personal belief based on their new experience.

This removes the "Measurement Problem" entirely, as there is no physical collapse since there was never a physical wave to begin with – only a *mental* update. QBism resolves famous paradoxes like Wigner's Friend or Schrödinger's Cat by insisting that facts are relative to the agent.

If a friend is inside a lab measuring a particle, and Wigner is outside, the friend has already experienced a result. For Wigner, who is outside, the lab is still in a superposition. In QBism, there is no conflict. The result is a fact for the friend, but it is not yet a fact for Wigner. A "fact" only exists once an agent experiences it. This suggests that the universe is "participatory" – reality is a combination of individual experiences rather than a single, objective, pre-existing script.

QBism is controversial because it strips away the idea of a "God's-eye view" of the universe. It suggests that science is not about discovering the objective "truth" of the world in itself, but about refining our ability to find our way and interact with the world. Critics argue that it borders on "solipsism" (the idea that only one's mind is certain to exist), but QBists counter that, while the wave function is subjective, the consequences of our actions are determined by a real, independent nature that we are constantly learning about.

Summary

Focusing on the foundational math and historical experiments that initiated the quantum revolution, Chapter Two opens with Max Planck's quantization of energy to resolve the "Ultraviolet Catastrophe" and Albert Einstein's explanation of the Photoelectric Effect.

It walks through the non-relativistic Schrödinger wave equation and contrasts it with relativistic variants, such as the Klein-Gordon and Dirac equations. From there, it explores the counterintuitive principles of the quantum world: wave-particle duality, Heisenberg's Uncertainty Principle, quantum tunneling, superposition (exemplified by Schrödinger's Cat), and quantum entanglement.

The latter half details various competing interpretations of these phenomena, including the Copenhagen interpretation, Superdeterminism, the Many-Worlds interpretation, Pilot-Wave (De Broglie-Bohm) theory, Objective Collapse theories, and Quantum Bayesianism (QBism).

Chapter Three: Quantum Physics

Ψ

"A wave function walks into a bar. The bartender looks at him and says, "How much have you had to drink?" The wave function replies, "I'm not certain...", and immediately collapses."

I'm using the term *Quantum Physics* to encompass Quantum Field Theory, Quantum Electrodynamics, and Quantum Chromodynamics. These combined theories study the universe at its most fundamental level, describing how energy and matter behave at atomic and subatomic scales. It departs from classical physics by showing that energy is quantized and that systems exist in probabilistic states – superpositions. We looked at some aspects of these theories under the four fundamental forces, so I'll try not to repeat myself.

The timeline is interesting in that it spans over 40 years, explaining and expanding upon basic quantum mechanics to describe particles and fields, leading to the Standard Model of Particle Physics – an incredible achievement.

Quantum Physics Timeline

The timeline I'm using begins in 1927 with Paul Dirac laying the foundation for quantum electrodynamics (QED), a complete theory developed in the late 1940s by Feynman, Schwinger, and Tomonaga. Quantum Field Theory (QFT) expanded to include other forces, with quantum chromodynamics (QCD) being developed in 1972 by Fritzsch, Gell-Mann, and Leutwyler to describe the strong nuclear force, leading to the development of the Standard Model of Particle Physics.

Quantum Field Theory (QFT)

The history of QFT began in the 1920s with Paul Dirac's combination of quantum mechanics and special relativity, which created a new framework for *fields*. In 1927, Paul Dirac published "*The quantum theory of the emission and absorption of radiation*," laying the foundation for QFT and coining the term "*quantum electrodynamics*". In the 1940s and 1950s, renormalized QED was introduced, resolving problems in earlier attempts and paving the way for the theory's success.

Quantum Electrodynamics (QED)

In 1927, Dirac's work provided the first systematic procedure for quantizing the electromagnetic field, which served as a model for subsequent field theories. In the late 1940s, Richard Feynman, Julian Schwinger, and Shin'ichirō Tomonaga independently developed the complete version of QED. In 1948, QED was unveiled as a complete theory that explains the interactions between atoms and light. In 1949, Freeman Dyson proved that the different versions of QED developed by Feynman, Schwinger, and Tomonaga are mathematically equivalent. Today, QED is one of the most accurately tested theories in all of science, with its predictions matching experimental results with high precision, for example, the electron's magnetic moment.

Quantum Chromodynamics (QCD)

In the early 1970s, the theoretical framework for the strong nuclear force, QCD, was developed to describe the interactions between quarks mediated by force-carrying particles called gluons. In 1972, the term "*Quantum Chromodynamics*" (QCD) was introduced, and the theory was published by Harald Fritzsch, Murray Gell-Mann, and Heinrich Leutwyler.[23] In the 1970s, QCD, along with the electroweak theory, became part of the Standard Model of Particle Physics. Today, QCD helps explain phenomena such as quark confinement, in which quarks are never observed in isolation, and the behavior of subatomic particles.

Quantum Field Theory (QFT)

Quantum Field Theory is the fundamental language physicists use to describe the universe at its most basic level. It's the framework that successfully merges quantum mechanics, special relativity, and the concept of forces into a single coherent picture. It's the parent theory of Quantum Electrodynamics and the foundation for the Standard Model of Particle Physics.

In quantum field theory, an "excitation" is a localized increase in energy within a quantum field, essentially representing a *particle* in that field. When a quantum field is disturbed or energized in a specific way, it creates a localized disturbance that behaves like a particle, with properties such as mass and energy, which we perceive as a fundamental particle, such as an electron or a photon.

The energy that causes an excitation is simply energy transferred from another source, which can be in the form of a collision with another particle, radiation, or even just fluctuations in the vacuum energy causing the field to move from its ground state to a higher energy level, manifesting as a particle excitation, or “wobble”; essentially, the excitation is a "wave" in the field with a discrete amount of energy, often referred to as a quantum of energy.

This elegant idea immediately solves a problem. In Einstein's special relativity, energy can be converted into mass and vice versa. This means particles can be created and destroyed. Traditional quantum mechanics struggled with this, but in QFT, it's natural: just add more energy to a field to create more ripples (particles), or a ripple can lose energy and disappear (annihilation).

QFT regards particles as excitations of underlying quantum fields that permeate all of space. These excitations are "quantized," meaning they can only occur in discrete packets of energy, like steps on a staircase, rather than at any arbitrary level. When a field is in its lowest energy state (no excitations), it is called the "ground state." An excitation represents a higher-energy state of the field, which we interpret as a particle. Imagine the electron field as a "field of possibilities" throughout space. An electron is a particle excitation within this field, meaning a localized disturbance with specific energy and momentum that behaves like a particle we can observe.

In QFT, operators mathematically describe the creation and annihilation of particles, which corresponds to exciting or de-exciting the underlying field. When particles interact, it is interpreted as the exchange of excitations between different quantum fields. Imagine the entire universe as a vast, invisible ocean. This "ocean" is the quantum field, which doesn't end at any boundary. It exists at every point in spacetime. Even in the vacuum of deep space, the field remains, just in its calm, lowest-energy state (its "ground state").

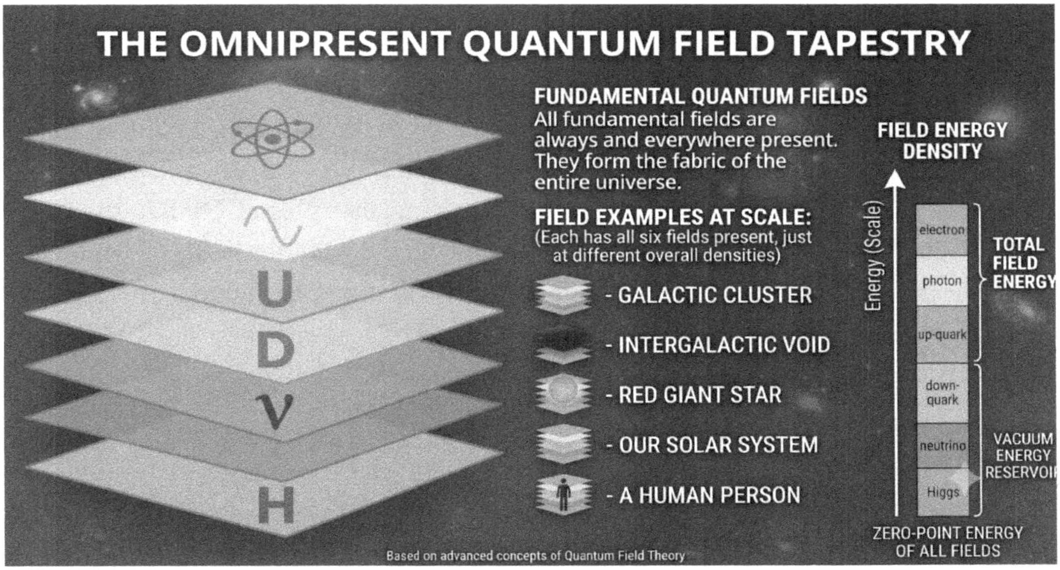

Think of the field as the still ocean and the particle (like an electron) as a ripple in that ocean. If you add a specific amount of energy to a point in the "electron field," a ripple appears, and we measure that ripple as a particle (an electron). If the ripple moves, we see the particle "moving." When the particle and its antiparticle meet (an electron and a positron), the two ripples cancel each other out, and the ocean at that spot becomes calm again (releasing the energy as a ripple in a *different* field, like the photon field). The "particle" is just the *manifestation* of the field vibrating. Fields are fundamental, and particles are not.

The theory is necessary because the old idea of particles as tiny billiard balls fails to explain reality. How can a particle, such as a photon, be "created" when you turn on a flashlight and "destroyed" when it hits a wall? In QFT, this is easy: The photon field is always on everywhere. Turning on the flashlight just "plucks" the field (adds energy) and creates a ripple (a photon). The wall then *absorbs* that ripple's energy, and the field at that location returns to a calm state.

To complete the picture, the universe isn't just *one* ocean. It's a whole set of fields, layered on top of each other and coexisting at every point in space. There is one field for each fundamental particle: an electron field, a photon field, quark fields, a Higgs field, etc. The Standard Model recognizes 17 distinct fundamental fields that permeate the universe, corresponding to 12 matter (fermion) fields and 5 force/Higgs (boson) fields.

All these fields permeate space and interact with one another. A ripple in the electron field can cause one in the photon field, which is just an electron emitting a photon of light. So, when you look at your hand, you're not seeing a collection of tiny, solid particles. You're seeing a complex, localized, interacting dance of vibrations in a dozen different quantum fields.

QFT is inherently quantum. You can never say with certainty what will happen in an interaction. You can only calculate the probability of a specific outcome. The Standard Model is the triumphant result of applying the rules of QFT to the known particles and forces. It's a "theory of *almost* everything." Matter fields (fermions) are what we think of as physical substance. Their excitations are quarks and leptons – such as electrons and neutrinos. Force Fields (bosons) mediate the fundamental forces. Their excitations are the force-carrying particles:

- ***Photon Field*** → Photons (Electromagnetism)
- ***Gluon Fields*** → Gluons (Strong Nuclear Force)
- ***W & Z Fields*** → W and Z Bosons (Weak Nuclear Force)
- ***Higgs Field*** → A unique scalar field that gives fundamental particles their mass by "dragging" on them as they move through it.

Theories built on QFT, like QED, have made the most precise and best-tested predictions in the history of science. For example, QED predicts the electron's magnetic moment to within about one part in a trillion. Experiments match perfectly, and QFT naturally predicted the existence of antimatter for every particle.

While QFT has successfully described three of the four fundamental forces, the glaring exception is gravity. When physicists try to apply the rules of QFT to gravity – to create a "graviton field" – the mathematics breaks down and spits out nonsensical infinities. Reconciling quantum field theory with Einstein's theory of gravity is seen as the single biggest challenge in theoretical physics today. The search for a "theory of quantum gravity," such as String Theory or Loop Quantum Gravity, is the quest for a framework more fundamental than QFT.

Quantum Electrodynamics (QED)

Quantum Electrodynamics is the quantum field theory that describes the interaction between light and matter. It is one of the most accurate and well-tested theories in all of physics. In essence, QED combines the principles of quantum mechanics with special relativity to provide a comprehensive framework for understanding electromagnetic interactions at the subatomic level.

QED explains how electrically charged particles, such as electrons, interact by exchanging photons, the quantum particles of light. This theory is fundamental to our understanding of a wide range of physical phenomena, from the emission and absorption of light by atoms to the scattering of electrons and positrons.

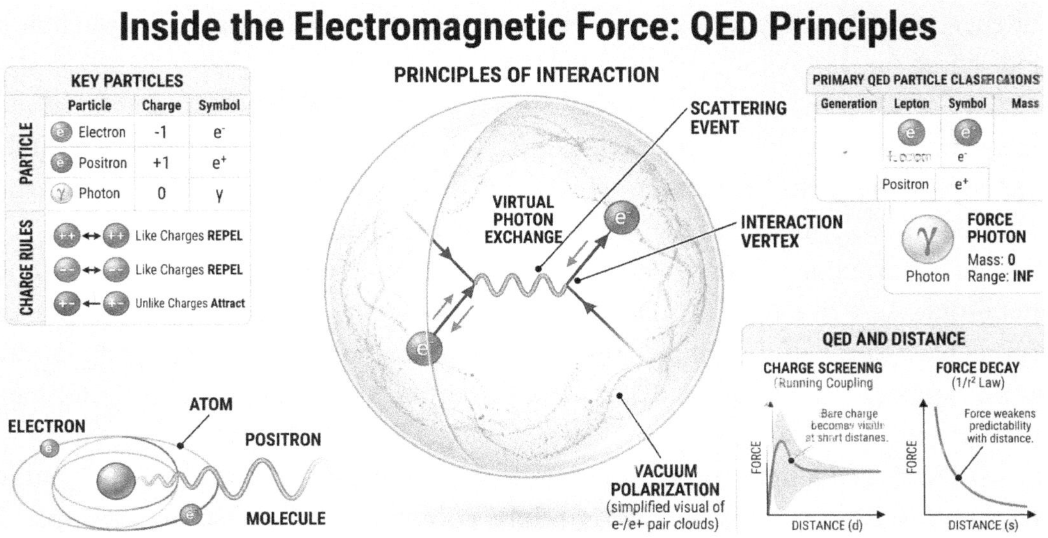

The fundamental entities that govern the world of QED are photons, electrons, and positrons. Electrons have a negative electric charge, while their antimatter counterparts, positrons, have a positive charge. Photons are the force-carrying particles of the electromagnetic field. They are massless and travel at the speed of light (c). Photons are exchanged between charged particles, mediating the electromagnetic force. Because photons are massless, the electromagnetic force has an infinite range, decreasing in strength as the square of the distance between objects. A revolutionary tool for understanding and calculating the interactions in QED is the Feynman diagram, developed by physicist Richard Feynman. These diagrams provide a simple and intuitive way to visualize complex particle interactions.

QED is a quantum *field* theory, meaning that particles are excitations of underlying fields that permeate all of space. For example, an electron is an excitation of the electron field, and a photon is an excitation of the electromagnetic field. Interactions are mediated by "virtual" particles – short-lived particles that pop into and out of existence, transferring energy and momentum between interacting particles.

Early in its development, calculations in QED were plagued by infinities, which rendered the theory nonsensical. A mathematical technique called renormalization was developed to address this issue. In simple terms, renormalization is a procedure that absorbs these infinities by redefining fundamental quantities like mass and charge, yielding finite, highly accurate predictions.

Key figures in QED's formulation include Paul Dirac, whose work in the late 1920s laid the foundation by developing a quantum theory of the electromagnetic field. In the 1940s, Richard Feynman, Julian Schwinger, and Sin-Itiro Tomonaga independently developed the QED framework and the renormalization technique, and shared the 1965 Nobel Prize in Physics.[24]

One of the most famous experiments supporting QED was Millikan's oil-drop experiment, which measured the elementary charge of the electron. The experiment was a pioneering 1909 study conducted by Robert Millikan and Harvey Fletcher, who successfully measured the elementary electric charge of a single electron.

The experiment comprises a sealed chamber containing two parallel horizontal metal plates. A fine mist of microscopic oil droplets is sprayed into the chamber's top section. Some of the droplets fall through a tiny hole in the upper plate into the space between the plates, and individual droplets are observed with a calibrated microscope.

Millikan used a light source positioned perpendicular to the microscope, making the illuminated oil drops appear as bright points against a dark background. To charge the droplets, he used X-rays to ionize the air in the chamber. The falling oil drops captured these free electrons and acquired a negative electric charge. The charge on a droplet is determined by measuring its velocity under different conditions.

The process involves two distinct phases. First, the electric field is turned off to allow a droplet to fall under the influence of gravity alone. The droplet reaches a terminal velocity when the downward gravitational force matches the upward forces of buoyancy and air resistance. A measurement is made for the time it takes the drop to pass between specific grid lines in the microscope. Using Stokes' Law, which relates viscous drag to a sphere's velocity and radius, the exact radius and mass of that specific droplet are calculated.

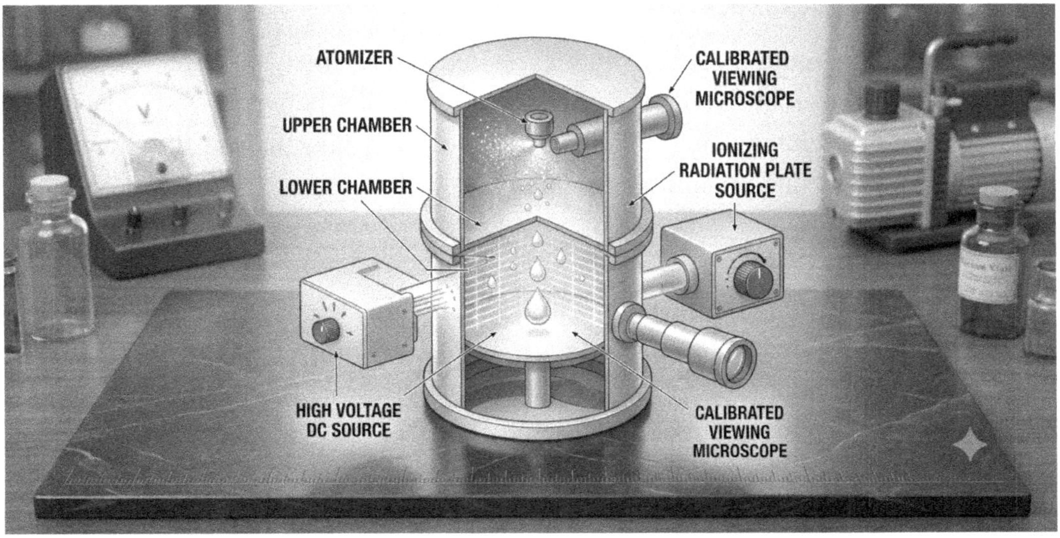

Second, a high-voltage direct current is applied to the metal plates to create a uniform vertical electric field. The upper plate carries a positive charge, and the lower plate carries a negative charge. Because the droplet is negatively charged, the electric field pulls it upward against gravity. The voltage is adjusted until the droplet hovers motionless in mid-air, or its upward terminal velocity is measured.

When a droplet hovers, the upward electric force equals the net downward force, and this balance is expressed with the equation:

$$qE = mg - F_b$$

Where q is the charge on the droplet, E is the electric field strength (voltage divided by the distance between the plates), m is the mass of the droplet, g is the acceleration due to gravity, and F_b is the buoyant force of the air.

Millikan repeated this measurement for thousands of individual drops. When he calculated the total charge q for each droplet, he discovered that every measured charge was an exact integer multiple of a fundamental base value:

$$e = 1.592 \times 10^{-19} \text{ Coulombs}$$

This result proved that electric charge is quantized. A droplet could capture one, two, or forty electrons, but it could never possess a fractional charge like 1.5 electrons. Modern experiments improved this fundamental value to 1.602×10^{-19} C. The small difference in Millikan's initial value stemmed from the use of an inaccurate air viscosity value available at that time.

QED is considered a cornerstone of modern physics: its predictions have been experimentally verified with astonishing precision, making it one of the most successful theories ever devised.

In summary, Quantum Electrodynamics is the theory that explains the quantum world of light and matter. It describes the interactions between charged particles through photon exchange, and its remarkable accuracy has made it an indispensable tool in our quest to understand the fundamental workings of the universe.

Feynman Diagrams

Feynman diagrams are graphical tools invented by physicist Richard Feynman in the late 1940s to visualize and calculate the probabilities of particle interactions in Quantum Field Theory (QFT). They are a pictorial shorthand for the complex mathematical expressions (integrals) used to describe particle scattering and decay.

They are also *different* in that Feynman bypasses what many thought was a sign that the theory was fundamentally wrong – infinities.

Renormalization

The need for Feynman diagrams arose from a crisis in physics: early QED calculations, particularly those describing higher-order processes (involving many virtual particles), yielded infinite expressions. This was known as the "problem of infinities."

In the late 1940s, physicists Julian Schwinger, Sin-Itiro Tomonaga, and Richard Feynman independently developed the renormalization technique. This provided a mathematically rigorous way to handle infinities by isolating and absorbing them into physical, measurable quantities, such as the electron's mass and charge.

Tomonaga’s Approach: Relativistic Wave Equations

Working in Japan during and after World War II, Sin-Itiro Tomonaga reformulated quantum field theory so that special relativity remained explicit at every step. Standard quantum mechanics used a single universal time coordinate, which conflicted with Einstein's relativity. Tomonaga resolved this by defining wave functions across arbitrary space-like surfaces in four-dimensional space-time. This mathematical frame allowed him to isolate and subtract infinite terms without violating relativistic invariance.

Schwinger's Approach: Operator Algebra and Field Equations

Julian Schwinger built a rigorous, operator-based formulation. He used variational principles and Green's functions to track field interactions directly, ensuring relativistic consistency throughout his derivations. Schwinger's calculations were massive and algebraically dense, but they produced exact predictions – such as the slight shift in the electron's magnetic moment – that matched precision laboratory experiments.

Feynman's Approach: Space-Time Diagrams and Path Integrals

Richard Feynman took an intuitive, geometric path. Rather than solving differential equations for changing wave functions, he treated quantum interactions as particles moving along every possible path through space-time. He visualized positrons as electrons moving backward in time and created simple drawings (Feynman diagrams) to track interactions. Each line and vertex in a drawing mapped to a specific mathematical factor, bypassing pages of algebra.

Freeman Dyson's Unification

As his startling contribution is often overlooked, and was derided by such greats as Oppenheimer, I wanted to ensure a brief account. In 1949, Freeman Dyson showed that these three seemingly unrelated approaches were mathematically identical.

Feynman's diagrams worked well in practice but lacked a formal mathematical derivation from standard quantum principles. Schwinger's and Tomonaga's operator mechanics had solid theoretical foundations, but were too complex for physicists to calculate higher-order interactions. Dyson resolved this by:

1. Deriving Feynman's diagram rules directly from the Schwinger-Tomonaga field equations using a time-ordered expansion (the Dyson series) of the scattering matrix (S-matrix).
2. Demonstrating that Feynman diagrams were the exact graphical equivalent of the perturbative terms in field theory.
3. Systematically classifying the infinities into three distinct types – electron self-energy, vacuum polarization, and vertex corrections. He proved that once these three terms were absorbed into the observed charge and mass, calculations at any level of precision remained completely finite.

While Dyson's contribution has been all but forgotten, the others were awarded a Nobel prize – perhaps Dyson should have been the fourth?

However, all credit to Feynman. His genius lay not only in deriving the renormalization procedure but also in creating a simplified graphical shorthand. His diagrams were an intuitive way to understand which terms to include and how to combine them, making QED calculations manageable.

Feynman diagrams are rooted in the path integral formulation of quantum mechanics, which Feynman also developed. The formulation states that a quantum particle does not travel a single, classical path between two points. Instead, it takes every possible path, and the probability amplitude for the particle's ultimate destination is the sum (or integral) over all these paths. A Feynman diagram represents a term in the path integral expansion.

Each diagram depicts a specific way the particles can interact and exchange virtual force carriers. The total probability for an interaction is obtained by summing the contributions of all relevant diagrams for that process. The lines connecting the vertices (interaction points) in a diagram represent virtual particles (force carriers such as photons, gluons, or W and Z bosons) that exist only briefly and mediate the force. They are "virtual" because they violate the standard energy-momentum relationship. This first example uses the relativistic energy-momentum relation (or the full mass-energy equivalence) equation:

$$E^2 = (pc)^2 + (m_0c^2)^2$$

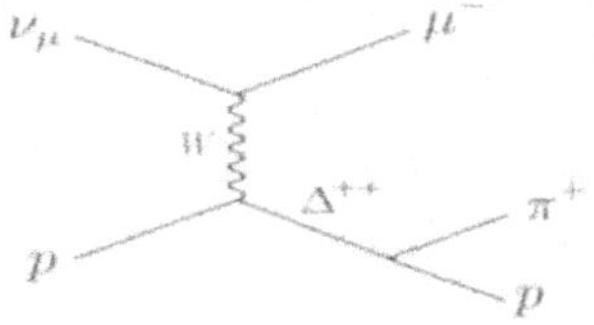

Every Feynman diagram is a schematic spacetime plot with specific elements.

- ***External Lines*** – Straight lines entering or exiting the diagram. Real, measurable particles (electrons, quarks, etc) before or after the interaction.
- ***Internal Lines (propagators)*** – Lines connecting two vertices. Virtual particles (force carriers) mediate the interaction.
- ***Vertex*** – The point where lines meet. Represents the fundamental interaction between particles, governed by a coupling constant.
- ***Arrows*** – Indicate the direction of particle flow. Denotes particle or antiparticle flow. An arrow pointing backward in time is an antiparticle.

Two electrons approach each other and repel (via Møller Scattering). The diagram below shows two straight lines (electrons) coming together at a vertex, exchanging a wavy line (a virtual photon), then separating.

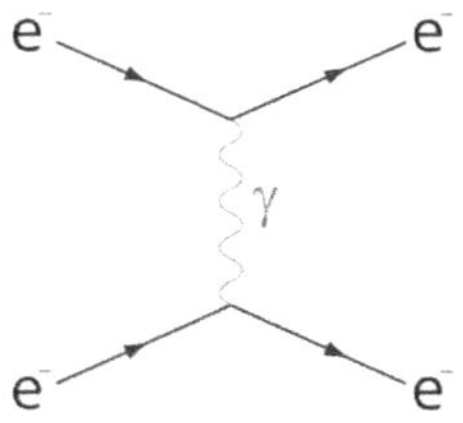

In the diagram below, an electron (e^-) and a positron (e^+) collide and annihilate, producing a photon, which then creates a new particle-antiparticle pair, such as a muon (μ^-) and an antimuon (μ^+). An electron (e^-) line (arrow forward in time) and a positron (e^+) line (arrow backward in time) meet at a vertex, annihilate to form a virtual photon (wavy line), which then splits into the final μ^- and μ^+ lines.

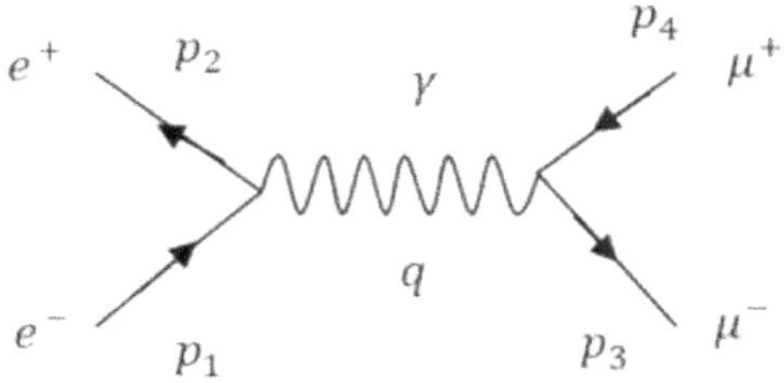

In Beta decay (via the weak nuclear force), a neutron decays into a proton, an electron, and an electron antineutrino. A down quark (d) line changes into an up quark (u) line, emitting a W^- boson (represented by a dashed or wiggly line). The boson immediately decays into an electron and an electron antineutrino ($\bar{v}_e$).

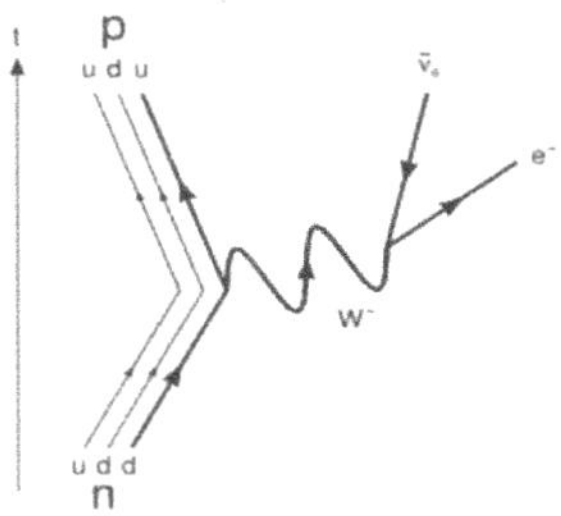

Unlike photons, gluons carry color charge and can interact with *other* gluons, which means QCD diagrams can involve vertices where three or four gluon lines meet – a complexity QED does not have.

Feynman Diagram Examples

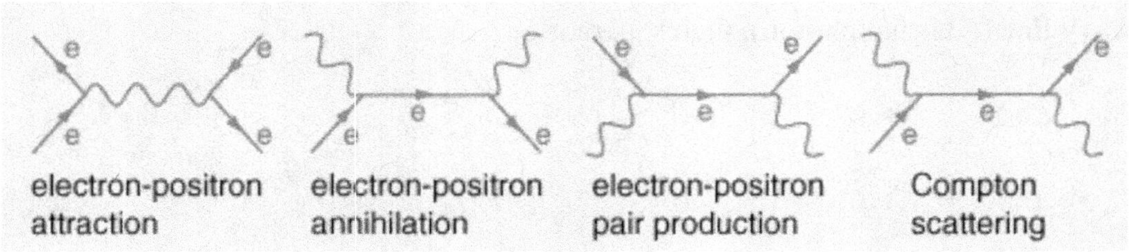

Above, a backward arrow represents the antiparticle; in these cases, a positron. Time moves upward, and a downward arrow is an antiparticle progressing upward (forward in time).

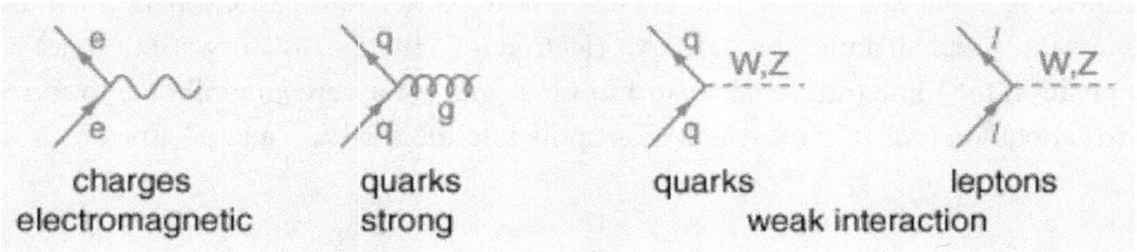

After being introduced in electromagnetic processes, Feynman diagrams were also developed for the weak & strong interactions. Forms of primitive vertices for these 3 interactions are shown above.

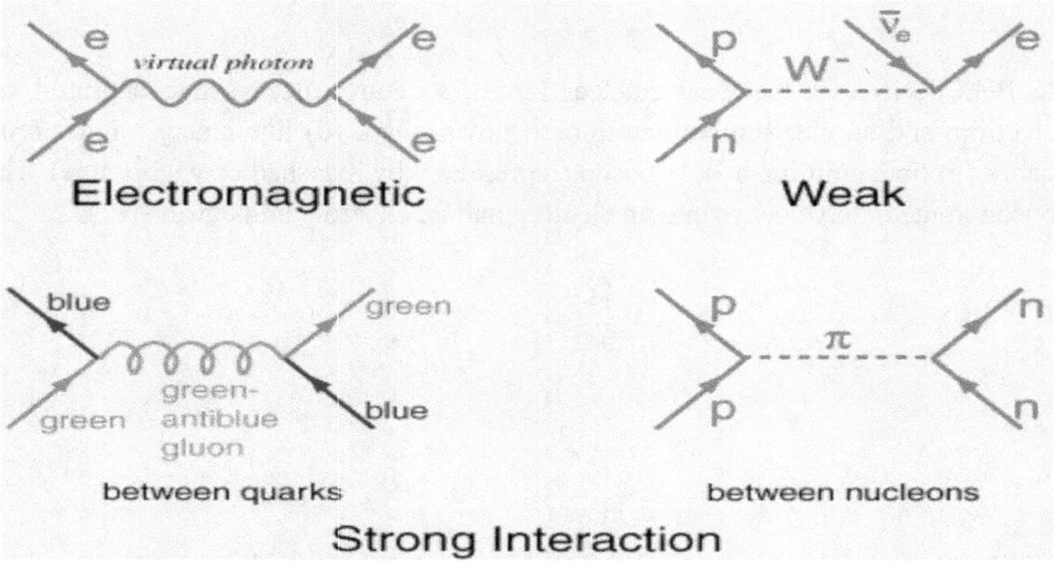

Particle interactions can be represented by diagrams with at least two vertices. They can be drawn for protons, neutrons, etc., even though they are composite objects. The interaction can be visualized as being between their constituent quarks.

Normalization Alternatives

While Richard Feynman, Julian Schwinger, and Sin-Itiro Tomonaga developed the original rigorous renormalization procedure to handle infinities in Quantum Electrodynamics (QED), several alternatives, refinements, and conceptually distinct approaches have since emerged.

Dimensional Regularization is the most widely used and practical alternative to Feynman's original method (which often used a rough "cutoff" to handle infinities before the formal subtraction). Instead of performing calculations in the familiar four spacetime dimensions (three space, one time), dimensional regularization mathematically generalizes the space to d dimensions, where. As long as ϵ (epsilon) is a small, non-zero number, the troublesome integrals remain finite. In these equations, ϵ represents the small shift away from the standard four-dimensional spacetime.

After performing the integration in d dimensions, the infinities appear as poles (terms like (1^{ϵ}). The regularization procedure isolates and subtracts the poles, and then the final result is taken in the limit as $\epsilon \to 0$. It preserves the fundamental symmetries of the theory (such as gauge invariance), which are crucial for QED and the Standard Model, making it the preferred method for modern calculations.

In the Standard Model, Quantum Chromodynamics describes how the strong force binds quarks together. However, at low energies, this force is incredibly strong, making it impossible to isolate quarks or use traditional mathematical approximations (like perturbation theory).

Invented by Nobel laureate Kenneth Wilson in the 1970s, Lattice QCD treats continuous spacetime as a discrete 4-dimensional hypercube grid. It's used to understand how the universe's visible mass is generated, the behavior of matter under extreme conditions (like inside neutron stars or the early universe), and to interpret particle accelerator experiments. Lattice QCD looks at systems as a whole for solving the QCD equations. It does so by converting spacetime into distinct, separate categories or values (for the technical among you, this is called discretization, and means that quarks are placed on the vertices of the hypercube grid, while the gluons are placed on the links connecting the points). Because these computations require massive resources, they are generally performed on world-leading supercomputer facilities.

Effective Field Theory (EFT) is not a direct replacement for renormalization, but a modern conceptual framework that explains *why* renormalization works and enables calculations of phenomena in which the fundamental theory is unknown or too complex. An EFT is built to describe phenomena only up to a particular maximum energy scale (Λ). which represents a theoretical energy threshold, often serving as a cutoff for effective field theories, usually ranging from TeV scales to the GUT scale ($\sim 10^{16}$GeV). Effects from physics at higher energies are systematically included as corrections to the theory. Instead of having to worry about infinite terms from arbitrarily high energies, the EFT framework treats them as coefficients that can be measured experimentally.

The Causal Perturbation Theory (Epstein–Glaser Method) is a mathematically rigorous alternative to the standard approach, emphasizing locality and causality from the beginning. Instead of regularizing and then renormalizing, this method starts with the required physical conditions of locality and causality. It iteratively builds the quantum fields, ensuring that all interactions are locally defined and that causality is obeyed at every step. This avoids introducing infinite terms or cutoffs altogether, offering a cleaner, more rigorous mathematical foundation for QFT.

Quantum Chromodynamics (QCD)

Quantum Chromodynamics is the fundamental theory of the strong nuclear force. It is a cornerstone of the Standard Model of Particle Physics. Where other forces describe how atoms and molecules hold together, QCD explains what keeps the very nucleus of the atom together. It details the interactions between the most fundamental particles (quarks) and the force-carrying particles that "glue" them together, known as gluons.

QCD, like all quantum field theories, describes a set of fundamental particles and a force that acts between them. Quarks are the basic particles of matter (fermions) that experience the strong force. They are the building blocks of composite particles called hadrons (such as protons and neutrons). Quarks come in six "flavors": up, down, charm, strange, top, and bottom. (All stable matter in the universe is composed of just up and down quarks.) The three current quark bare masses sum to only ~ 9–$12\ \mathrm{MeV}/c^2$ (1% of the proton's $938\ \mathrm{MeV}/c^2$ mass), with the remaining 99% generated dynamically via chiral symmetry breaking and QCD binding energy. They possess familiar properties such as mass and electric charge, but they also carry a unique new type of charge.

Gluons are the force-carrying particles (gauge bosons) of the strong force. They are the "messengers" of the strong force, "gluing" quarks together. Crucially, unlike photons (which carry the electromagnetic force but have no electric charge), gluons *also* carry the strong force's charge. This single fact is key to all QCD's unique properties.

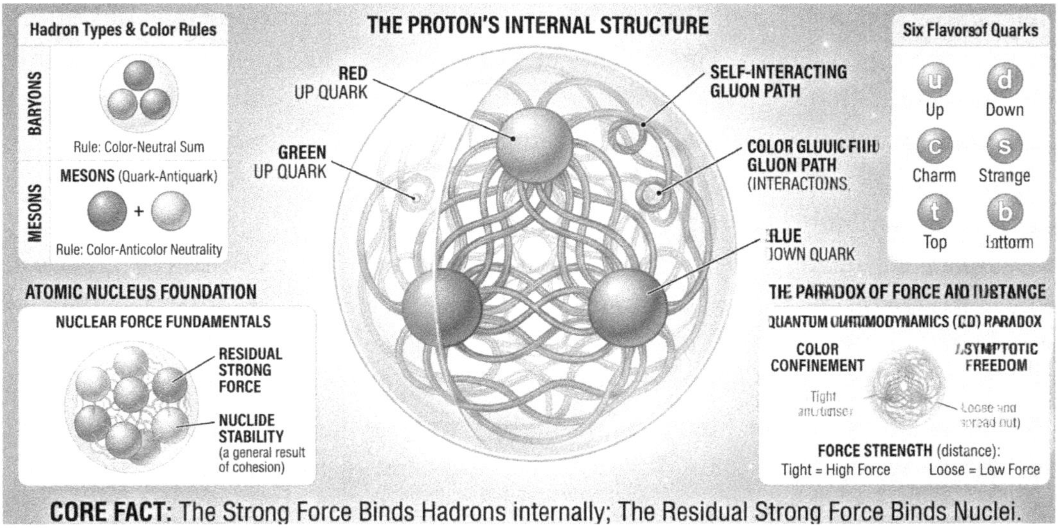

The strong force's charge is called "color charge." This name is purely an analogy; it has absolutely nothing to do with visible light or color. It's a label for a quantum property. Where electromagnetism (Quantum Electrodynamics, or QED) has one type of charge (positive and negative), QCD has three types: Red, Green, and Blue. Just as positive and negative electric charges attract, these three color charges do as well. For matter to be stable, it must be "color neutral" – a central QCD rule. A particle can be color-neutral in two ways:

1. Baryons (like protons) can form a combination of three quarks, one of each color (Red + Green + Blue = "White" or neutral).
2. 1. Mesons are composed of a quark and an antiquark. The antiquark carries an "anti-color" (e.g., Red + Anti-Red = "White" or neutral).

This three-charge system is described mathematically as an $SU(3)$ gauge theory. The "3" refers to the three colors, and "SU" (Special Unitary group) is the mathematical description of the symmetry (see the background materials).

The most famous property of QCD is color confinement. This principle states that no particle with color charge (such as a single quark or gluon) can ever be isolated and observed. They are permanently *confined* inside their color-neutral hadrons. This happens because of the unique way the strong force behaves at short distances, which is a direct result of gluons carrying their own charge.

In electromagnetism, the force is carried by photons, which are chargeless. The force lines spread out, and the force gets *weaker* with distance (following an inverse-square law). In QCD, the force is carried by color-charged gluons. Because gluons carry color, they interact *with each other*. This self-interaction "pinches" the force field into a narrow "string" between two quarks. The energy in this flux tube is *constant* along its length. This means the force required to separate the quarks does not decrease with distance. It remains constant and incredibly strong.

If you try to pull a quark out of a proton, you stretch this energy string. You have to put in more and more energy. Eventually, you put *so much* energy into the string that, via $E=mc^2$, it becomes more energetically favorable for the universe to create a new quark-antiquark pair from the vacuum. The string "snaps," but you don't get a free quark. You get a *new* color-neutral hadron (a meson). You started with one proton and ended with a proton and a meson. The quark is still confined.

The second, and equally bizarre, property of QCD is the exact opposite of confinement. While the force is infinitely strong at large distances, it becomes shockingly weak at very short distances (or, equivalently, at very high energies). This discovery earned David Gross, David Politzer, and Frank Wilczek the Nobel Prize in Physics in 2004.[25]

In QED, a "bare" electron is surrounded by a cloud of virtual electron-positron pairs that pop in and out of existence. These virtual pairs are polarized, and they *screen* the electron's charge. From a distance, this cloud makes the electron's charge appear *weaker* than it actually is. In QCD, a quark is also surrounded by virtual quark-antiquark pairs (which screen the charge). But it is *also* surrounded by a cloud of virtual gluons. Because gluons carry color, their self-interactions create an "anti-screening" effect. This gluon cloud *amplifies* the color charge, making it appear stronger at a distance. The result is that when you are far from a quark, you see the full effect of this "anti-screening" gluon cloud, and the force appears immensely strong (confinement). When you get incredibly close to a quark (like by smashing particles together at the LHC), you "penetrate" this cloud. You see the "bare" quark, and the force becomes very weak.

This is why quarks *inside* a proton can rattle around almost as if they were free particles, only to be yanked back by an incredible force if they stray too far. The fundamental difference between electromagnetism and the strong force lies in their mathematical structures.

The Strong CP Problem

The Strong CP Problem is one of the deepest unsolved puzzles in particle physics. It asks a simple question: Why does Quantum Chromodynamics (QCD), the fundamental theory of the strong nuclear force, preserve CP symmetry when the mathematics of the theory allows it to be violated?

What is CP Symmetry?

- ***C (Charge Conjugation):*** Swapping a particle with its antiparticle (e.g., reversing the sign of electric charge).
- ***P (Parity):*** Inverting spatial coordinates (x, y, z $\rightarrow -x, -y, -z$), equivalent to looking at a physical process in a mirror.
- ***CP Symmetry:*** The rule that the laws of physics should remain identical if you simultaneously swap all particles with their antiparticles and view the interaction in a spatial mirror.

The electroweak force (the unified electromagnetic and weak interaction) is known to violate CP symmetry. This violation is observed experimentally in the decays of neutral kaons (K-mesons) and B-mesons, and it is parameterized naturally within the Standard Model by a complex phase in the Cabibbo–Kobayashi–Maskawa (CKM) matrix. Because the strong force operates within the same Standard Model, one would naturally expect CP violation to occur in strong interactions as well.

In QCD, the non-trivial vacuum structure of the gluon fields introduces an extra topological term into the QCD Lagrangian:

$$\mathcal{L}_\theta = \theta \frac{g_s^2}{32\pi^2} G^a_{\mu\nu} \tilde{G}^{a\mu\nu}$$

Where:

- g_s is the strong coupling constant.
- $G^a_{\mu\nu}$ is the gluon field strength tensor.
- $\tilde{G}^{a\mu\nu} = \frac{1}{2}\epsilon^{\mu\nu\alpha\beta} G^a_{\alpha\beta}$ is its dual tensor.
- θ is a bare fundamental parameter (an angle between 0 and 2π).

Because the term contains the Levi-Civita symbol $\epsilon^{\mu\nu\alpha\beta}$, it breaks both **P** (parity) and **T** (time reversal), which implies an automatic violation of **CP** (by the CPT theorem – see *Chapter Four*).

Furthermore, when physicists perform chiral transformations to diagonalize the quark mass matrices, an additional CP-violating phase enters from the quark sector:

$$\theta_{\text{quark}} = \arg(\det M_q)$$

The physically measurable parameter governing CP violation in the strong force is the sum of these two unrelated contributions:

$$\bar{\theta} = \theta + \arg(\det M_q)$$

Mathematically, $\bar{\theta}$ is a free parameter. There is no requirement in the Standard Model for $\bar{\theta}$ to be zero; natural values would be somewhere on the order of ~ 1. If $\bar{\theta}$ were non-zero, it would give the neutron an Electric Dipole Moment (EDM) – a slight separation of positive and negative charge along its spin axis. The predicted EDM of the neutron as a function of $\bar{\theta}$ is approximately:

$$|d_n| \approx 3 \times 10^{-16}\bar{\theta} \quad e \cdot \mathrm{cm}$$

Physicists have spent decades measuring the neutron EDM using ultra-cold neutron traps. Current experimental bounds place the neutron EDM at:

$$|d_n| < 1.8 \times 10^{-26} \quad e \cdot \mathrm{cm}$$

Plugging this bound back into the formula constrains the value of $\bar{\theta}$:

$$\bar{\theta} \lesssim 10^{-10}$$

The Strong CP Problem is not that CP violation cannot happen in QCD, but that $\bar{\theta}$ is measured to be zero to at least ten decimal places. For $\bar{\theta} = \theta + \arg(\det M_q)$ to be less than 10^{-10}, the bare vacuum angle θ from the gluon sector and the complex phase$\arg(\det M_q)$ from the electroweak quark mass matrices would have to cancel each other out with uncanny precision – despite originating from two completely distinct sectors of the Standard Model.

In theoretical physics, when a parameter is extraordinarily close to zero without an underlying symmetry forcing it to be zero, it is considered an unnatural fine-tuning. Physicists have proposed three main avenues to resolve the problem:

1. **The Peccei–Quinn Mechanism and the Axion (Most Popular):** In 1977, Roberto Peccei and Helen Quinn proposed an additional global symmetry, denoted $U(1)_{\mathrm{PQ}}$. In this framework, $\bar{\theta}$ is no longer a static, fixed constant. Instead, it becomes a dynamic quantum field. The field naturally settles into the minimum of its potential energy, which dynamically drives $\bar{\theta} \to 0$ and restores CP symmetry.
   Frank Wilczek and Steven Weinberg realized that the spontaneous breaking of this symmetry generates a new pseudo-Goldstone boson: the axion. Because axions would be light, neutral, and weakly interacting, they are currently one of the leading candidates for cold dark matter.
2. **A Massless Up-Quark ($m_u = 0$):** If the lightest quark (the up-quark) had zero bare mass, the phase $\arg(\det M_q)$ could be rotated away completely through a chiral transformation without altering physical predictions, rendering $\bar{\theta}$ unphysical and eliminating CP violation.
   However, high-precision lattice QCD calculations and meson mass ratios have established that the up-quark has a non-zero mass ($m_u \approx 2.2\ \mathrm{MeV}/c^2$), largely ruling out this solution.
3. **Spontaneous CP Violation at High Energies:** In these models, CP is an exact fundamental symmetry of nature at very high energy scales ($\bar{\theta} = 0$ natively). CP

symmetry is then broken spontaneously at lower energies in a way that generates the CKM phase observed in weak decays, while keeping the strong $\bar{\theta}$ parameter suppressed at high loop orders.

QCD is built on the $\mathrm{SU}(3)$ gauge symmetry group. Unlike the single-charge type in electromagnetism, QCD requires eight different gluons to act as the force carriers that mediate interactions between these colored quarks.

Gluons are unique because they carry color charge themselves, which allows them to interact with other gluons. This self-interaction causes the strength of the strong force to vary with the distance between particles. If you try to separate two quarks, the energy in the gluon field increases until it becomes energetic enough to create a new quark-antiquark pair. This behavior leads to color confinement, a physical principle that ensures that you never find a quark or a gluon in isolation.

At very high energy scales or extremely short distances, the coupling constant of the strong force decreases – known as asymptotic freedom. It treats quarks as independent, non-interacting particles during high-energy experiments. Perturbative mathematics is used to calculate these high-energy results. However, at lower energies, the force becomes so strong that you must rely on non-perturbative methods like Lattice QCD to calculate the mass and structure of hadrons. This energy dynamic explains why nearly all the visible mass in the universe arises from gluon fields rather than from the quarks' masses.

Lattice QCD

Quantum Chromodynamics (QCD) is the mathematical framework that describes the strong force, the interaction that binds quarks into protons and neutrons. In the physical universe, spacetime is smooth and continuous. The equations governing these quark interactions are too complex to solve using standard analytical mathematics because the force grows stronger as the distance between the particles increases.

Lattice QCD is a computational technique for solving these equations by altering the geometry of spacetime. Physicists replace continuous spacetime with a rigid, four-dimensional grid of points called a lattice. In this framework, quarks exist only on the individual points of the grid, while gluons, the force-carrying particles, occupy the links connecting those points.

Replacing a continuum with a finite grid transforms complex differential equations into a massive array of discrete numerical operations. Supercomputers evaluate these equations arithmetic step by step. To produce data that reflects the real world, researchers run the calculations multiple times using progressively smaller grid spacings, then mathematically calculate the trajectory of the results as the grid spacing shrinks to zero. Well, that's the simple explanation – what follows gets a bit complex, so feel free to skip it.

You define the theory through a Lagrangian density that governs the interactions of color-charged particles.

$$\mathcal{L}_{QCD} = -\tfrac{1}{4}F^a_{\mu\nu}F^{a\mu\nu} + \sum_f \bar{\psi}_f(i\gamma^\mu D_\mu - m_f)\psi_f$$

In this expression, $F^a_{\mu\nu}$ represents the gluon field strength tensor and ψ_f represents the quark fields of flavor f. While this formulation works well for high-energy processes where the coupling constant is small, it fails at low energies. At the scale of approximately 1 GeV, the strong coupling constant α_s becomes large. This prevents you from using perturbative methods such as Feynman diagrams to calculate hadron masses or decay constants.

Lattice QCD provides a non-perturbative method to solve QCD equations by discretizing spacetime. Instead of a smooth continuum, you use a four-dimensional Euclidean grid with a finite spacing, denoted as a. This discretization turns the infinite-dimensional path integral of the continuum theory into a finite-dimensional integral that you can evaluate using numerical methods.

You place the quark fields $\psi(x)$ on the lattice sites. You represent the gluon fields as gauge links $\mathrm{U}_\mu(x)$, which are elements of the $\mathrm{SU}(3)$ color group. These links reside on the edges connecting the sites. This structure preserves gauge invariance on the grid. To perform calculations, you replace the derivatives in the continuum Lagrangian with finite difference operators. The Wilson action is a common choice for the gluon part of the lattice: $S_G = \beta \sum_P (1 - \tfrac{1}{3}\mathrm{Re\ Tr}\ U_P)$. The Wilson action is a foundational mathematical formula created by physicist Kenneth Wilson in 1974. It serves as a discrete, mathematical model used to approximate the Yang-Mills action on a spacetime grid (lattice gauge theory). It allows physicists to study complex quantum interactions using powerful computer simulations.

Here, U_P is the "plaquette," which is the product of four link variables around the smallest possible square on the grid. Or, more simply, a plaquette is the smallest closed loop on a spacetime lattice grid used to calculate physical equations. The variable β relates to the bare coupling constant g through the relation $\beta = 6/g^2$.

Lattice QCD relies on the Path Integral formulation in Euclidean time. You calculate the expectation value of a physical observable O by integrating over all possible configurations of the fields:

$$\langle \mathcal{O} \rangle = \tfrac{1}{Z}\int \mathcal{D}U\mathcal{D}\bar{\psi}\mathcal{D}\psi\mathcal{O}e^{-S[\psi,\bar{\psi},U]}$$

Because the number of degrees of freedom is massive, you use Monte Carlo importance sampling. Computers generate thousands of gauge field configurations that follow the probability distribution e^{-S}. The observable is then averaged over these configurations to obtain the physical result.

The lattice serves as a regulator for the theory. The spacing a acts as an ultraviolet cutoff with a momentum scale proportional to $1/a$. To get results that apply to the real world, you must take the continuum limit. This involves two steps:

1.The Continuum Limit: The simulation is repeated at smaller and smaller values of a while keeping physical quantities like the proton mass constant. The results are then extrapolated to $a = 0$.

2. The Infinite Volume Limit: The physical size of the lattice L is large enough that the particles do not feel the effects of the periodic boundaries. This requires L to be several times larger than the Compton wavelength of the pion.

Lattice QCD has successfully predicted the mass of the proton and other hadrons with a precision of about 1%. It remains the only systematic way to calculate the effects of confinement and chiral symmetry breaking directly from the fundamental QCD Lagrangian.

Understanding Subatomic Particles and Interactions

Atoms are the basic units of matter and consist of a central nucleus containing protons (p^+) and neutrons (n^0) surrounded by electrons (e^-). Protons and neutrons are composed of quarks held together by gluons, which transmit the strong force.

The strong force binds quarks into protons and neutrons and holds the nucleus together. Nucleons are depicted as three simple quarks, but there is actually a complex hierarchy of "dressed" particles and quantum fluctuations from center to cloud, dynamically generating mass and form. In this context, "dressed" means that a particle's observable properties (like its mass and charge) are altered by the dense, continuous cloud of virtual particles and field fluctuations surrounding it.

The Core Atomic Structure

- *Atom:* The fundamental unit of matter.
- *Nucleus:* The dense, positively charged center containing protons and neutrons.
- *Protons (p^+):* Positively charged particles; the number of protons determines the element's identity.
- *Neutrons (n^0):* Neutral particles in the nucleus that provide stability and affect atomic mass.

- ***Electrons (e^-):*** Negatively charged particles in a cloud surrounding the nucleus, governing chemical bonding.

Subatomic Components (The Standard Model)

- ***Quarks:*** Fundamental particles that constitute protons and neutrons (specifically "up" and "down" quarks). They carry "color charge".
- ***Gluons:*** Massless particles that carry the strong force, acting as "glue" to bind quarks together.
- ***Strong Nuclear Force:*** The most powerful force, holding quarks together within protons/neutrons and protons/neutrons together within the nucleus.
- ***Weak Nuclear Force:*** Involved in radioactive decay, changing quarks from one type to another (e.g., in beta decay).

Summary of Subatomic Particles

- ***Protons & Neutrons (Nucleons):*** Comprised of three quarks each; mass of ≈ 1 amu (atomic mass unit).
- ***Electrons:*** Fundamental particles ≈ 0.0005 amu.
- ***Quarks & Gluons:*** Fundamental, indivisible building blocks.
- ***Force Particles (Bosons):*** Gluons (strong force) and W/Z bosons (weak force).

Other Quantum Field Theories

Topological Quantum Field Theory (TQFT)

A Topological Quantum Field Theoryis a quantum field theory that calculates and describes topological invariants. These have the property of a geometric shape or space that remains unchanged under continuous deformations – such as stretching.

This makes TQFTs useless for describing phenomena we usually care about, such as particle scattering at specific energies or the force of gravity. Instead, they provide a mathematical framework for studying the "shape" of space and time, with applications in mathematics and condensed-matter physics. twisting, or bending – provided that you do not tear, cut, or glue the shape. The overall idea is that TQFT is independent of geometry and, unlike standard quantum field theories, doesn't care about distances, angles, or the curvature of spacetime.

Chern-Simons theory is a three-dimensional TQFT that is widely used in condensed matter physics for Fractional Quantum Hall (FQH) states and in mathematics for computing knot and three-manifold invariants, such as the Jones polynomial. Briefly, the Jones polynomial is a mathematical invariant in knot theory. It assigns a unique polynomial, allowing both positive and negative powers, to each oriented knot or link, helping mathematicians determine if two knots are structurally the same.

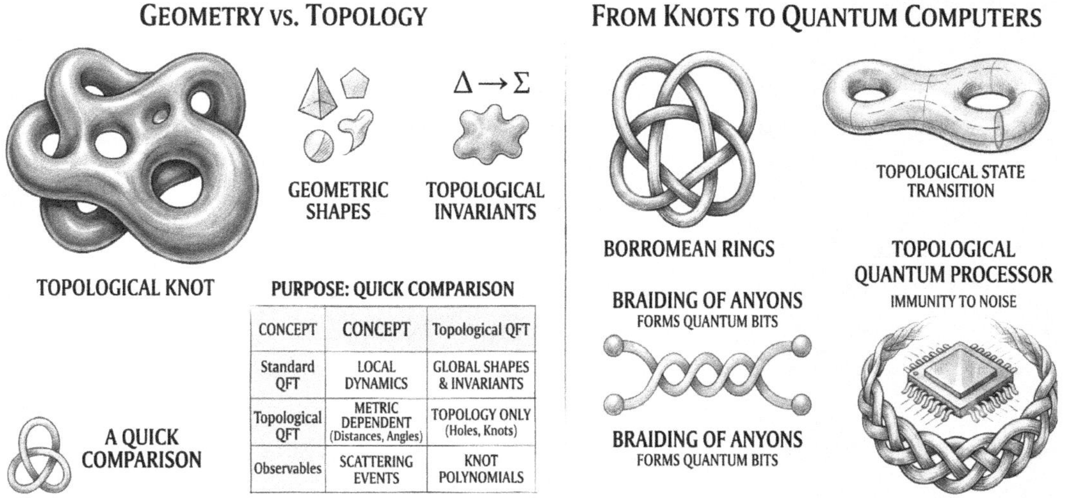

Chern-Simons theory was discovered by Albert Schwarz and named after Shiing-Shen Chern and James Harris Simons.[26] The TQFT vs. "Normal" QFT differences are best seen by comparing their core properties:

Feature	Standard QFT (QED, QCD)	TQFT (e.g., Chern-Simons)
Depends on	Metric (geometry, distances, angles)	Topology (shape, holes, knots)
What it describes	Local dynamics. Particles and forces propagating and interacting in spacetime.	Global properties. An invariant describing the entire spacetime manifold.
Observables	Depend on coordinates, energy, and momentum (e.g., scattering cross-sections).	Topological Invariants (e.g., knot polynomials, manifold invariants).
Dynamics	A complex "Hamiltonian" that describes how states evolve in time.	A trivial Hamiltonian (H=0). There are no locally propagating particles or "dynamics" in the usual sense.

Because a TQFT's Hamiltonian is zero (i.e., it has exactly zero total energy), its states do not "evolve" in time. The only way a system can change is if the topology of the space itself changes.

The best way to understand a TQFT is through something called the Atiyah-Segal axioms. These axioms define it as a mathematical "functor" – broadly speaking, a functor represents a mapping or relationship.

While it sounds a bit complex, the idea is actually not that difficult to understand: a TQFT is a rulebook that can map geometric shapes to algebraic equations, allowing mathematicians to study and solve complex topology problems using algebra. In traditional physics, a quantum field theory describes how particles and fields behave across space and time by tracking their changes second by second. The Atiyah–Segal axioms change this approach entirely. Instead of focusing on the continuous passage of time, they treat physics geometrically by looking at the starting point, the ending point, and the region of spacetime that connects them.

Imagine a physical pipe. The pipe represents a region of spacetime. The two open ends of the pipe represent boundaries in time – for example, the left opening is the universe at 1:00 PM, and the right opening is the universe at 2:00 PM. In this geometric category, the shapes are the boundaries (the circles at the ends), and the paths connecting them (the pipe itself) are called cobordisms.

The magic is in how it "stitches" spacetimes together; If you physically glue two spacetimes (bordisms) together, the TQFT calculation is just the composition (multiplication) of their linear maps. The most famous example is the "pair of pants" bordism. This is a 2D surface that starts with *one* circle and ends with *two* circles. The TQFT assigns a linear map that algebraically describes a state on one circle "splitting" into a state on two circles.

By using these simple rules, any complex, knotted, or twisted spacetime can be "chopped up" into simple pieces (like cylinders and pants), calculated algebraically, and then "glued" back together to get a number – a topological invariant – that describes the entire complex shape.

TQFTs are effective low-energy theories for exotic states of matter called topologically ordered states. The most famous example is the Fractional Quantum Hall effect (FQH). This occurs when a 2D sheet of electrons is cooled to near absolute zero and placed in a powerful magnetic field. The electrons stop behaving as individual particles and enter a collective, strongly correlated quantum "liquid." This state is "topological" because its properties (like its precisely quantized electrical conductance) are robust and don't depend on the material's shape or imperfections. They only rely on the system's global topology.

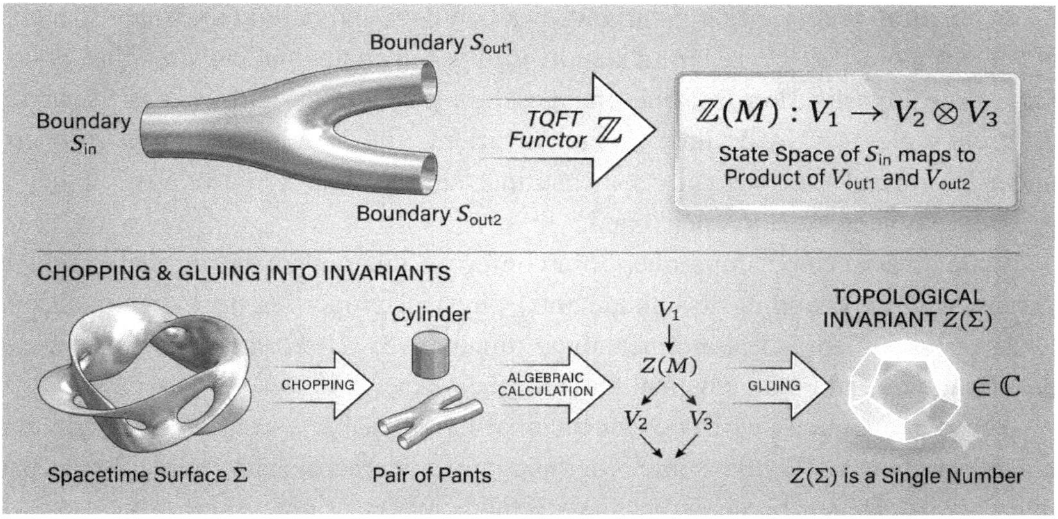

The excitations (quasiparticles) in this state are not electrons, but bizarre particles called anyons. In our 3D world, all particles are either fermions or bosons. Anyons, which can only exist in 2D, are neither. When you swap two anyons, their quantum wave function gets a phase that is *between* that of a fermion and a boson. The TQFT describes the behavior of these anyons.

Conformal Field Theory (CFT)

A Conformal Field Theory (CFT) is a quantum field theory that possesses an extra layer of symmetry: conformal symmetry. This symmetry makes CFTs remarkably well-behaved and often exactly solvable, unlike many more complex QFTs. At its heart, a conformal transformation is a coordinate change that locally preserves angles. Think of zooming in on a map – the shapes of small regions are preserved, even though their sizes change.

The important lesson is that conformal symmetry is stronger than scale invariance alone. While scale-invariant theories look the same at all zoom levels, conformal theories also look the same after these more complex angle-preserving transformations.

CFT is invariant under conformal transformations – those that preserve angles but not necessarily lengths, encompassing rotations, translations, dilations (scaling), and special conformal transformations. CFTs are especially useful in two dimensions, where their infinite-dimensional symmetry algebras allow for exact solutions and classification. They are a key component of string theory, which describes the behavior of fundamental strings.

In quantum theory, scale invariance is a common and natural symmetry because any fixed point of the renormalization group – a mathematical tool that allows systematic investigation into the changes in a physical system as it is viewed at different scales – is, by definition, scale invariant. The defining feature is symmetry under conformal transformations – ensuring that the shapes and relative angles of geometric objects remain unchanged.

Scale invariance is a consequence of conformal invariance: the ability to scale the system without changing its fundamental physical properties. In two dimensions. CFTs possess an infinite-dimensional algebra of conformal transformations, providing powerful mathematical tools for their exact solution.

The rich symmetries of two-dimensional CFTs enable exact solutions, yielding precise predictions. Unlike some other quantum field theories formulated using path integrals, CFTs can be constructed with mathematical rigor. Their critical role in applications, from phase transitions to string behavior, highlights their importance in describing fundamental aspects of physics.

Scalar Quantum Electrodynamics and Chromodynamics

These two concepts are theoretical models in quantum field theory used to study how particles with zero spin (scalar particles) interact via fundamental forces. They are extensions of the standard QFT models (QED and QCD), in which the charged matter particles (such as electrons or quarks) are treated as scalar fields (θ) rather than spinor fields (which have spin ½).

Both theories are based on the principle of gauge invariance, which governs how matter fields interact with force-carrying particles (gauge bosons).

- ***Gauge Group:*** $U(1)$ (the Abelian group of electromagnetism).
- ***Gauge Boson:*** The photon (A_μ), which is massless and uncharged.
- ***Matter Field:*** A complex scalar field (θ) that carries electric charge.

Scalar Quantum Electrodynamics (SQED) is the model describing the interaction between a charged scalar field and the photon. It is the scalar analogue of Quantum Electrodynamics. Scalar Quantum Chromodynamics (SQCD) is a model describing the interaction between a scalar field carrying color charge and gluon fields. It is the scalar analogue of Quantum Chromodynamics.

The major distinction between SQED and SQCD (and the key feature of all non-Abelian gauge theories) is that the gauge bosons (gluons) themselves carry color charge. SQCD is used as a theoretical model to study the complex behavior of non-Abelian gauge theories, particularly to understand confinement and the quantum vacuum in the presence of color charge. It provides a simpler mathematical framework for calculations than full QCD, especially for studying phenomena such as bound-state formation (analogous to mesons or baryons).

The Yukawa Model

The Yukawa Model is a theory in particle physics, proposed by Japanese physicist Hideki Yukawa in 1935, that explained the short-range nature of the strong nuclear force by hypothesizing that it is mediated by a massive particle (the pion, or π-meson).[27] It was the first successful attempt to apply principles of quantum field theory to nuclear forces, demonstrating that the mass of a force-carrying particle determines the range of the force.

In the 1930s, physicists struggled to explain why the atomic nucleus, packed with positively charged protons, did not fly apart due to strong electromagnetic repulsion (the Coulomb force). The force holding the nucleus together (the strong nuclear force) had to be immensely powerful, yet it was known to act only over extremely short ranges (about 1-2 femtometers, or 10^{-15} meters).

Yukawa proposed that the nuclear force arises from the exchange of a virtual particle between nucleons (protons and neutrons), in direct analogy to the electromagnetic force, which is mediated by the massless photon. He reasoned that the short range of the nuclear force meant its mediating particle could not be massless like the photon. He named this hypothetical particle the meson.

Yukawa's predicted particle, the π-meson, was discovered in 1947, validating his theoretical model. While the Yukawa Model was essential for understanding the nuclear force, it has been superseded as the fundamental theory of the strong force by Quantum Chromodynamics. In QCD, the force acts between quarks and is mediated by massless gluons. The Yukawa nuclear force is now understood as a residual effect of the strong force (analogous to the weak van der Waals forces between neutral molecules that arise from transient shifts in electron density).

Despite being superseded, the concept of a Yukawa coupling is a fundamental interaction in the Standard Model of Particle Physics. It describes how the Higgs field (a scalar field) interacts with fermions (like quarks and leptons). When the Higgs field acquires its vacuum expectation value, the Yukawa coupling mechanism is precisely what gives fundamental fermions their mass.

Yang-Mills Theory

The Yang-Mills Theory is the fundamental mathematical framework underlying the Standard Model, which describes three of the four known fundamental forces: the electromagnetic, weak nuclear, and strong nuclear forces. It is a revolutionary generalization of QED and is the blueprint for all modern gauge theories.

The entire Yang-Mills theory is built on extending the concept of gauge invariance from the simple Abelian group of QED to a more complex non-Abelian group. (see *Abelian* and *Gauge Theory* in the background materials). In QED, the force is mediated by the photon. The theory remains physically unchanged (invariant) if you "rotate" the electron field by a phase angle (θ) that can vary locally in spacetime. This is an Abelian symmetry – the order of two consecutive transformations doesn't matter (like $(A + B = B + A)$.

In 1954, physicists Chen Ning Yang and Robert Mills proposed generalizing this idea to a non-Abelian symmetry group.[28] That is, the order in which you perform two symmetry transformations *does* matter, like matrix multiplication $(AB \neq BA)$. To maintain invariance under this more complex non-Abelian symmetry, the theory requires the existence of new force-carrying particles (gauge bosons). This led to the mathematical necessity that symmetry dictates the *exact form* of the forces and interactions, thereby providing a powerful method for constructing fundamental theories.

The non-Abelian nature has two consequences that distinguish the Yang-Mills field from the electromagnetic field. Instead of just one photon, the non-Abelian symmetry requires multiple gauge bosons.

The gauge boson self-interaction (color change) is the most crucial consequence of the non-Abelian symmetry. The gauge bosons themselves carry charge. In QED, the photon does not have an electric charge, so it cannot directly interact with other photons. In the Yang-Mills Theory, gluons carry color charge and can directly mediate interactions with other gluons. This results in three-gluon and four-gluon interaction vertices in the Feynman diagrams, which are responsible for the complex phenomena of the strong force, specifically confinement (trapping quarks inside protons) and asymptotic freedom (the force weakening at high energies).

The Yang-Mills framework provides a precise description of the interactions between massless particles. However, to maintain gauge invariance, the theory required that all gauge bosons must be massless. While the photon and gluons are massless, the W and Z Bosons of the weak force are known to be massive. This difficulty was solved by introducing the Higgs mechanism. This mechanism allows the underlying Yang-Mills symmetry to be maintained while the W and Z bosons acquire mass through interactions with the ubiquitous Higgs field. This final piece completed the Standard Model, which is built entirely on the mathematical foundation of Yang-Mills theory, using the combined gauge group $SU(3) \times SU(2) \times U(1)$.

Source Theory

The Nobel Prize-winning physicist Julian Schwinger developed Source Theory in the late 1960s and 1970s (detailed in his 1970 book, *Particles, Sources, and Fields)*.[29] Schwinger created source theory as an alternative to mainstream quantum field theory. Instead of dealing with complex operators and fields, his approach focused on physically observable "sources" that emit and absorb particles. A source field is a mathematical construct that acts like a "tap," generating particles at a specific location, much as a current source generates an electric current in a circuit. By introducing a source field, physicists can calculate a particle's propagation through space and time.

Source fields are instrumental in the path-integral formulation of QFT, where the behavior of a field is computed by integrating over all possible paths a particle could take, with the source term influencing the probability of each path. Imagine wanting to study the creation of an electron-positron pair. You could introduce a source field at a specific point in spacetime that represents the location where the pair is created, then calculate how the electron and positron propagate away from that point using the propagator derived from the source field.

The source field is not a physical object, but rather a mathematical tool used to simplify calculations. Source fields are often used alongside perturbation theory to introduce small perturbations to the system and to study complex interactions. (Perturbation theory in quantum mechanics is a set of approximation schemes used to find the energy levels and wavefunctions of a complex, unsolvable system by building upon a similar, exactly solvable one. It relies on treating the difference between the two systems as a small "perturbation").

The source field appears on the right-hand side of the field equations, acting as a driving force for the field dynamics, much like a current source drives an electromagnetic field. Source theory is often associated with the path integral formulation of quantum field theory, in which the source field can be used to compute correlation functions and vacuum expectation values. The advantage is simplicity: focusing on the source simplifies calculations for specific interactions. Source theory can sometimes avoid complex renormalization procedures, which are often necessary in other approaches to quantum field theory. Imagine studying electron-photon interactions: the source field could represent the electron's position and motion, and the resulting electromagnetic field would be calculated from it.

Quantum Theory Experiments

The Double-Slit Experiment

I would be amazed if you hadn't heard of the double-slit experiment, the primary example of quantum wave-particle duality and the measurement problem.

Thomas Young first described this type of experiment in 1801, when he argued for the wave nature of visible light. In 1927, Clinton Davisson and Lester Germer (see later in this chapter), and, independently, George Paget Thomson and his research student Alexander Reid, demonstrated that electrons exhibit the same behavior, a finding later extended to atoms and molecules. This experiment provided the crucial confirmation of wave-particle duality for matter (specifically, the electron). Louis de Broglie hypothesized in 1924 that if light (a wave) can behave like a particle, then matter (a particle) must also be able to act like a wave.

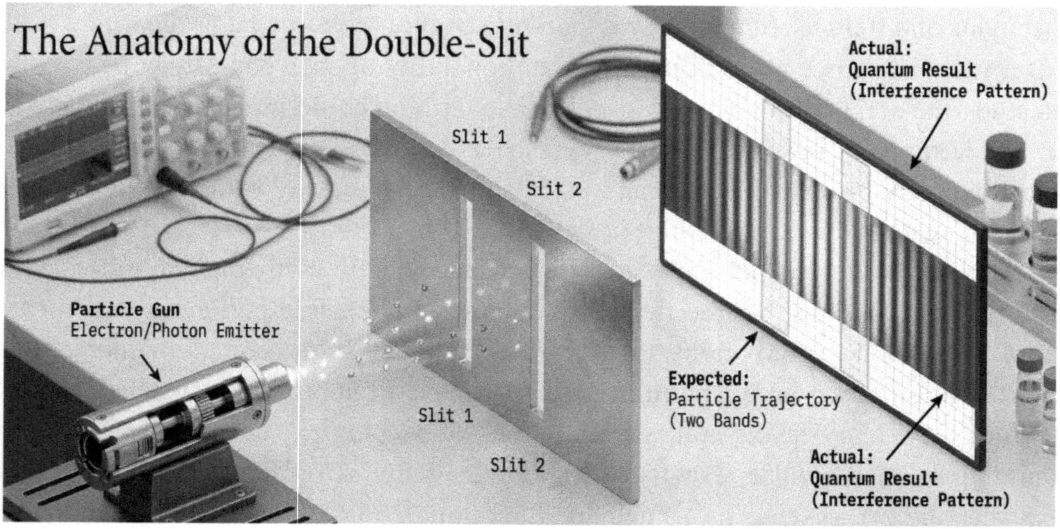

Electrons were fired at a nickel crystal, and the pattern of scattered electrons was observed. The electrons were observed to produce a diffraction pattern – a phenomenon characteristic of waves (such as water or light). This confirmed the electron's wave nature, demonstrating that matter, like light, exhibits wave-particle duality and is governed by quantum mechanics.

The Double-Slit Experiment is the most famous in physics because it dramatically demonstrated the strange, non-intuitive nature of quantum mechanics and the wave-particle duality of matter. Richard Feynman famously called it "the single mystery" of quantum mechanics, as it contains all the peculiarities of the quantum world. The experiment was first performed not with quantum particles but with light to settle a centuries-old debate. Young shone a light source through two adjacent slits onto a screen and observed an alternating pattern of bright and dark bands, known as an interference pattern. This pattern is the signature of waves combining: bright bands occur where wave crests meet crests (constructive interference), and dark bands occur where crests meet troughs (destructive interference). Young's experiment proved light behaves like a wave. To understand the quantum paradox, we first look at the expected results for classical objects. The experiment involves a source, a barrier with two slits, and a detector screen behind the barrier. In the first scenario, the waves pass through both slits simultaneously. The waves interfere with each other, creating a distinctive interference pattern.

In the second scenario, the individual photons must pass through either slit. The photons strike the screen directly behind the slits, creating two distinct bands. The total pattern is simply the sum of the particles that went through the first slit and those that went through the second slit.

In the 20th century, physicists repeated the experiment using individual quantum particles, such as electrons, and found results that confounded the measurement problem. Electrons were fired one at a time, ensuring that each electron could only interact with itself (if it were a wave). Even when fired one by one, the electrons gradually build up the interference pattern on the detector screen, exactly like waves, and not like classical particles. The only way to create an interference pattern is for the electron to somehow pass through both slits simultaneously and interfere with itself – a behavior expected only of a wave, even though the source sends discrete particles. The electron is in a superposition of having gone through both slits.

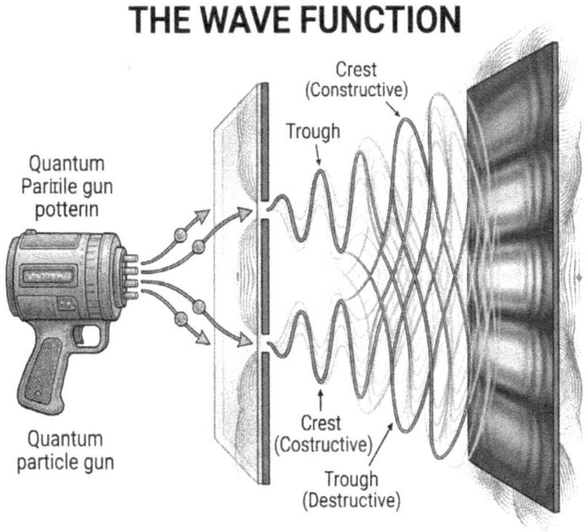

The Wave Function: Particles exist in a blurry superposition of all possible paths, interfering with themselves to create an interference pattern.

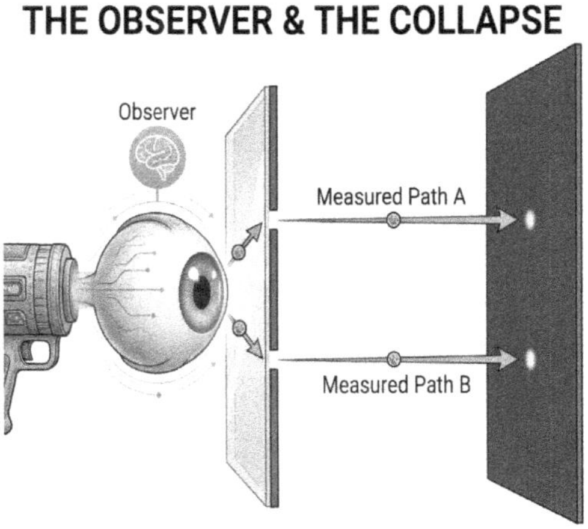

The Collapse: The moment path information is explicitly measured (observed), the superposition vanishes. The wave collapses into a definite particle, traveling through a single slit, destroying the interference pattern.

Next, physicists placed a detector next to the slits to measure which path the electron took. A measuring device records whether the electron passes through the left or the right slit. The moment the observer attempts to measure the electron's path, the interference pattern disappears. The electrons behave exactly like classical particles, creating only two distinct bands. The act of measurement forces the electron's wave function (superposition) to "collapse" into a definite state – it is forced to choose just one slit.The double-slit experiment demonstrates the fundamental principle of wave-particle duality. When not observed, the electron behaves like a wave, existing in a superposition of states, and interferes with itself.

When observed, the electron behaves like a particle, forcing it to commit to a single location or path. The experiment shows that simply gaining knowledge of a quantum system's path fundamentally changes its reality, illustrating the Measurement Problem–the inability of quantum mechanics to explain the transition from an indefinite quantum state to a definite classical state.

When discussing this experiment, you may find some people with "criticisms", but none can contradict the basics of an experiment that has been performed so often. The experiment has been repeated millions of times with photons, electrons, and even large molecules (and even a sodium metal with over 6000 atoms), and the data is mathematically bulletproof. Instead, the criticisms are almost entirely directed at how the experiment is interpreted, taught, and sometimes misrepresented in popular science.

The most frequent criticism from the physics community is directed at the pop-science extrapolation that a *conscious observer* is required to change the outcome of the experiment. When physicists say that "observing" or "measuring" which slit the particle goes through collapses the wave function (destroying the interference pattern), "observation" does not mean a human mind looking at it. It means a physical interaction. Any macroscopic measuring device placed at the slits must interact with the quantum particle (usually by hitting it with a photon or electromagnetically entangling with it) to record a data point. This physical interaction causes decoherence, effectively forcing the particle out of its delicate quantum state.

In most classrooms, the double-slit experiment is taught strictly through the lens of the Copenhagen Interpretation (championed by Niels Bohr and Werner Heisenberg). This interpretation asserts that the particle exists in a "superposition" of probabilities until it is measured, at which point the wave function "collapses" to a single point. Critics argue that teaching this as the only explanation is fundamentally flawed, as it ignores other mathematically valid interpretations that do not require a magical "collapse". Theories such as the De Broglie-Bohm (Pilot-Wave) Theory and the Many-Worlds Interpretation offer alternative explanations. Another criticism targets the assumption that what happens at the quantum scale in the double-slit experiment easily translates to the macroscopic world.

Some theoretical extensions suggest that, because an electron can be in a superposition of passing through both slits, the universe as a whole is composed entirely of blurry probabilities. Critics note that this ignores environmental decoherence. In the real world, macroscopic objects (like a baseball or Schrödinger's Cat) are constantly being bombarded by billions of photons, air molecules, and thermal radiation. This constant environmental "measurement" ensures that macroscopic objects cannot exist in a double-slit-style superposition; they are forced into definite classical states almost instantaneously. Decoherence explains the *loss of interference* (turning pure states into improper mixed states) but does *not* select a single unique outcome without an underlying interpretation (the measurement problem remains).

The Quantum Eraser Experiment

The Quantum Eraser Experiment is a variation of the double-slit experiment that shows that by "erasing" information about which path a particle took, you can restore an interference pattern that had previously disappeared.

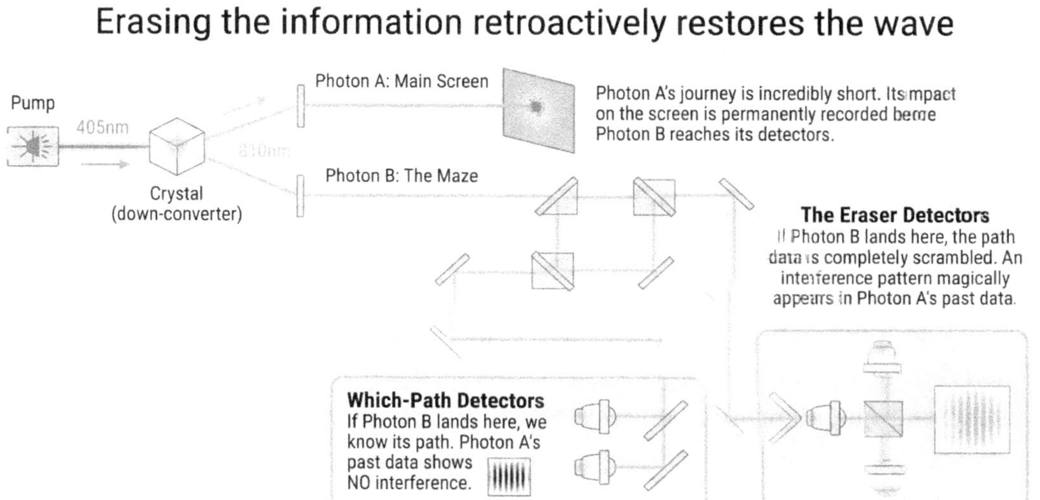

Marlan Scully and Kai Drühl first proposed the quantum eraser experiment in 1982. They published their theoretical concept in a seminal paper titled "*Quantum eraser: A proposed photon correlation experiment concerning observation and 'delayed choice' in quantum mechanics*".[30] This initial design laid the groundwork for testing wave-particle duality and quantum complementarity by manipulating path information. While the original idea was born in 1982, it evolved through subsequent variations and physical laboratory testing. They suggested that "which-path" information could be stored in a quantum system and later erased, causing an interference pattern to reappear.The most famous variation, the delayed-choice quantum eraser, was experimentally realized by Yoon-Ho Kim, R. Yu, S.P. Kulik, Y.H. Shih, and Marlan Scully.[31] Submitted in 1999 and published in 2000, this laboratory experiment used entangled photons to prove that path data could be scrambled or erased after a detector had already recorded a photon.

In a standard double-slit experiment, particles such as electrons or photons produce an interference pattern, showing that they behave like waves. However, if you place a detector at the slits to determine which slit the particle passes through, the interference pattern vanishes, and the particles behave like little "billiard balls," creating two distinct piles on the back screen.

In the Quantum Eraser experiment, instead of a standard detector, physicists "tag" the particles. For example, they might use a light source and place different polarizers over each slit – one that polarizes the light vertically and one that polarizes it horizontally. Because the light passing through the left slit is vertically polarized and the right horizontally polarized, it is theoretically possible to know which slit each photon went through. This is called "Which-Path" information. The wave functions of the two paths can no longer interfere with each other because they are now "distinguishable." The interference pattern on the screen disappears, replaced by a simple, non-interfering clump of light.

The "Eraser" part of the experiment involves placing a third polarizer in front of the final detection screen. This third polarizer is set at a 45-degree angle. When the vertically and horizontally polarized light hits this 45-degree polarizer, both become polarized at that same diagonal angle. Once they are both polarized the same way, they are no longer distinguishable. The "Which-Path" information has been erased – there is no longer any way to tell which slit the photon originally passed through. When this information is erased, the interference pattern reappears on the detection screen. The quantum superposition depends entirely on whether the information about the particle's path is available to the rest of the universe.

The most famous version is the Delayed-Choice Quantum Eraser. In this setup, the "erasing" of the information happens *after* the photon has already hit the detection screen. Physicists use entangled pairs of photons. One photon (the "signal") goes to the screen, while its entangled twin (the "idler") goes through a series of mirrors and sensors that can either preserve or erase the path information. Paradoxically, the pattern formed by the signal photons on the screen changes depending on whether the information about their twins was eventually erased or not – even if that erasure occurred later.

While this leads to sensationalist claims about "rewriting the past," most physicists interpret it through the lens of quantum entanglement. The results of the two photons are correlated in a way that remains consistent regardless of the order of measurement, illustrating that at the quantum level, the concepts of "here" and "now" are far more interconnected than they appear in the macroscopic world.

The Franck-Hertz Experiment (1914)

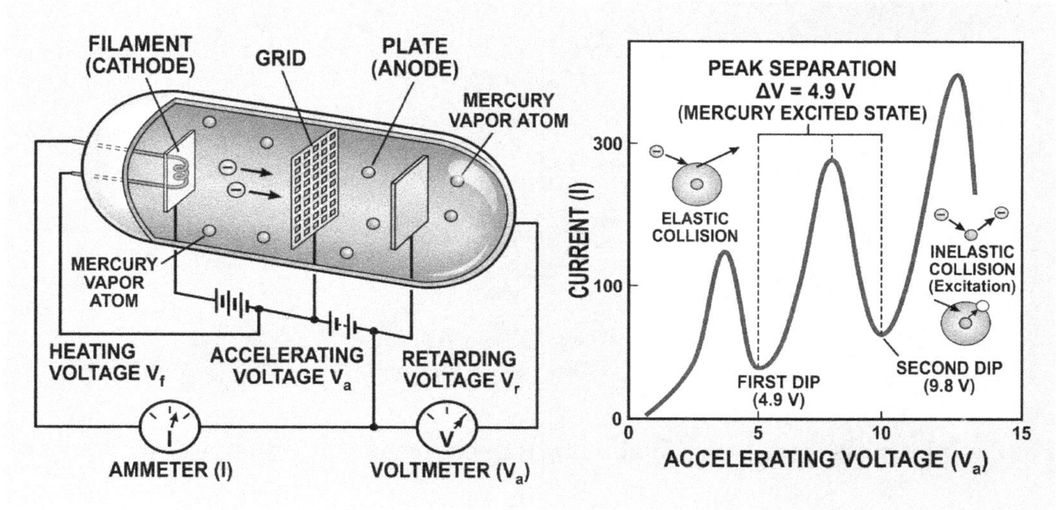

This experiment demonstrated the quantization of atomic energy levels, supporting Niels Bohr's model. Electrons were fired through mercury vapor, and their energy loss was measured. Instead of continuously losing energy, electrons only lost energy in discrete amounts when they reached specific energy thresholds.

This proved that mercury atoms could absorb energy only in discrete packets, confirming that electrons occupy fixed, quantized energy levels within an atom rather than a continuous range of energies.

The Compton Effect

The Compton Effect provided definitive, smoking-gun evidence for the photon's momentum (its particle nature); the experiment confirmed that photons carry measurable momentum $p = h^{\lambda}$ and, in this context, behave exactly like billiard balls, thereby validating the particle view of light.

In the experiment, an X-ray tube generates a high-energy beam that is directed at a scattering target, such as a Perspex (acrylic) disc. Within the material, X-ray photons collide with weakly bound or "free" electrons. During these collisions, the photons transfer some of their energy and momentum to the recoiling electrons. As a result, the scattered X-ray photons have lower energy and, consequently, a longer wavelength (A') compared to the incident ones (A). The change in wavelength depends on the scattering angle (0). This experiment provides strong evidence for the particle-like behavior of light (photons) with definite energy and momentum.

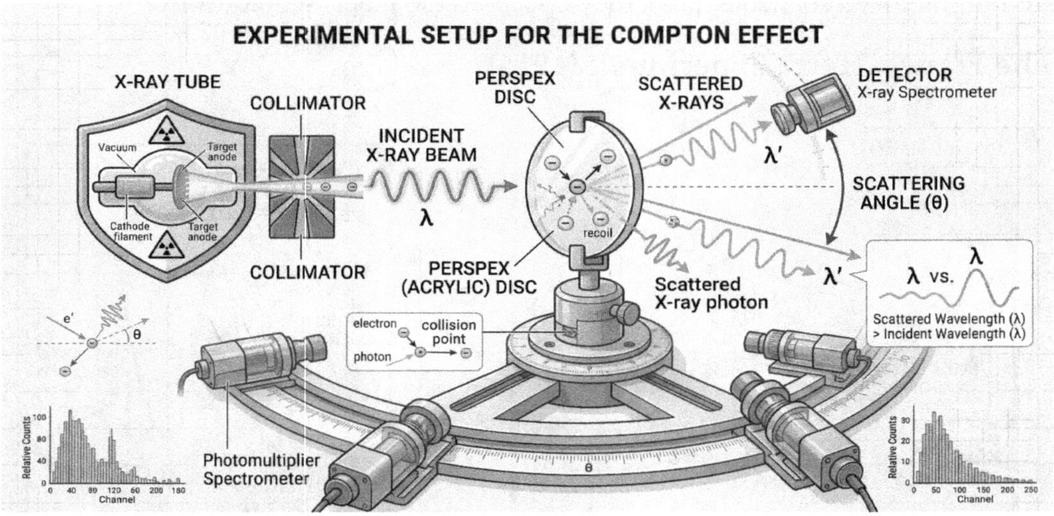

For completeness, here is the final mathematical result of Compton's work:

$$\lambda' - \lambda = \frac{h}{m_e c}(1 - \cos\theta)$$

Where:

- λ is the initial wavelength of the incoming photon
- λ' is the wavelength after scattering
- h is Planck's constant (6.626×10^{-34} Joule-seconds)
- m_e is the rest mass of the electron (9.109×10^{-31} kilograms)
- c is the speed of light (3.00×10^{8} meters per second)
- θ represents the scattering angle of the photon
- $\frac{h}{m_e c}$ is a constant known as the Compton wavelength of an electron, which equals 2.426×10^{-12} meters.

To illustrate with a concrete example, if a photon scatters at an angle of 90 degrees, the term $\cos(90°)$ equals 0. The factor (1 - $\cos\theta$) becomes 1. The resulting shift in wavelength ($\lambda' - \lambda$) equals exactly 2.426×10^{-12} meters.

The Davisson-Germer experiment (1927)

The Davisson-Germer experiment, conducted in 1927 by American physicists Clinton Davisson and Lester Germer, was a landmark study in the history of physics. It provided the first direct experimental evidence for wave-particle duality, specifically proving that electrons exhibit wave-like behavior.

In 1924, Louis de Broglie proposed a revolutionary hypothesis: if light (a wave) could behave like a particle (a photon), then matter (particles) should be able to behave like waves, and Germer set out to test this idea by observing how electrons scattered off the surface of a solid.

The experiment used a vacuum chamber to prevent electrons from colliding with air molecules. The apparatus consisted of three main components:

1. ***Electron Gun:*** A tungsten filament was heated to emit electrons via thermionic emission. These electrons were then accelerated through a known potential difference (V), giving them a specific kinetic energy and momentum.
2. ***Target (Nickel Crystal):*** The beam of electrons was directed at a target made of a single crystal of nickel.
3. ***Movable Detector:*** A detector was mounted on an arc, allowing it to move and measure the intensity of electrons scattered at various angles (θ).

The physicists bombarded the nickel crystal with the electron beam and measured the number of scattered electrons at different angles. Initially, the results were unremarkable. However, after an accidental break in the vacuum system caused the nickel target to oxidize, they heated it to clean it, which inadvertently turned the polycrystalline nickel into a few large single crystals.

When they resumed the experiment, they observed a striking phenomenon: at a specific accelerating voltage (54 volts) and a specific scattering angle $(50°)$, there was a distinct "peak" in the intensity of the scattered electrons. This peak could not be explained if electrons were merely tiny "billiard balls" bouncing off a surface. Instead, it behaved exactly like the interference patterns seen in X-ray diffraction.

The researchers realized that the rows of atoms in the nickel crystal were acting as a diffraction grating. They applied Bragg's Law, which describes how waves interfere when reflected from a crystal lattice: $n\lambda = 2d\sin\theta$. By calculating the wavelength from the observed diffraction pattern observed $(1.65\AA)$ and comparing it with the wavelength predicted by de Broglie's formula for electrons at 54 volts $(1.67\AA)$, they found the results were in nearly perfect agreement.

The Davisson-Germer experiment was a significant moment in the development of quantum mechanics, as it demonstrated that matter exhibits wave properties, confirming de Broglie's hypothesis. It provided experimental grounding for the Schrödinger equation and the wave-based descriptions of the subatomic world.

The discovery earned Clinton Davisson the Nobel Prize in Physics in 1937 (shared with George Paget Thomson, who independently observed electron diffraction through thin films).

The discovery that electrons can be diffracted led directly to the development of the electron microscope, which uses the extremely short wavelengths of high-speed electrons to see objects far smaller than those visible with light.

Bell Test Experiments (late 20th Century)

Experimental tests of John Stewart Bell's Theorem were performed to measure correlations between entangled particles separated by large distances. These Bell test experiments determine whether the universe obeys local realism or follows the predictions of quantum mechanics. These tests measure correlations between entangled particles to determine whether they exceed the limits allowed by classical physics.

In 1935, Albert Einstein, Boris Podolsky, and Nathan Rosen proposed the EPR paradox. They argued that quantum mechanics was incomplete because it appeared to allow "spooky action at a distance." They suggested that "local hidden variables" determined a particle's properties before measurement.

John Bell published a theorem in 1964, creating a mathematical inequality – known as Bell's inequality – that any theory based on local hidden variables must satisfy.[32] He proved that quantum mechanics predicts a violation of this inequality. If the correlations between two particles exceed a specific numerical limit, then local hidden variables cannot explain the results.

Physicists John Clauser, Michael Horne, Abner Shimony, and Richard Holt refined Bell's work into the CHSH inequality.[33] This version is easier to test in a lab using photon polarization or electron spin of. In this model, a variable S represents the correlations between measurements. According to local realism, the absolute value of S cannot exceed 2. Quantum mechanics predicts that S can reach $2\sqrt{2}$ (approximately 2.82).

Timeline of Major Bell Experiments

In 1972, Freedman and Clauser performed the first successful Bell test.[34] They used entangled photons and found a clear violation of the inequality, supporting quantum mechanics. However, critics identified "loopholes," such as the possibility that the detectors were too inefficient to capture a representative sample of particles.

In 1982, Alain Aspect conducted a series of experiments in Paris that closed the "locality loophole."[35] His team changed the orientation of the polarizers while the photons were in flight. Because the change happened in less than 30 nanoseconds, no signal traveling at the speed of light could reach the other detector. The results violated Bell's inequality again.

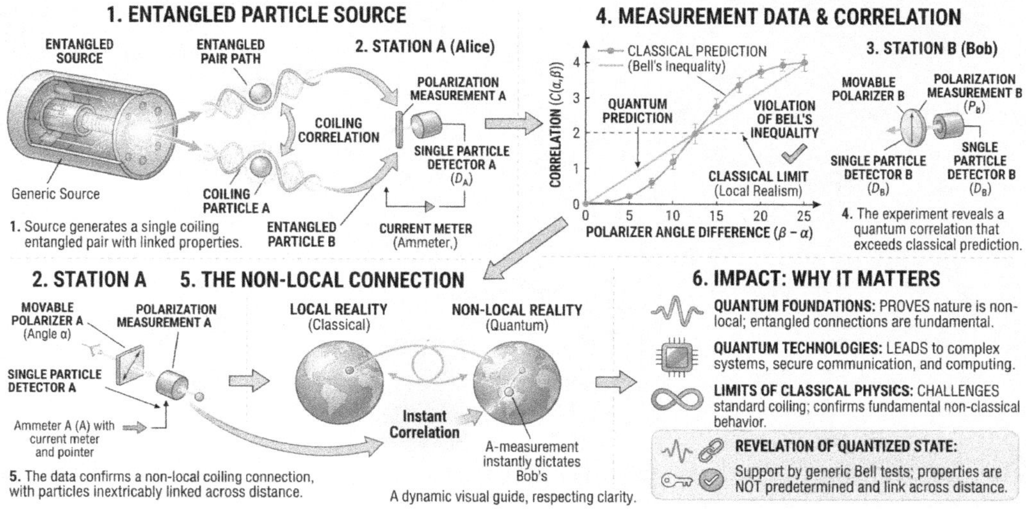

In 2015, Several groups achieved "loophole-free" Bell tests. Ronald Hanson and his team used entangled electrons in diamonds separated by 1.3 kilometers. They completed 245 trials over 220 hours. By ensuring high detection efficiency and sufficient distance, they simultaneously closed both the locality and detection loopholes. The experiment produced a result of $S = 2.42$, which significantly exceeds the classical limit of 2.

The results of these experiments demonstrate that the universe is not locally real. You must accept that either the properties of particles do not exist until they are measured (abandoning realism) or that particles can influence each other instantaneously across any distance (abandoning locality). These findings provide the experimental proof required for modern quantum technologies.

The Mach–Zehnder Interferometer Experiment (1994)

The Mach–Zehnder interferometer was independently developed by physicists Ludwig Zehnder in 1891 and Ludwig Mach (son of the famous philosopher and physicist Ernst Mach) in 1892. Ludwig Zehnder designed it to observe and measure variations in the refractive index of transparent media. Ludwig Mach improved Zehnder's design, improving the structural stability and alignment mechanisms, making it much easier to achieve and measure precise interference fringes.While it was created as a classical optical instrument for measuring phase shifts, density, or length, the Mach–Zehnder Interferometer Experiment is widely known for its modern application in basic quantum mechanics.

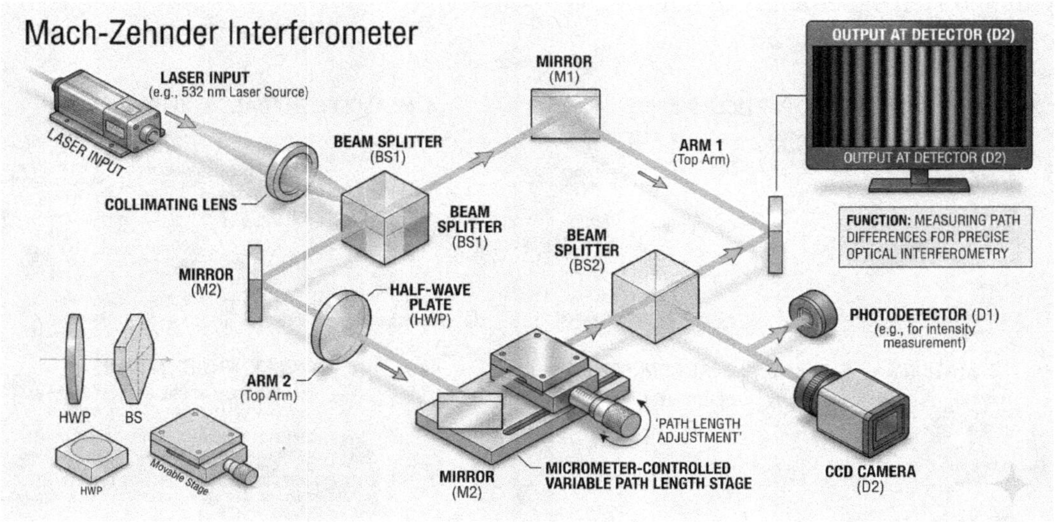

When used in the "Bomb Tester" thought experiment, it illustrates one of the most counterintuitive aspects of quantum mechanics: Interaction-Free Measurement.

A standard Mach–Zehnder Interferometer consists of:

- ***A light source:*** Sending out single photons.
- ***Two beam splitters:*** Half-silvered mirrors that reflect 50% of the light and transmit 50%.
- ***Two fully reflecting mirrors.***
- ***Two detectors (D1 and D2).***

In a standard configuration, a single photon enters the first beam splitter and enters a "superposition" of two paths (Path A and Path B). The mirrors reflect these paths toward a second beam splitter. Because the paths are calculated in this way, the waves interfere with each other. In a perfectly aligned system, constructive interference ensures the photon always hits Detector 1, while destructive interference means the photon never hits Detector 2.

Imagine you have a collection of bombs. Some are "duds" and do nothing when a photon hits them. Others are "live" and are so sensitive that if even a single photon hits the trigger, the bomb explodes. The goal is to find a live bomb without setting it off. In classical physics, this is impossible. To know if the trigger works, it is touched, which causes an explosion. Quantum mechanics provides a loophole. The bomb's trigger is placed in Path B of the interferometer.

If the bomb is a dud: The photon behaves as if nothing is there. It travels both paths simultaneously, interferes with itself at the second beam splitter, and, due to destructive interference, always hits Detector 1. If you see a hit on D1, you haven't learned anything; it could be a dud or a live bomb that just didn't explode this time.

If the bomb is live: It acts as a "measuring device." For the bomb to be live, it must be capable of being triggered. This "possibility" of interaction forces the photon to choose a path – often described as the collapse of the wave function.

- ***Scenario A*** (50% chance): The photon takes Path B, hits the trigger, and the bomb explodes.
- ***Scenario B*** (50% chance): The photon takes Path A. Because the superposition was broken, the photon reaches the second beam splitter as a particle rather than a wave. It now has a 50/50 chance of being reflected or transmitted at that final splitter.

Because the live bomb destroys the interference pattern, the probabilities for the detectors change:

1. ***Explosion*** (50%): You lost the bomb.
2. ***Detector 1 hits*** (25%): This result is inconclusive (the same as a dud).
3. ***Detector 2 hits*** (25%)**:** This is the "Eureka" moment.

In a standard, empty interferometer, it is physically impossible for a photon to hit Detector 2. Therefore, if Detector 2 clicks, you know with 100% certainty that there is a live bomb in Path B, even though the photon never touched it.

The Bomb Tester experiment demonstrates that, in the quantum realm, we can learn about an object's properties without ever physically interacting with it. It highlights the "Measurement Problem" in physics: the mere act of measurement (the trigger being present) alters the physical outcome of the experiment. This suggests that the universe is not just composed of objects and forces, but also of information and possibilities. Even if a photon doesn't "touch" the bomb, the fact that it could have touched it is enough to alter the reality of the photon's path.

Recent Experiments in Quantum Theory

Science and technology never stop moving forward, sometimes extremely rapidly. It's impossible to publish a book like this and stay completely current.

Experiments in quantum theory have focused heavily on both foundational physics and advanced technological applications, especially in computing and sensing. Researchers have developed ultra-small optical phase modulators, nearly

100 times smaller than a human hair, to efficiently control the lasers used to operate trapped-ion and trapped-neutral-atom qubits. This represents a major step toward practical quantum computers.

A team successfully trapped and entangled ultra-cold polar molecules (sodium-cesium) for the first time, using their electric dipole-dipole interactions to perform quantum operations. This opens new pathways for quantum computing by leveraging the complexity of molecular structures.

Quantum engineers have achieved quantum entanglement between two distant atoms in silicon, using electrons as a bridge. Entangled states are the essential resource for the computational power of quantum computers. Scientists have successfully teleported information between two separate light-emitting devices at wavelengths compatible with ordinary fiber-optic internet cables, demonstrating the potential for future quantum communication networks that use existing infrastructure. (Only quantum information is teleported; regular information must still obey the speed of light law).

John Clarke, Michel H. Devoret, and John M. Martinis awarded the Nobel Prize in Physics 2025 for experiments, earning this prestigious recognition "*for the discovery of macroscopic quantum mechanical tunnelling and energy quantization in an electric circuit*".[36] Their 1984 and 195 experiments were carried at UC Berkeley in 1984 and 1985. These electrical circuits were large enough to be held in the hand.

Researchers discovered that a single-atom-thick layer of tungsten ditelluride, a semi-metal, can exhibit novel quantum behavior, including transitions between insulating and superconducting phases, challenging current theories of superconductivity.

In January 2026, researchers achieved a major milestone by demonstrating quantum interference in sodium nanoparticles containing over 7,000 atoms, setting a new record for the largest object in a "Schrödinger's Cat" state.[37]

Researchers produced a beam of sodium clusters, which were sent through three diffraction gratings (using ultraviolet lasers) to observe interference. The clusters comprised 5,000 to 10,000 atoms and weighed over 170,000 atomic mass units, making them larger than most proteins. To prevent decoherence, the experiment was conducted under cryogenic conditions and in an ultra-high vacuum. The results demonstrated a superposition spread of 133 nanometers, challenging the assumption that macroscopic objects cannot exhibit quantum behavior.

They have also experimentally proven, for the first time, that angular momentum is conserved even when a single photon splits into two, thereby validating a fundamental principle of physics at the quantum level.

Summary

Chapter Three closes the gap between quantum mechanics and the Standard Model by explaining Quantum Field Theory. In contrast to earlier chapters, which approached particles as discrete or solid-like entities within classical or early quantum frameworks, this chapter shifts to viewing the universe as a collection of layered, fluid-like fields, in which particles are understood as localized, quantized vibrations – ripples within these fields. This builds upon and combines earlier discussions, moving from the foundations of quantum mechanics toward a unified field perspective.

The chapter then breaks down the two pillars of the Standard Model's field interactions: Quantum Electrodynamics, which describes how light and matter interact via photon exchange, and Quantum Chromodynamics, which describes the strong force that holds quarks together via gluons.

We highlight Richard Feynman's invention of Feynman diagrams and renormalization used to tame infinities in field calculations. Ongoing experiments continue to prove that the quantum theories, while so odd to our classical way of thinking, are our best representation of the universe we inhabit.

Chapter Four: Quantum-Related Subjects

Ψ

"A quantum physicist can know how fast they do it, or where they do it, but not both."

Conservation of Quantum Information

What is conservation of quantum information, and why is it a fundamental pillar of quantum mechanics? The short answer is that information can't be destroyed because the underlying laws of quantum physics are reversible. We should start with this question: what exactly is quantum information?

I wanted to be sure I understood what it is in relation to the topic at hand. A basic definition of *normal* information is any processed, structured, or contextualized data that resolves uncertainty and carries meaning. It's the bridge between raw facts and usable knowledge, acting as the foundation for communication, decision-making, and understanding systems.

In quantum mechanics, information refers to the exact configuration, quantum states, and fundamental properties of the particles that form an object. The principle that information cannot be destroyed, known as quantum unitarity, means that the universe always maintains a perfect record of its own past.

Unitarity is a foundational principle in quantum mechanics that states the total probability of all possible outcomes for any physical process must always sum to exactly 1 (or 100%). It ensures that quantum systems evolve in a perfectly conserved, predictable, and mathematically reversible way over time. A simple definition is this: If you have a complete, perfect description of any physical system right now, the laws of physics allow you to mathematically reverse time to reconstruct its exact past state with 100 percent certainty.

Think about burning a book (not this one!). You've destroyed the *classical* information. You can't reverse combustion to get the book back. The information is gone forever, converted into random heat and ash. Quantum mechanics doesn't work that way. For any closed system (like the entire universe), its current state is described by a quantum wave function. Unitarity dictates that if you know the *exact* final state of the system, you can (in principle) apply the reverse of the quantum-mechanical "rules" to find its *exact* initial state. The exact content of the text is encoded in the precise directions of the rising smoke particles, the atomic arrangement of the ash, and the wavelengths of the infrared light emitted by the flame. If you possessed an instrument capable of measuring every single atom and photon involved in the fire, you could run the quantum equations backward. You would successfully reconstruct the original paper and read the text.

If you could *truly* destroy quantum information (like the book), it would be impossible to rewind the "movie" of the universe, because you wouldn't know what to rewind *to*. The final state (ash and heat) would have "forgotten" the initial state (the book). This would violate unitarity.

So, what *does* happen when information seems to be lost? It gets scrambled or entangled with its environment. The No-Hiding Theorem states that if quantum information is "lost" from one system, it *must* have moved to another system (the environment). It cannot simply vanish, nor can it be "hidden" purely in the correlations *between* the two systems.

Imagine tearing a diary into millions of tiny pieces, then scattering them in a hurricane. For all practical purposes, the information is "lost" *to you*. You can't read it. However, the information remains *fundamentally* intact on those tiny pieces, which are now part of the wider environment. It hasn't been destroyed; it's just become difficult to access. In quantum mechanics, this scrambling occurs through decoherence, in which a quantum state interacts with its surroundings and becomes entangled with them. The information isn't gone; it's just "leaked" into the environment. Like entropy, it has moved from a more ordered state to a less ordered one.

This principle faced its biggest challenge when Stephen Hawking calculated that black holes evaporate over time by emitting Hawking radiation. His calculations suggested that this radiation was "thermal," meaning it was completely random and contained no information about what had fallen into the black hole.

If the book (a complex quantum state) fell into a black hole, and the black hole slowly evaporated into random thermal radiation, the information *inside* the book would be well and truly destroyed. This destruction would be a catastrophic violation of unitarity.

This paradox sparked a 40-year debate. Today, the overwhelming consensus among physicists (including Hawking himself, who later conceded) is that unitarity holds and information *is* conserved. The information that falls into a black hole must somehow be "imprinted" or "encoded" in the outgoing Hawking radiation in an incredibly subtle, scrambled form. Some theories suggest that information about the object may be encoded at the black hole's event horizon, rather than being destroyed. Another idea is that black holes might have a kind of "quantum hair" that stores information about what fell into them. Some physicists speculate that black holes may be connected to other regions of spacetime via wormholes, allowing information to escape in a manner not yet fully understood. The "Black Mirror" model proposes that, rather than an interior leading to a singularity, the event horizon is a boundary that connects our region of spacetime to its own CPT-reversed mirror image. In this mirror region, time flows in reverse – the temporal direction is the opposite of ours. Matter is replaced by antimatter – charge and parity are inverted. The curvature of spacetime remains smooth and bounded, removing the mathematical "dead end" of a singularity.

This model addresses the black hole information paradox – the conflict between general relativity (which says information is lost) and quantum mechanics (which says it must be preserved) - by ensuring that the system remains unitary.

The Pusey–Barrett–Rudolph (PBR) Theorem

The Pusey–Barrett–Rudolph theorem is a crucial result in the foundations of quantum mechanics that places strict limits on how we can interpret the quantum wave function (ψ).[38] It strengthens the argument that the wave function is a real, physical state of the system, rather than merely a representation of an observer's knowledge. Physicists divide historical models of the wave function into two main categories:

- ψ-ontic **(Realist):** The wave function is a concrete physical property of reality. It exists objectively, much like the position or momentum of a planet in classical physics.
- ψ-epistemic **(Informational):** The wave function is not a physical object. Instead, it serves as a statistical tool that represents an observer's uncertainty or lack of information about the system's true underlying state.

If a model is epistemic, two distinct quantum states can potentially overlap. This means that a single underlying physical configuration could occasionally be compatible with more than one quantum state. The theorem addresses the long-standing philosophical question about the nature of the wave function. In the first, the wave function is an ontological entity – it corresponds directly to the system's physical state, much like an electron's mass or charge. In the second, the wave function is an epistemic entity -- it merely represents the knowledge, belief, or uncertainty an observer has about the actual, underlying physical state (often called "hidden variables").

To properly formulate the result, we need to know how quantum states are represented within a hidden-variable theory. In such a theory, quantum systems are assumed to have real, pre-existing properties that determine what happens when we make a measurement. A full specification of these properties constitutes an ontic state of the system. In general, we don't have precise control over the ontic state, so a quantum state corresponds to a probability distribution over the ontic states. This framework is illustrated below.

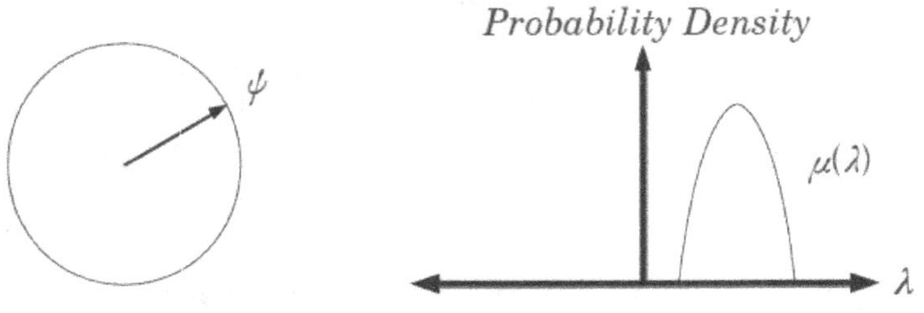

In an ontic model, a quantum state (indicated heuristically on the left as a vector in the Bloch sphere) is represented by a probability distribution over ontic states, as indicated on the right. The Bloch sphere is a geometric 3D unit sphere representation of a qubit's quantum state, mapping pure states to its surface using spherical coordinates.

A hidden-variable theory is psi-ontic if knowing the ontic state of the system allows you to determine the (pure) quantum state that was uniquely prepared. Equivalently, the probability distributions corresponding to two distinct pure states do not overlap, as illustrated below.

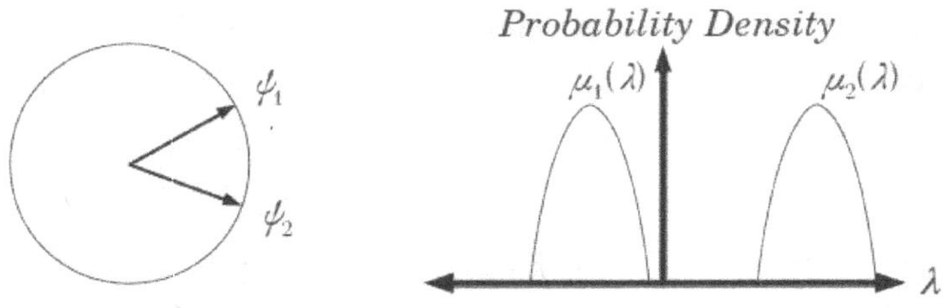

A hidden variable theory is psi-epistemic if it is not psi-ontic, i.e., there must exist an ontic state that is possible for more than one pure state, or, in other words, there must exist two non-orthogonal pure states with corresponding distributions that overlap. This is illustrated below.

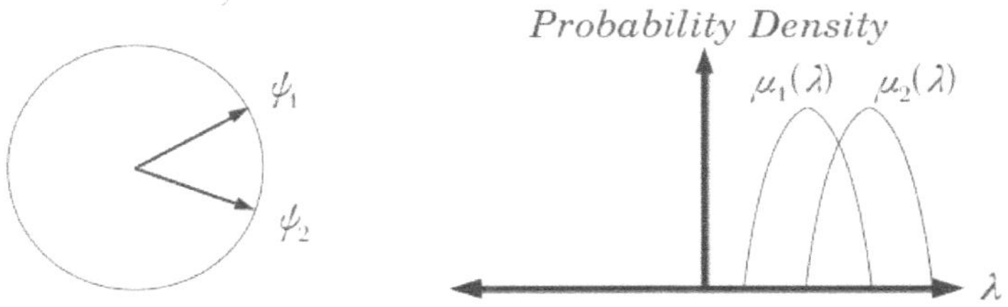

These definitions of psi-ontology and psi-epistemicism may seem abstract, so a classical analogy may be helpful. In Newtonian mechanics, a particle's ontic state is a point in phase space, i.e., a specification of its position and momentum. Other ontic properties of the particle, such as its energy, are given by functions of the phase-space point; i.e., they are uniquely determined by the ontic state. Likewise, in a hidden variable theory, anything that is a unique function of the ontic state should be regarded as an ontic property of the system, and this applies to the quantum state in a psi-ontic model. The definition of a psi-epistemic model as the negation of this is very weak; e.g., it could still be the case that most ontic states are possible only in a single quantum state, while just a few are compatible with more than one. Nonetheless, PBR rules out even this very weak notion.

The PBR theorem provides strong mathematical evidence against the broad class of epistemic theories. The theorem mathematically proves that any theory in which the wave function is purely epistemic will yield statistical predictions that contradict the known, verified predictions of standard quantum mechanics. The theorem targets epistemic theories that satisfy a very reasonable constraint called λ-preparation independence. When two quantum systems are prepared independently in separate pure quantum states (ψ_1 and ψ_2), their underlying physical states (λ, the hidden variables) must also be statistically independent.

PBR examined a scenario in which the joint preparation and measurement of two or more quantum systems (qubits) are performed independently, each in a known quantum state. The mathematical proof showed that if the wave functions (ψ_1 and ψ_2) were merely epistemic, the statistics resulting from specific joint measurements would violate the predictions of standard quantum mechanics. Since standard quantum mechanics is verified by decades of experimental data, the initial premise (that the wave function is merely epistemic) must be false. The wave function (ψ_1) must be ontic; it must correspond to a state of physical reality.

The equations and mathematical foundations of the PBR theorem involve two main components. The theorem begins with two distinct, non-orthogonal quantum states, usually represented in a two-dimensional Hilbert space (like the spin of an electron):

$$|\psi_0\rangle = |0\rangle$$
$$|\psi_1\rangle = |+\rangle = \frac{1}{\sqrt{2}}(|0\rangle + |1\rangle)$$

The inner (dot) product of these two states is non-zero, meaning they have a finite overlap:

$$\langle\psi_0|\psi_1\rangle = \frac{1}{\sqrt{2}}$$

The theorem considers two independent systems prepared in combinations of these states:

$$|S_1\rangle = |\psi_0\rangle \otimes |\psi_0\rangle$$
$$|S_2\rangle = |\psi_0\rangle \otimes |\psi_1\rangle$$
$$|S_3\rangle = |\psi_1\rangle \otimes |\psi_0\rangle$$
$$|S_4\rangle = |\psi_1\rangle \otimes |\psi_1\rangle$$

PBR proved there exists a specific entangled measurement (a projective measurement) with four orthogonal outcomes, designed such that exactly one outcome is strictly forbidden (yielding a Born rule probability of 0 for each of the four states above.

For example, the measurement operators (P_k) are designed such that the probability of a certain outcome for the first state is:

$$P(\text{Outcome } 1 \,|\, |\psi_0\rangle \otimes |\psi_0\rangle) = 0$$

By assuming that quantum states are merely statistical knowledge about a "real underlying physical state" (the ψ-epistemic view) and applying an assumption of preparation independence, PBR showed that these distributions cannot overlap. If they did, there would be a non-zero probability of getting measurement results that quantum mechanics strictly dictates must be zero, leading to a logical contradiction.

The PBR theorem forces a choice: one must either reject a very natural assumption about independence (which is hard to justify) or accept that the wave function is real. By ruling out local epistemic theories, the PBR theorem emphasizes the necessity of bizarre quantum phenomena such as non-locality. It supports the view that the oddities of quantum mechanics are fundamental features of nature rather than mere flaws in our knowledge. While PBR confirms the wave function is real, it does not solve the Measurement Problem – it still doesn't explain *how* or *why* the wave function collapses upon observation.

The Quantum Hall Effect (QHE)

The Quantum Hall Effect, discovered by Edwin H. Hall, is a fascinating phenomenon observed in two-dimensional electron systems at low temperatures and in strong magnetic fields.[39] It's characterized by the quantization of the Hall resistance, meaning it takes on precise, discrete values. When a current flows through a conductor in a magnetic field, a voltage develops perpendicular to both the current and the magnetic field. This is the classical Hall effect. As the magnetic field increases, the Hall resistance doesn't increase smoothly. Instead, it forms plateaus at specific quantized values. This quantization is incredibly precise and is related to fundamental constants of nature: Planck's constant (h) and the elementary charge (e).

The QHE requires electrons to be confined to a two-dimensional plane, most often at the interface between two different semiconductor materials. The strong magnetic field causes electrons to move in circular orbits (cyclotron orbits). Low temperatures are crucial for minimizing thermal excitations and preserving electron quantum coherence.

In a magnetic field, the energy levels of electrons in a two-dimensional electron system (2DES) become quantized into discrete levels called Landau levels. Each Landau level can hold a specific number of electrons. The Filling Factor (ν) is a crucial parameter in the QHE and represents the number of filled Landau levels (Think of it as a way to measure how crowded the electron "parking lot" is). When an integer number of Landau levels is filled, the system exhibits the Integer Quantum Hall effect.

Quantization arises because, when Landau levels are filled, electrons in the bulk of the 2DES are localized and cannot contribute to conduction. However, at the edges of the sample, electrons can "skip" along the boundaries in one-dimensional, chiral edge states. These edge states are topologically protected, meaning they are well protected against imperfections and scattering. These edge states carry current without dissipation, leading to perfectly quantized Hall resistance and a vanishing longitudinal resistance.

So why is it of interest? Mostly because it could provide a superhighway for electrons. Inside a Quantum Hall material, the middle acts like an insulator (no electricity flows), but the edges act like perfect one-way streets. Electrons move in one direction and can't "bounce back" or scatter off defects. Because they don't scatter, they don't lose energy as heat. This "topological" protection is a holy grail for making electronics that don't get hot and use almost no power.

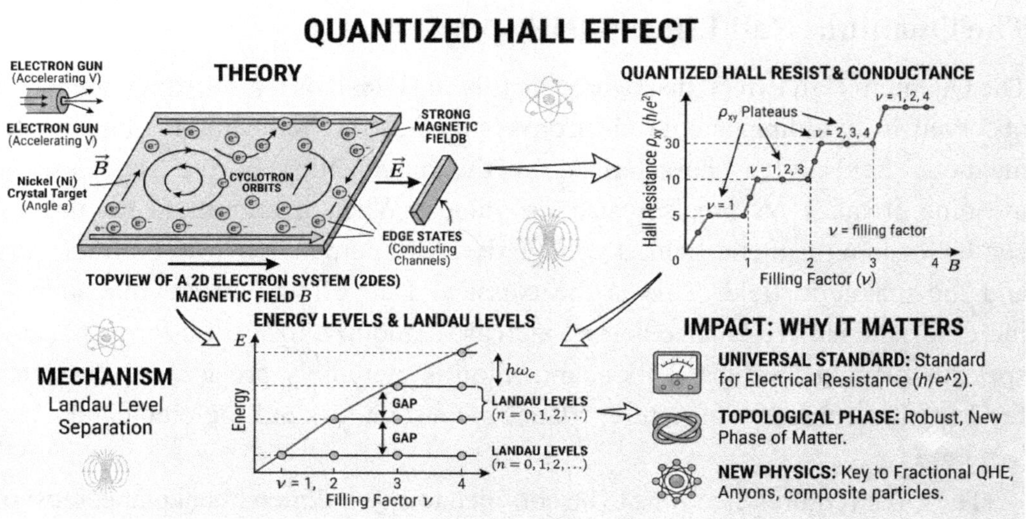

The Fractional Quantum Hall Effect (FQHE) is when ν can be a fraction (e.g., $\frac{1}{3}$, $\frac{2}{5}$). This is a more complex phenomenon involving strong electron-electron interactions that cause to exotic *quasiparticles* (anyons) with fractional elementary charges (such as $\frac{1}{3}$ of an electron). Studying these helps us understand how groups of particles can behave in ways a single particle never could.

Fractional Excitons

The theory of fractional excitons describes a fascinating and complex quasiparticle that can emerge in certain condensed-matter systems, particularly those with strong electron-electron interactions and often in the presence of a strong magnetic field.

To understand fractional excitons, we first need to understand regular excitons. An exciton is a quasiparticle that represents an excited electronic state in a solid. It's formed when an electron in a material absorbs energy (e.g., from a photon) and jumps from a lower-energy band (the valence band) to a higher-energy band (the conduction band), leaving behind an empty state called a hole.

Due to the attractive electrostatic (Coulomb) force between the negatively charged electron in the conduction band and the positively charged hole in the valence band, they can bind together to form a composite particle – the exciton. This exciton is electrically neutral and can move through the material, carrying energy.

In condensed matter physics, fractionalization is a remarkable phenomenon that can occur in strongly correlated electron systems, especially in low dimensions (like two-dimensional electron gases). In these systems, electrons can behave as if they have been broken down into smaller, independent quasiparticles with fractional quantum numbers, such as fractional charge (as seen in the fractional quantum Hall effect). These fractionalized quasiparticles are not simply fractions of individual electrons but rather collective excitations of the many-body system.

Fractional excitons are essentially excitons where either the electron, the hole, or both are themselves fractionalized quasiparticles. This can occur in systems that already host fractionalized excitations, such as those that exhibit Fractional Quantum Hall (FQH) states. As we saw earlier, FQH states that the strong interactions between electrons in a two-dimensional electron gas under a strong magnetic field lead to the formation of exotic quasiparticles with fractional charge (e.g., charge $\frac{e}{3}$, $\frac{e}{5}$). A fractionalized electron and a fractionalized hole could form an exciton in such a system. The properties of this fractional exciton would then be governed by the fractional charges and statistics of its constituents.

Fractionalization can also occur in different systems with strong interactions and specific symmetries, such as certain topological insulators or Mott insulators. In these cases, excitons formed from fractionalized excitations would also be considered fractional excitons. In low-dimensional materials (e.g., two-dimensional materials or quantum wires), reduced dimensionality often enhances electron-electron interactions, making them potential hosts of fractional excitons.

The overall charge of the fractional exciton is still an integer multiple of the elementary charge (usually zero, as with regular excitons), but its constituent quasiparticles (the fractionalized electron and hole) carry fractional charges. The fractionalized quasiparticles that make up the fractional exciton can obey different exchange statistics compared to fermions (like electrons). They can be anyons, whose statistics are neither purely fermionic nor purely bosonic.

Quantum Computing

Quantum computing uses the strange behaviors of quantum mechanics, like superposition and entanglement, to perform calculations faster than traditional computers by using "qubits," which can be both a zero and a one at the same time, allowing them to explore many possibilities simultaneously.

Unlike regular computers, which use bits (0 or 1), quantum computers use qubits, which can be in a superposition of 0 and 1. A qubit's ability to exist in multiple states simultaneously (superposition) allows quantum computers to explore many potential solutions to a problem at once. When two qubits are entangled, they are linked so that knowing the state of one instantly tells you the state of the other, regardless of distance.

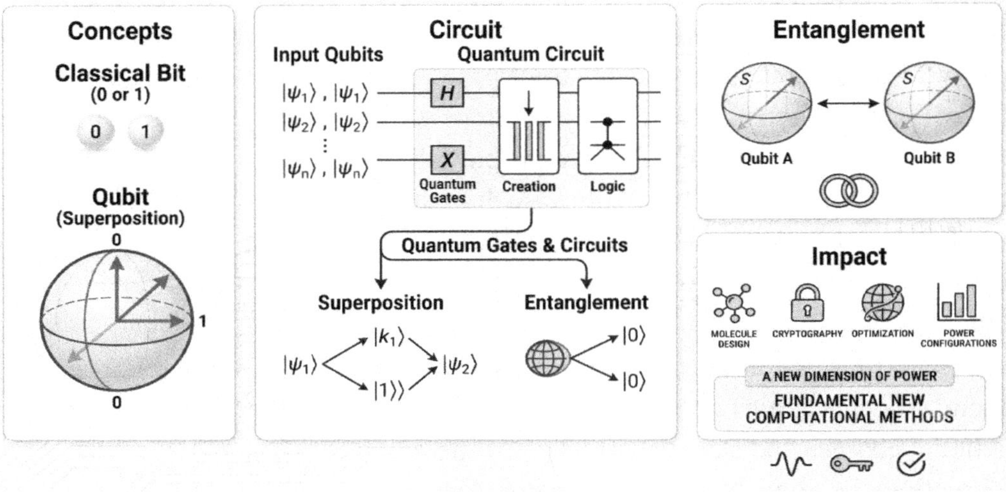

To understand why a quantum computer is faster, it is helpful first to realize that it isn't just "faster" in the way a sports car is faster than a sedan. Instead, it is faster because it takes a shorter path to the answer. While a regular (classical) bit is like a light switch that is either on (1) or off (0), a qubit uses the laws of quantum mechanics to process information in a fundamentally different way.

The first reason for the speed is superposition. A regular bit can only be in one state at a time. If you have 2 bits, they can represent 4 possible combinations (00, 01, 10, 11), but they can only "be" one of those at any given moment. A qubit, however, can exist in a state that is a mathematical combination of both 0 and 1 simultaneously. To check 4 combinations, a regular computer usually has to check them one by one (or use 4 separate processors). 2 qubits in superposition can represent and process all 4 combinations simultaneously. As you add more qubits, this power grows exponentially:

- 10 qubits = 1,024 states at once.
- 30 qubits = over 1 billion states at once.
- 300 qubits = more states than there are atoms in the observable universe.

The second reason is entanglement. Qubits can be linked so that the state of one is instantly entangled with the state of another, no matter how far apart they are.

In a classical computer, if you double the number of bits, you double the amount of work the computer can do (linear growth). In a quantum computer, because of entanglement, adding just one more qubit doubles the entire computational space (exponential growth). This allows the computer to manipulate massive, complex data structures simultaneously rather than piece by piece.

You might wonder: *"If the computer is doing all calculations at once, don't you just get a messy pile of random answers?"* This is where Interference comes in. Quantum algorithms (like those used to crack encryption) are structured like complex wave patterns. The "wrong" paths in the calculation cancel each other out (like noise-canceling headphones) – this is destructive interference. The "right" path – the correct answer – is amplified. This is constructive interference.

When the scientist finally "measures" the qubits, the probability of getting the correct answer approaches 100%, and the wrong answers disappear. Imagine a massive maze with a prize at the end.

A classical computer is like a mouse. It runs down a path, hits a wall, turns around, and tries another. It eventually finds the prize, but it takes a long time. In a quantum computer, it's like a mist that enters the maze. It floats down every path simultaneously. When the mist reaches the exit, it "solidifies" into a mouse exactly at the prize's location.

While quantum systems are often depicted as pristine and isolated, in reality, they interact with their surroundings, leading to a loss of quantum coherence. Understanding and mitigating this process is crucial for the development of quantum computing and communication. This is not a trivial issue; it explains why most modern systems must be supercooled to minimize decoherence. Error correction is improving, and the number of viable qubits continues to grow, but progress seems very slow. That's why there are very few algorithms that can be shown to outpace classical-based solutions by any significant amount.

The technology could become viable for solving problems currently intractable even for the most powerful supercomputers; it might redefine industries and accelerate scientific discovery, so the question is not when, but "if" and "what".

To date, experts believe that quantum computing has potential applications in various fields, such as designing new pharmaceutical molecules or novel materials for manufacturing, which is an incredibly complex task that involves simulating molecular interactions. Global financial markets are a web of intricate variables.

Quantum algorithms appear to be uniquely suited to optimizing investment portfolios, assessing risk, and pricing complex financial derivatives.

Internet security, which relies on the difficulty of factoring large numbers, is vulnerable to future quantum computers. This creates a "quantum threat" that requires new, "quantum-resistant" cryptography. Quantum communication promises unhackable networks, and quantum computing can supercharge specific machine learning models by enabling them to analyze vast and complex datasets, leading to breakthroughs in areas such as pattern recognition and data analysis.

Current quantum processors are incredibly delicate. They must be kept in highly controlled environments, shielded from vibrations and electromagnetic interference, and often cooled to temperatures colder than those found in deep space to maintain their quantum state. Qubits are notoriously fragile and prone to "decoherence," which causes them to lose their quantum properties and introduce errors into calculations. A major challenge in the field is developing strong quantum error correction.

In a paper published in the Proceedings of the National Academy of Sciences, Emeritus Professor Tim Palmer FRS considers quantum mechanics from a different perspective.[40] His theory of Rational Quantum Mechanics (RaQM) suggests that, above a critical number of entangled qubits, there isn't enough information in the quantum state to allocate even one bit of information to each dimension of Hilbert Space. When this happens, quantum algorithms that use all of Hilbert Space will no longer have a quantum advantage over classical algorithms. An example is Shor's algorithm for factoring integers, thereby enabling the decryption of RSA-encrypted messages.

Professor Palmer predicts that Shor's algorithm will start to fail in this way when a few hundred (error-corrected) qubits are entangled. This is bad news for those looking to develop quantum computers for practical applications. However, Professor Palmer views things more positively:

"*If quantum computers provide the experiments not only to find a successor theory to QM, but more importantly to find the theory which synthesizes quantum and gravitational physics, that would surely be an extraordinarily good outcome for all the work that has been put into quantum computing over the years.*"

The most likely near future is hybrid solutions where classical and quantum computers work in tandem. Classical computers will continue to handle the vast majority of computational tasks, while offloading specific, complex problems to specialized quantum co-processors, which will likely be accessed via the cloud.

Photonic Computing

Photonic computing (optical computing) is an emerging technology that uses photons rather than electrons for data processing, storage, and transmission. This shift from electronics to photonics promises to overcome fundamental limitations of modern semiconductor technology, particularly in speed, heat dissipation, and power consumption. Given the explosion of data centers to keep the AI boom going, the latter two are becoming increasingly urgent.

The history of using light for computation can be divided into three main phases: theoretical foundations, early analog efforts, and the digital transition.

Claude Shannon's work in information theory established the theoretical limits of information transfer, which hinted at the immense bandwidth potential of light compared to electrical signals. The invention of the laser by Theodore Maiman in 1960, provided the necessary high-quality, coherent light source required for optical processing. This immediately spurred research into optical devices.

From the 1960s to the 1980s, efforts focused on using light to solve complex, parallel problems rather than a direct electronic replacement. Researchers used lenses and mirrors to perform large, simultaneous mathematical operations, such as Fourier transforms (essential for image processing), almost instantaneously.

From the 1990s to the 2010s, the focus shifted from large analog systems to miniaturized digital components integrated directly onto silicon chips, drawing on semiconductor industry techniques.

Today, the field is moving toward true co-integration and the use of photons for computation itself, not just communication. Optical Neural Networks (ONNs) for Artificial Intelligence are among the most impactful current developments.

Photonic platforms are proving ideal for building scalable quantum computers. The polarization or path of a single photon can serve as a qubit. Photons are robust against decoherence and can transmit quantum information over long distances at the speed of light, making them excellent candidates for quantum networks.

Photonic computation inherently solves the two biggest problems facing Moore's Law: 1. Photons generate virtually no waste heat as they travel through optical waveguides, enabling far denser integration of logic circuits without thermal throttling. 2. Photonic signals can overlap without interference (unlike electrical signals), allowing a massive increase in the number of channels within a small chip area. This is essential for highly parallel computing architectures.

Quantum Technologies

Cryptography

Quantum cryptography has transitioned from a niche laboratory experiment to a commercially available, critical technology for securing data. While not yet a mainstream replacement for all forms of encryption, it is developing rapidly, driven by the looming threat of quantum computers that will one day be able to break much of the cryptography that protects our digital world today.

The current state of quantum cryptography can be summarized as a race to build a secure communications infrastructure for the quantum era. Fundamentally, the most mature application of quantum cryptography is Quantum Key Distribution (QKD).

Unlike traditional encryption, which relies on mathematical complexity, the security is guaranteed by the fundamental laws of physics. It enables two parties to securely exchange a secret, random key that can then be used with classical encryption algorithms to protect data. Any attempt by an eavesdropper to intercept and measure the quantum particles (photons) used to create the key inevitably disturbs their state, immediately alerting the legitimate users to the security breach.

Dedicated fiber optic QKD networks have been established in several countries. The primary adopters are organizations with extremely high security needs, such as government agencies, the military and defense sectors, financial institutions securing high-value transactions, and critical infrastructure, including power grids.

To overcome the limitations of fiber-optic distance, researchers have successfully demonstrated QKD via satellite. Despite this progress, several challenges are preventing the widespread adoption of quantum cryptography:

1. In fiber-optic cables, signal loss limits the range of a single QKD link to a few hundred kilometers. While trusted nodes can extend this range, they also introduce a potential security vulnerability.
2. The rate at which secret keys can be generated is still relatively slow compared to classical methods, limiting their use in applications that require very high-speed data encryption.
3. The hardware required for quantum cryptography remains expensive and requires specialized knowledge to operate and maintain, making it inaccessible for most consumer and small business applications.
4. Integrating quantum security solutions with existing classical network infrastructure is a complex, ongoing effort.

The primary driver of the increased urgency to develop quantum cryptography is the rapid advancement in quantum computing. It is widely predicted that a sufficiently powerful quantum computer will be able to break widely used public-key encryption algorithms, such as RSA and Elliptic Curve Cryptography. This would render much of the world's secure data, from financial records to state secrets, vulnerable to unauthorized access.

In response, the field of Post-Quantum Cryptography (PQC) is developing new classical encryption algorithms believed to be resistant to attacks by both classical and quantum computers.[41] The prevailing view is that the future of secure communications will involve a combination of PQC for its ease of deployment and QKD for its unparalleled physical security in high-stakes environments.

Quantum Navigation

Quantum navigation is an emerging technology that promises a revolutionary, standalone method for determining an object's precise position, orientation, and velocity, without relying on external signals such as those from the Global Positioning System (GPS). Instead of looking to satellites, it looks inward, using quantum-mechanical principles to sense motion with extraordinary precision.

Although the technology is still in its early stages of development and faces engineering challenges, the underlying science is sound.

Quantum navigation is a highly advanced form of inertial navigation. All Inertial Navigation Systems (INS) use accelerometers and gyroscopes to track changes in motion. By starting from a known position and continuously measuring accelerations, turns, and changes in direction, an INS can calculate its new position.

Your smartphone and commercial airliners use this technology today, but it suffers from a critical flaw: drift. Tiny errors in the sensors accumulate over time, causing the calculated position to "drift" further and further from the true position. Quantum navigation seeks to address the drift problem by utilizing quantum sensors, primarily cold-atom interferometers. Here's a simplified explanation of the process:

A cloud of atoms is trapped and cooled by lasers to a temperature just fractions of a degree above absolute zero. At these extreme temperatures, the atoms behave like a single quantum object, a wave of matter. A laser pulse splits the atomic cloud, causing the atoms to enter a quantum superposition in which they travel along two different paths simultaneously. The device's movement affects the two quantum waves slightly differently. After a set time, another laser pulse recombines the two waves. The way they interfere with each other–either reinforcing or canceling each other out–creates a measurable pattern.

This interference pattern is exquisitely sensitive to the motion the device experienced while the atoms were separated. By measuring these interference patterns with extreme accuracy, quantum accelerometers and gyroscopes can detect motion with a precision theoretically thousands of times better than that of their classical counterparts. This dramatically reduces drift, allowing a vehicle to maintain its position for days, weeks, or even longer without needing to correct its course with an external signal.

It's a credible future technology because modern society is dangerously reliant on GPS. The satellite signals are weak, easily jammed by adversaries, "spoofed" with false information, and can be disrupted by solar weather or blocked in underwater, underground, or dense urban environments. Quantum navigation offers a path to "GPS-free" navigation, a strategic advantage for military and commercial applications.

Quantum Optics

Quantum optics is the study of how photons interact with matter at the quantum level, examining the behavior of light when treated as particles rather than waves, and allowing scientists to explore phenomena such as entanglement and coherence, which are key aspects of quantum mechanics. Unlike classical optics, which treats light as a wave, quantum optics looks into the individual behavior of photons. This field examines how photons interact with atoms and molecules, focusing on absorption, emission, and scattering phenomena. Quantum optics enables scientists to study quantum-mechanical concepts, such as superposition and entanglement, using photons.

Quantum optics sits at the heart of one of the most famous paradoxes in science: light acts like a wave when it travels through space, but acts like a particle when it hits a detector. Light can interfere with itself, creating patterns of light and dark (interference). Light can knock electrons off a metal surface (the Photoelectric Effect), causing them to behave like tiny billiard balls. When a photon meets an atom, one of three things usually happens:

1. ***Absorption:*** The atom "eats" the photon and gains energy.
2. ***Spontaneous Emission:*** An atom randomly emits a photon to lose energy.
3. ***Stimulated Emission:*** A passing photon "nudges" an excited atom, forcing it to release an identical photon, and this is the basic principle of lasers.

Without understanding the quantum nature of light, we would not have:

- ***Fiber Optic Cables:*** The high-speed backbone of the internet.

- ***Quantum Computers:*** Which use "photonic qubits" to process information.
- ***Precision Medicine:*** Using lasers for surgery and ultra-sensitive imaging.
- ***Cryptography:*** Using the "indivisibility" of a photon to create unhackable communication.

Research in quantum optics is crucial for developing quantum computers, which use the quantum properties of light to perform complex calculations. Secure communication systems can be developed using quantum properties of light to ensure data privacy. Due to their high sensitivity, quantum-optical techniques are used in advanced measurement devices.

Quantum Sensors

Quantum sensors are a new class of ultra-precise measurement devices that exploit the bizarre and delicate properties of quantum mechanics to detect tiny changes in their surroundings. Think of the difference between a simple mercury thermometer and a modern infrared thermal camera. Both measure temperature, but the camera offers a level of detail and sensitivity orders of magnitude greater. Quantum sensors represent a leap forward in measuring phenomena such as gravity, magnetic fields, and time.

The key to understanding is to recognize that individual quantum systems – such as single atoms, electrons, or photons–are incredibly fragile. Their quantum states (such as their energy levels or "spin") are exquisitely sensitive to the slightest disturbance from their environment. In quantum computing, this fragility is a major problem known as "decoherence," which engineers work diligently to prevent. In quantum sensing, this same fragility is their greatest strength. They are designed to be so sensitive that even a minuscule change in gravity or a faint magnetic field from a brain's neurons will predictably and measurably disturb their quantum state.

They achieve this by using quantum phenomena such as superposition, entanglement, and Quantum Interference, where the wave-like nature of particles is used to measure tiny shifts. It isn't just theory; several types of quantum sensors are already in use or in advanced stages of development.

Atomic clocks are the most mature and well-known quantum sensors. They work by using the perfectly consistent, quantized energy transitions of an atom (such as cesium) as a hyper-accurate pendulum. The clock's "tick" is the frequency of the microwave radiation needed to make the atom's outer electron jump between two specific energy levels. They are the backbone of GPS, global financial networks, and the internet.

Atom Interferometers measure gravity, acceleration, and rotation with mind-boggling precision. Lasers cool a cloud of atoms to near absolute zero. Another laser then places the atoms in a superposition, effectively allowing them to travel two different paths at once. Gravity and motion affect these two paths differently. When the paths are recombined, the way the atom's waves "interfere" reveals tiny variations in the forces it experiences.

The incredible sensitivity of quantum sensors is opening new possibilities across many fields, including, as we saw earlier, GPS-free navigation. Using highly accurate atom interferometers, submarines, aircraft, and drones could navigate for weeks or months without a GPS satellite signal.

In healthcare and brain imaging, quantum sensors can detect the incredibly weak magnetic fields produced by the human brain (magnetoencephalography) or even a single neuron firing. This could lead to a revolution in understanding and diagnosing neurological disorders.

For geology and civil engineering, vehicles equipped with quantum gravity sensors could map underground structures with incredible detail. This would allow us to find underground water reserves, detect hidden tunnels or sinkholes before they collapse, and monitor magma chambers beneath volcanoes.

In fundamental science, the giant LIGO observatories that first detected gravitational waves are essentially massive-scale quantum interferometers. Future quantum sensors are being designed to find dark matter and other undiscovered particles.

Quantum Teleportation

Clear the image of science fiction from your mind, and any of the sensational "teleportation achieved!" headlines. Quantum teleportation does not transport regular matter, such as "beaming" a person from one place to another as in *Star Trek*. Instead, quantum teleportation is the faithful transfer of a particle's quantum state from one location to another, without the particle physically traveling between them. Think of it like a "quantum fax machine." You aren't physically sending the original piece of paper. You scan it, send the information, and create a perfect replica at the destination while destroying the original. With quantum teleportation, you do this with the precise quantum information of a particle.

The process relies on the phenomenon of quantum entanglement. Imagine there are two "entangled" particles, let's say two electrons. We'll refer to them as Particle A and Particle B. When they are entangled, they are linked especially, no matter how far apart they are – whether it's across a lab bench or across the galaxy. Their properties, like "spin", are perfectly correlated. If an experiment is set up in which one is"spin-up," the other is guaranteed to be "spin-down," and vice versa. The weird part is this: before measurement, neither particle has a definite state. They are in a quantum superposition of all possibilities. This instantaneous correlation is what makes quantum teleportation possible.

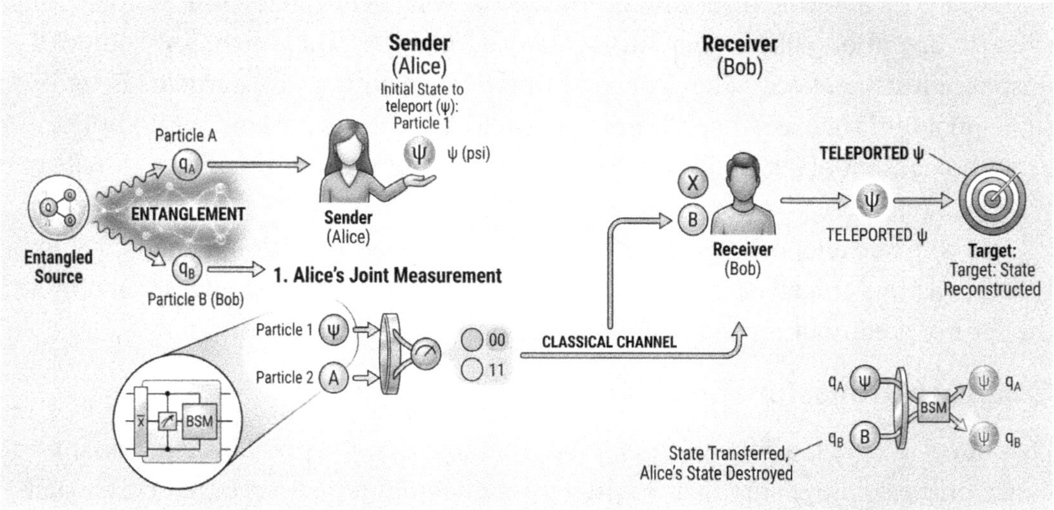

Let's imagine a classic scenario with three characters: Alice, Bob, and a quantum third party, "Charlie." Alice wants to teleport the quantum state of Particle C (Charlie) to Bob. Particle C is in an unknown quantum state. Alice doesn't know its state, and if she tries to measure it, she will destroy the very information she wants to send. This is a key rule in quantum mechanics. Alice and Bob each hold one particle from an entangled pair (particles A and B, respectively). They could have generated these years ago and then moved far apart. Bob might even be here on Oak Island, while Alice is on a spaceship orbiting Mars.

Alice takes her entangled particle (A), and Bob takes his (B). They are now physically separated but still quantum-ly linked. Alice brings together the particle she wants to teleport (C) and her half of the entangled pair (A). She performs a special joint measurement on them called a Bell-state measurement. This measurement forces particles A and C to interact.

The moment Alice completes her measurement, the link between her particle (A) and Bob's particle (B) does its magic. The measurement she performed on A and C instantly changes the state of Bob's particle (B) in a very specific way. Bob's particle now holds the state of the original particle (C), but it's slightly scrambled. Crucially, at this exact moment, the original state of Particle C is destroyed.

Alice's measurement gives her one of four possible outcomes. This result shows that Bob's particles are scrambled. She must then send this information to Bob. This part is not instantaneous; she has to use a classical communication channel, such as a phone call or an email. The information is just two classical bits (e.g., "00," "01," "10," or "11"). Bob receives both bits of information from Alice. This information is like a key or a set of instructions. Based on what Alice sent him, he performs a specific operation (like a rotation) on his particle (B). This operation "unlocks" or unscrambles the state. After Bob performs the operation, his particle, B, is in the same quantum state as Alice's original particle, C. The teleportation is complete. The exact process involves Bell measurements and unitary transformations – I will spare you the equations.

We won't be teleporting people anytime soon (the information content of a human being is astronomically large and complex). But it's a real and vital technology for the future of computing and communications.

Quantum Consciousness

Quantum theory offers a potential, though controversial, explanation for consciousness, proposing that it arises from quantum processes in the brain, such as quantum entanglement or wave function collapse.

Proposed by Roger Penrose and Stuart Hameroff, the Orchestrated Objective Reduction (Orch-OR) theory suggests that consciousness occurs due to the objective reduction (OR) of quantum superpositions within microtubules. The microtubule protein structures within neurons are thought to provide the necessary environment for maintaining quantum coherence long enough to contribute to consciousness.

In quantum mechanics, measurement causes a system's wave function (which represents multiple possible states) to collapse into a single state. Penrose and Hameroff suggest that this collapse, occurring within microtubules, gives rise to a moment of consciousness. There are crucial issues with the theory, including environmental instability and decoherence.

The brain is considered a "warm and noisy" environment, which is generally expected to lead to rapid decoherence, destroying delicate quantum states. Estimates suggest that quantum coherence in microtubules would collapse too quickly (on the order of 10^{-13} seconds) to be meaningful for consciousness, which requires much longer durations.

Researchers are designing experiments to investigate the role of microtubules and quantum processes in consciousness, including studies on the effects of anesthesia on microtubule function. Some approaches re-evaluate the foundations of quantum theory to find connections with consciousness, moving beyond simple particle-based models. Other researchers are exploring alternative models, such as the idea that consciousness arises when quantum superpositions form rather than when they collapse, to avoid potential paradoxes.

Superradiance is a quantum phenomenon in which a group of atoms or emitters (an ensemble of excited molecules, such as amino acids) synchronizes to emit light collectively as a single, intense flash, rather than radiating independently. Because their emissions constructively interfere, the intensity of this burst scales quadratically with the number of emitters, producing a much brighter, faster pulse.

First predicted by physicist Robert Dicke in 1954, this occurs when excited emitters are confined to a region smaller than the wavelength of the light they emit. Instead of decaying randomly and independently (which results in a standard glowing emission), the atoms act as a single, giant antenna.

A major criticism of Orch-OR is that the warm, wet brain causes "decoherence" – destroying fragile quantum states before computation can occur. Proponents suggest that superradiance may actively protect quantum states. The collective synchrony of molecules such as tryptophan within microtubules can shield the system from environmental noise, enabling stable, large-scale quantum integration. According to Orch-OR, conscious moments happen when this massive, unified quantum state reaches a critical gravitational threshold and collapses (Objective Reduction). Superradiance acts as the "orchestration" mechanism. It provides a way for billions of tubulin proteins to synchronize their energy states into a unified, coherent "musical" vibration across the brain, eventually triggering a collective collapse that manifests as a conscious thought.

If superradiance is how the brain generates consciousness, disrupting it should render a person unconscious. Proponents point to studies showing that general anesthetics inhibit exciton movement and fluorescence in microtubules. By interfering with collective energy states, anesthetics prevent superradiance, effectively halting the proposed quantum computations of consciousness. While superradiance is a widely accepted, well-studied principle of quantum optics, applying it – and the Orch-OR theory – to human consciousness remains a highly speculative, intensely debated concept in neuroscience and quantum biology.

So although there is interest in experimental testing, the precise mechanisms linking quantum activity to subjective consciousness remain unproven. Key predictions of the Orch-OR theory are subject to debate.

I am slightly amused that, when the theory was first suggested, it was laughable, and now it might actually be real. Perhaps another quantum theory that may prove very difficult to prove – but one that is likely to stir the imagination of science fiction. Even I can think up some fun ways to exploit this "functionality" in a story.

The Multiverse

The multiverse is a mind-bending and exciting idea in modern physics. While it's a staple of science fiction, the scientific concept of a multiverse isn't just a wild guess; it's a set of related hypotheses that naturally follow from our current best theories of the universe.

First, it's important to understand that there is no single "multiverse theory." Instead, it's a collection of different models that propose the existence of other universes besides our own. These ideas stem from attempts to address fundamental problems in cosmology and quantum mechanics. The idea of a multiverse isn't just pulled from thin air; it is a potential consequence of well-established scientific theories. When physicists push theories such as cosmic inflation and quantum mechanics to their logical conclusions, they often find that the existence of other universes is not just possible but, in some cases, almost unavoidable.

The most widely accepted cosmological concept of a multiverse is the Inflationary Multiverse (Bubble Universes). Cosmic inflation is our leading theory for what happened in the first fraction of a second after the Big Bang. It posits that the universe underwent an incredibly rapid, exponential expansion, smoothing out the cosmos and seeding the structures we see today, such as galaxies.

The theory of *eternal inflation* suggests that once started, this expansion process never completely stops. While small patches of space can stop inflating and form stable universes (like our own), the vast expanse of space between them continues to expand at a breakneck pace, constantly spawning new "bubble" universes.

Imagine a vast pot of water that is perpetually boiling. Our universe is like a single bubble that formed, expanded, and cooled. But the "water" itself – the fabric of spacetime–is still boiling and constantly creating countless other bubbles. Each of these bubble universes could have different fundamental laws of physics and different physical constants. One might have stronger gravity, another might have only two dimensions of space, and most would be completely sterile and incapable of supporting any form of complexity or life.

The second is the Quantum Multiverse model, which comes not from cosmology but from the rules of quantum mechanics, which state that a particle (such as an electron) can exist in a "superposition" of multiple states at once. It's only when we measure it that it "chooses" one definite state. The Many-Worlds Interpretation (MWI) is one way to explain what happens during this "choice."

The MWI, proposed by Hugh Everett III in the 1950s, suggests that the particle doesn't choose one state. Instead, the entire universe splits into multiple copies. In each copy, one of the possible outcomes becomes reality. Think of it like a *"Choose Your Own Adventure"* book. Every time a quantum event happens, you don't just turn to one page. The universe itself branches, and you follow every possible path simultaneously, with a different version of "you" living in each parallel timeline.

These parallel worlds would have the same laws of physics as our own, but their histories would differ based on the outcomes of quantum events. In one universe, you took that job in California; in another, you stayed home.

The String Theory Multiverse (Brane-Worlds) theory is based on String Theory, a candidate for a "theory of everything" that seeks to unify all the forces of nature. String Theory requires the existence of extra spatial dimensions beyond the three we experience. It could be 9, 10, or even 25 dimensions of space. In this framework, our entire three-dimensional universe might be a membrane, or "brane," floating within a higher-dimensional space, often called the "bulk." The theory allows for many such branes to exist, each a self-contained universe.

Picture a single loaf of sliced bread. Our universe is just one slice (a brane). But there are other slices parallel to ours, potentially just millimeters away, in a higher dimension we cannot access directly. These brane-worlds could have different physical laws. One intriguing possibility is that gravity from other branes could "leak" into ours, potentially explaining why gravity is so much weaker than the other fundamental forces.

It's important to understand that there is currently no direct, verifiable evidence for the existence of any multiverse. Scientists have looked for indirect evidence, such as distortions or "bruises" in the Cosmic Microwave Background (the afterglow of the Big Bang) that could have been caused by a collision with another bubble universe in the distant past. To date, no conclusive findings have been established. Another idea is to try to detect gravitational anomalies, the influence of gravity leaking from another brane in a higher dimension.

The multiverse is a fascinating and mathematically plausible concept that arises from our most fundamental physical theories. It offers potential explanations for questions, such as why the physical constants of our universe seem so perfectly "fine-tuned" for life to exist (the anthropic principle).

Matter-Antimatter Asymmetry

A matter-antimatter asymmetry means that in our universe, there is much more matter than antimatter, even though they should have been created in equal amounts during the Big Bang; essentially, the universe is made mostly of "regular" matter, and not much antimatter, which is a big puzzle because when matter and antimatter meet, they annihilate each other, leaving only energy behind.

Current theories of matter-antimatter asymmetry involve mechanisms that violate fundamental symmetries, such as Leptogenesis and CP violation. Leptogenesis suggests that a breach of lepton-number conservation generates the asymmetry, while CP violation proposes that matter and antimatter behave differently in certain processes. Experiments at the Large Hadron Collider (LHC) are actively searching for these differences, particularly in the behavior of baryons and the Higgs boson, which could provide evidence to support these theories.

Cosmological models suggest the Big Bang created equal amounts of matter and antimatter, which should have annihilated each other, leaving behind only energy. The fact that a small excess of matter particles survived suggests a process occurred that tipped the balance in favor of matter. This is a key concept explaining the imbalance. CP (Charge-Parity) symmetry is a combination of two distinct symmetries:

1. ***C (Charge):*** A particle is swapped with its antiparticle (e.g., an electron with a positron).
2. ***P (Parity):*** A particle's mirror image is considered (e.g., its "handedness" or spin is flipped).

If CP symmetry is violated, matter and antimatter particles would not have behaved the same. This slight difference could have led to a tiny surplus of matter before most particles annihilated.

Leptogenesis suggests that the matter-antimatter asymmetry was produced by a violation of lepton-number conservation in the early universe. The properties of neutrinos, particularly their potential to violate CP symmetry, are being studied as a possible source of the asymmetry. Briefly, leptogenesis involves extremely heavy, "sterile" neutrinos. These are hypothetical right-handed counterparts to the light, left-handed neutrinos we observe in nature. In the extreme heat of the infant universe, these heavy neutrinos were produced in vast quantities. As the universe expanded and cooled, these particles became unstable. Because they were so massive, they decayed into lighter particles, specifically standard neutrinos (leptons) and Higgs bosons.

For leptogenesis to create an imbalance, it must satisfy conditions first outlined by physicist Andrei Sakharov. The most critical part is CP violation (Charge-Parity violation). If CP symmetry were perfect, a heavy neutrino would decay into a lepton and an anti-lepton at the same rate. However, in leptogenesis models, these heavy neutrinos decay slightly more often into leptons than into antileptons. This creates a "lepton asymmetry" – a surplus of leptons over anti-leptons.

The mystery of the missing solar neutrinos

For three decades, observations of the Sun consistently conflicted with the theoretical models of how stars produce energy. To understand how we have reached our current understanding of neutrinos and the problem with the Sun, we need to review the timeline:

- **1964:** John Bahcall calculates the theoretical flux of neutrinos from the Sun using the Standard Solar Model, while Raymond Davis Jr. begins designing a detector to capture them.
- **1968:** The Homestake experiment in South Dakota reports its first results. It detects only about one-third of the electron neutrinos (ν_e) predicted.
- **1980s:** The Kamiokande-II experiment in Japan confirmed the deficit and provided the first directional data proving they originate from the Sun.
- **1990s:** Radiochemical experiments GALLEX and SAGE observe the same deficit in lower-energy neutrinos, ruling out certain astrophysical explanations for the missing flux.
- **1996:** Super-Kamiokande (Super-K) begins operations in Japan with a 50,000-ton tank of ultrapure water.
- **1998:** Super-K announces the first definitive evidence of neutrino oscillation based on atmospheric neutrino data.
- **2001–2002:** The Sudbury Neutrino Observatory (SNO) in Canada detected all three flavors of neutrinos (ν_e, ν_μ, ν_τ), confirming that the total flux matches Bahcall's predictions. The "missing" neutrinos had changed flavor.

The Standard Solar Model describes nuclear fusion in the solar core, where hydrogen nuclei fuse into helium. This process releases massive amounts of energy and electron neutrinos (ν_e). Raymond Davis used a tank filled with 100,000 gallons of perchloroethylene (dry-cleaning fluid) to detect these particles. When an electron neutrino hits a chlorine-37 atom, it transforms the atom into radioactive argon-37. Because Davis only found one argon atom every few days – representing one-third of the expected count – physicists faced a dilemma. Either the solar model was wrong about the Sun's core temperature, or neutrinos had properties not described by the Standard Model of particle physics.

I remember hearing about this before the solution had been found – I was actually worried that our sun's fusion reactions in the core might have stopped. Luckily, and much more interestingly, the sun was fine, and the real answer led to a whole new world of neutrino physics.

Super-Kamiokande detected neutrinos via Cherenkov radiation. When a neutrino interacts with water in the tank, it can produce a high-speed charged particle, such as an electron or a muon. This particle travels faster than the speed of light in water, creating a "shockwave" of blue light.

- ***The Detector:*** About 11,200 photomultiplier tubes line the tank walls to detect these faint rings of light (Kajita, 2010).
- ***Atmospheric Anomaly:*** In 1998, Super-K researchers studied neutrinos created by cosmic rays hitting Earth's atmosphere. These reactions produce both muon neutrinos (ν_μ) and electron neutrinos (ν_e) in a 2:1 ratio.
- ***The Findings:*** The number of muon neutrinos coming from directly above matched the predictions. However, the number of muon neutrinos coming from "down" – having traveled 12,000 kilometers through the Earth – was only half the expected amount. This proved that the neutrinos were oscillating into a different flavor, specifically tau neutrinos (ν_τ), during their flight.

In the Standard Model, neutrinos were assumed to be massless. However, quantum mechanics dictates that for a neutrino to oscillate (change flavor), it must have a non-zero mass. Neutrinos exist in two different sets of states: flavor states $(\nu_e, \nu_\mu, \nu_\tau)$ and mass states (ν_1, ν_2, ν_3). A single flavor, such as a ν_μ, is actually a superposition of the three different mass states.

As a neutrino travels, its mass states propagate as waves. If the masses are different, these waves travel at slightly different velocities. This creates a phase shift between the waves over distance. This shift changes the "mixture" of the mass states, which in turn changes the flavor the detector sees. The probability of a neutrino changing flavor depends on the difference between the squares of the masses, expressed as:

$$P(\nu_\alpha \to \nu_\beta) = \sin^2(2\theta)\sin^2\left(1.27\frac{\Delta m^2 L}{E}\right)$$

In this equation, Δm^2 represents the mass difference, L is the distance traveled, and E is the energy. If neutrinos had no mass (or identical masses), Δm^2 would be zero, the sine function would remain zero, and no oscillation would occur. Therefore, the discovery of oscillations at Super-K is direct proof that neutrinos are massive particles.

The Mass Ordering Problem

Neutrinos are created in specific "flavor" states (electron, muon, or tau). However, these flavors are not the same as their "mass" states (m_1, m_2, m_3). As a neutrino travels, its flavor state evolves because each mass component propagates as a wave with a slightly different frequency. Oscillation experiments (such as T2K, NOvA, and Super-Kamiokande) measure the probability of a neutrino changing flavor over a distance (L) at a given energy (E). These experiments are sensitive to the difference between the squares of the masses, but not the absolute mass of the particles themselves. Because oscillation depends on the *difference* between masses, we know there are two distinct gaps:

- ***The "Solar" gap*** (Δm_{21}^2)***:*** A small difference between m_1 and m_2. We know m_2 is heavier than m_1 due to how neutrinos interact with matter in the Sun (the MSW effect).
- ***The "Atmospheric" gap*** ($|\Delta m_{31}^2|$)***:*** A much larger difference involving m_3.

The "Mass Ordering" problem is that we do not know the sign of this larger gap, and this leads to two possible scenarios:

- ***Normal Ordering (NO):*** $m_1 < m_2 < m_3$. The two light neutrinos are at the bottom, and m_3 is the heaviest (the "light" hierarchy).
- ***Inverted Ordering (IO):*** $m_3 < m_1 < m_2$. The two heavy neutrinos are at the top, and m_3 is the lightest.

We cannot identify the heaviest neutrino yet because oscillation experiments are "blind" to the absolute scale. They tell us that m_3 is roughly 0.05 eV away from the others, but they don't tell us if it is 0.05 eV *above* them (making it the heaviest) or 0.05 eV *below* them (making it the lightest). To find the heaviest neutrino, we need to determine the sign of Δm_{31}^2. Experiments like JUNO and DUNE are specifically designed to observe how neutrinos interact with matter over long distances, which will eventually reveal the correct ordering.

In neutrino physics, "degeneracy" refers to a mathematical problem in which different sets of physical parameters yield the same experimental result. This makes it difficult to pinpoint the truth. The most common is the Hierarchy-CP Degeneracy:

- A certain oscillation result could be explained by Normal Ordering combined with one value of CP violation.
- The *same* result could also be explained by Inverted Ordering combined with a *different* value of CP violation.

There is also the Octant Degeneracy, where we don't know if the mixing angle θ_{23} is slightly less than or slightly greater than 45 degrees. Until we have more precise data, these "clashing" interpretations remain equally valid.

CP violation is the phenomenon where physics behaves differently for particles than for antiparticles. In neutrinos, this is represented by a phase factor (δ_{CP}) in the PMNS mixing matrix.

- ***The Goal:*** If we find that neutrinos oscillate at a different rate than antineutrinos, it means δ_{CP} is not zero or 180 degrees.
- ***The Significance:*** This theory suggests that an asymmetry in the early universe's neutrinos eventually led to the dominance of matter over antimatter.

Experiments, such as T2K, have hinted at maximal CP violation, but we are still waiting for the "5-sigma" level of certainty that would officially constitute a discovery.

While creating an excess of leptons is the first step, the universe we see today is dominated by "baryons" (protons and neutrons). The theory relies on a high-temperature process called a sphaleron interaction. Sphalerons are non-perturbative effects in the Standard Model that occur only at very high energies. They have the unique ability to transform lepton numbers into baryon numbers. Through these interactions, the initial excess of leptons was partially converted into an excess of quarks, which eventually cooled and bound together to form the protons and neutrons that make up all visible matter. Physicists highly regard leptogenesis because it ties together several unexplained phenomena:

Neutrino Mass: The same heavy neutrinos used in leptogenesis are part of the "see-saw mechanism," which explains why the neutrinos we detect in labs are so incredibly light compared to other particles. Neutrinos, spin-$\frac{1}{2}$ particles may not be affected by the Higgs Field. Many physicists suspect neutrinos are Majorana particles, meaning they are their own antiparticles. If this is true, neutrinos can gain mass through the See-Saw mechanism. In this mathematical "see-saw," the more massive the right-handed neutrino is, the lighter the left-handed neutrino becomes. This would naturally explain why the neutrinos we detect have such incredibly tiny masses.

Matter Dominance: It provides a mathematically consistent explanation for the matter-antimatter imbalance without requiring new, undiscovered forces. This is definitely an area that I hope bears more fruit.

Dark Matter: Some variations of the theory suggest that a specific type of sterile neutrino could also explain the missing Dark Matter in the universe.

Experiments are searching for CP asymmetry in the Higgs boson's interactions with other particles, which could offer further clues to the origin of matter.

In July 2025, the LHCb experiment at CERN observed CP violation for the first time in the decay of baryons, which are fundamental to the visible matter in the universe. Scientists at CERN continue to collect data from the Large Hadron Collider to more precisely measure CP-violating effects and search for new sources of this asymmetry. Let's look at CPT-Symmetry a bit deeper.

CPT-Symmetry

In the Standard Model, the neutron's electric dipole moment is predicted to be a factor of ten billion larger than our observational limits show. The only explanation is that somehow, something beyond the Standard Model is protecting this CP symmetry in the strong interactions. If the C symmetry is violated, so is PT; if P is violated, so is CT; if T is violated, so is CP. CPT symmetry is a fundamental principle of physics that describes a deep, underlying balance in the laws of nature.

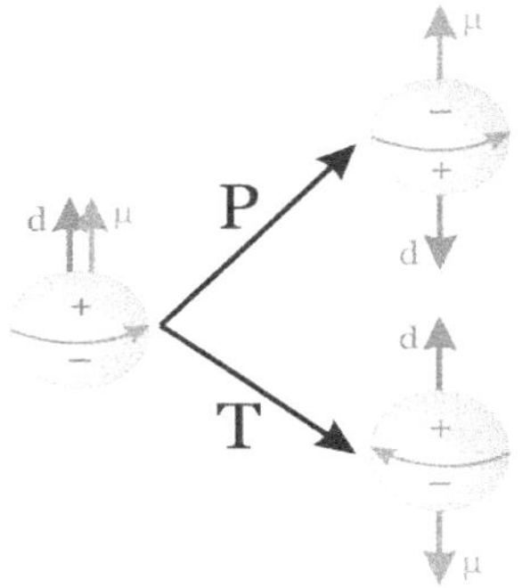

It states that the laws of physics remain unchanged (invariant) if you simultaneously flip three specific properties: Charge, Parity, and Time. The CPT Theorem is a cornerstone of the Standard Model and quantum field theory. It suggests that a universe where matter is replaced by antimatter, everything is reflected in a mirror, and time runs backward would behave exactly like our own.

To understand the combined symmetry, it is best to look at each operation individually:

1. The Charge Conjugation operation involves swapping every particle with its corresponding antiparticle. For example, an electron (a negatively charged particle) becomes a positron (a positively charged particle). C-symmetry suggests that the laws of physics should be the same for matter and antimatter. We know C-symmetry is violated on its own, particularly in the weak nuclear force.

2. Parity is a spatial reflection, similar to looking in a mirror. It involves flipping the coordinates of space (swapping "left" for "right" and "up" for "down"). If a process obeys P-symmetry, its mirror image should be physically possible. In 1956,

Chien-Shiung Wu famously proved that the weak force violates P-symmetry–it has a "handedness" and prefers left-handed particles.

3. The Time Reversal (T) operation involves replacing time (t) with ($-t$), effectively "rewinding the movie" of a physical process. T-symmetry suggests that if a motion is allowed by physics, the exact reverse of that motion is also allowed. While most classical laws are T-symmetric, certain rare subatomic processes (involving kaons and B-mesons) have been shown to violate T-symmetry.

While nature violates C, P, and T symmetries individually (and even violates the combination of CP), the CPT Theorem states that the combination of all three must be an exact symmetry of any local quantum field theory that respects Lorentz invariance (the principle that the laws of physics are the same for all observers moving at constant velocity). If you find a violation of CP symmetry in a laboratory, the laws of physics *must* compensate by violating T-symmetry in the same way to ensure the total CPT remains balanced.

CPT symmetry is important because it predicts that every particle must have the same mass and lifespan as its antiparticle. So far, every experiment (measuring the mass of protons vs. antiprotons, for example) has confirmed this to incredible precision. It provides a mathematical anchor for why time flows in one direction in our macroscopic world, even if the subatomic universe is technically reversible when all other symmetries are flipped.

Some modern theories use CPT symmetry to explain the Big Bang. They suggest that the birth of our universe also created a "mirror" universe on the other side of the starting point, where time runs in the opposite direction and matter is replaced by antimatter. This model helps explain why our universe is so orderly and why we don't see massive amounts of antimatter in our own sector of space.

In essence, CPT symmetry is the universe's ultimate consistency check. It ensures that, no matter how we "view" the system – whether mirrored, inverted, or reversed – the fundamental mathematical logic governing particles and forces remains perfectly synchronized. Any violation of CPT would require a total rewrite of our understanding of relativity and quantum mechanics.

Aharonov–Bohm (AB) effect

The Aharonov–Bohm (AB) effect shows that the quantum world behaves differently than the classical world we see every day. In classical physics, an electric or magnetic field must touch a particle directly to change its path. In quantum mechanics, a magnetic field can alter a particle's behavior even if the particle never enters the field.

To understand this, consider how magnetic fields work. When you place a magnet near iron filings, the filings line up along invisible lines of force. This is the magnetic field. In the mid-nineteenth century, physicists developed amathematical tool called a "potential" to help calculate these fields. For decades, scientists viewed potentials as mere mathematical shortcuts, while the fields themselves were the only real, physical entities. The AB effect proved that potentials have a physical reality of their own.

In 1959, Yakir Aharonov and David Bohm proposed an experiment using a long, tightly wound coil of wire called a solenoid. When an electric current runs through a solenoid, it creates a magnetic field trapped entirely inside the coil. Outside the coil, the magnetic field is zero.

They envisioned firing a beam of electrons toward this solenoid. The experiment splits the electron beam into two paths: one path goes to the left of the solenoid, and the other goes to the right. Both paths stay completely outside the coil, traveling only through regions where the magnetic field is zero.

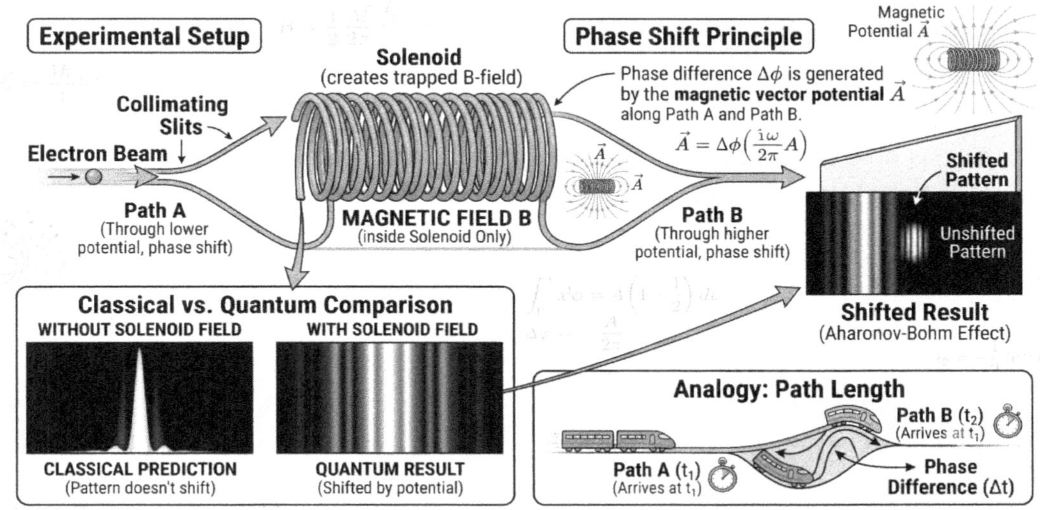

According to traditional physics, since the electrons never pass through the magnetic field, the magnet should have no impact on them. However, quantum mechanics treats electrons as both particles and waves. As the electron waves travel along their separate paths, they experience the magnetic potential, which still exists outside the coil. This potential alters the phase of the waves. It speeds up the wave on one side and slows it down on the other.

When the two paths recombine on a screen behind the solenoid, they create an interference pattern – a series of bright and dark lines caused by the overlapping wave crests and troughs. When technicians turn on the current inside the solenoid, this interference pattern shifts, even though the electrons never touched the magnetic field.

A real-life analogy is a train track that splits into two routes around a mountain and recombines on the other side. If the left route is slightly downhill and the right route is slightly uphill, two identical trains starting at the same time will arrive at the destination at different times. The mountain itself did not push the trains, but the landscape altered their arrival times.

In the quantum experiment, the magnetic potential alters the arrival of the electron waves, proving that the vacuum outside the magnet is not truly empty, and that quantum particles sense fields from a distance.

Summary

Chapter Five surveys a diverse array of advanced concepts and technological applications stemming from quantum mechanics. It explores the conservation of quantum information, the Pusey–Barrett–Rudolph (PBR) theorem, and the properties of the Quantum Hall Effect (including its integer and fractional forms).

A major portion is dedicated to explaining the mechanics of quantum computing (how qubits manipulate superposition and entanglement to vastly outpace classical computers) and adjacent quantum technologies. It discusses fractional excitons and the speculative physics surrounding the Quantum Multiverse.

Then we looked at the matter-antimatter asymmetry problem and the strange nature of neutrinos (the "ghost particle"), leading to the issue of CPT symmetry, which predicts that every particle must have the same mass and lifespan as its antiparticle.

Finally, we finished with the Aharonov–Bohm effect that shows the quantum world behaves differently than the classical world we see every day. In quantum mechanics, a magnetic field can alter a particle's behavior even if the particle never enters the field.

Chapter Five: General Relativity

In which we learn that space and time are one entity, mass and energy curve spacetime, creating gravity, nothing with mass can travel faster than the speed of light, space itself expands faster than light, and that black holes swallow everything - including light.

$T_{\mu\nu}$

"Einstein developed a theory about space, and, boy, it was about time, too!"

Main Relativity Theories Timeline

In 1905, Einstein published his theory of Special Relativity, which explains how space and time are related for objects moving at constant speed – but didn't consider gravity. In 1915, building on Special Relativity, Einstein published his General Relativity theory, which explains how gravity affects the fabric of spacetime, including the curvature it causes from massive objects. In 1917, Einstein applied his theory of relativity to the universe for the first time, thereby founding the field of relativistic cosmology. He introduced a cosmological constant into his theory to produce a static universe – he later called this his "biggest blunder".

The first observational proof came in 1919, when a solar eclipse expedition led by astronomer Arthur Eddington confirmed Einstein's prediction that the Sun's gravitational field deflects light, making Einstein world-famous. In 1927, the Belgian astronomer and priest Georges Lemaître found non-static cosmological solutions to the Einstein equations (without the cosmological constant) that describe an expanding universe. He derived Hubble's law theoretically from these solutions, which await experimental confirmation.

Hubble's Law was formulated in 1929 by the American astronomer Edwin Hubble, who observed that extragalactic objects are receding from Earth. This occurs at a speed proportional to the object's distance from Earth. Hubble's law is the first experimental confirmation that the universe is expanding.

In 1960, American physicist Joseph Weber reported detecting gravitational waves, but his results could not be reproduced. The search for gravitational waves continued until a binary pulsar was discovered, providing evidence that matched predictions. The LIGO Laser Interferometer Gravitational-Wave Observatory directly detected gravitational waves from the merger of two black holes, and this detection has been observed multiple times.[42]

In 1965, American astronomers Robert Wilson (1936) and Arno Penzias (1933) discovered the cosmic microwave background radiation (CMB). This radiation is interpreted as the afterglow of the Big Bang, providing evidence that the universe had a beginning and is not expanding forever, contrary to the steady-state hypothesis.

In 1974, String theory, originally developed to explain nuclear forces, was proposed as a possible unifying theory of general relativity and quantum mechanics. British physicist Stephen Hawking (born 1942) published a quantum theory of gravity that predicts black holes emit a type of radiation known as Hawking radiation.

The Standard Model of Cosmology

As we did in Chapter One for particle physics, let's look at the best current theory of the cosmos – the Standard Model of Cosmology – and then look into how we got to where we are today, and how it relates to General Relativity. We could go straight to Mr. Einstein and his theories, but a quick tour of this model will be helpful, and (I think) interesting.

The Standard Model of Cosmology, also known as the λCDM (Lambda Cold Dark Matter) model, is the current working framework that astrophysicists use to explain the origin, evolution, and composition of the universe. It is often referred to as the "concordance model" because it successfully accounts for the vast majority of cosmological observations, from the afterglow of the Big Bang to the large-scale distribution of galaxies.

To understand the λCDM model, it is best to break it down into its core components (the "ingredients" of the universe) and its timeline (how those ingredients evolved). According to the Standard Model, the universe is made of three distinct components:

1. Ordinary (Baryonic) Matter (~5%)

This is the matter that makes up stars, planets, gas clouds, and us. It consists of protons, neutrons, and electrons. Despite being the foundation of the visible universe, it accounts for about 5% of the total mass-energy density of the cosmos.

2. Cold Dark Matter (CDM) (~27%)

"Dark matter" is a hypothetical form of matter that neither emits, absorbs, nor reflects light, making it completely invisible to electromagnetic observations. However, its presence is inferred through its gravitational effects on visible matter, such as the rotation speeds of galaxies and the bending of light (gravitational lensing). The "Cold" in Cold Dark Matter refers to these particles moving slowly compared to the speed of light. Because they are slow, they can clump together under the force of gravity, forming the invisible "scaffolding" or "halos" upon which ordinary matter gathers to build galaxies.

3. Lambda / Dark Energy (λ) (~68%)

The Greek letter Lambda (λ) represents the Cosmological Constant, a concept originally introduced by Albert Einstein. In the Standard Model, it represents "Dark Energy," a mysterious, repulsive force that permeates all of space. While gravity pulls matter together, dark energy pushes space itself apart. Because dark energy is a property of space itself, as the universe expands and creates more space, the total amount of dark energy increases, leading to the accelerated expansion we observe today.

Dark Energy is still the subject of extensive investigation and speculation. In 2026, mathematicians showed that Einstein's equations can account for the accelerated expansion if coupled with the assumption that the universe is not as homogeneous as astrophysicists had assumed.

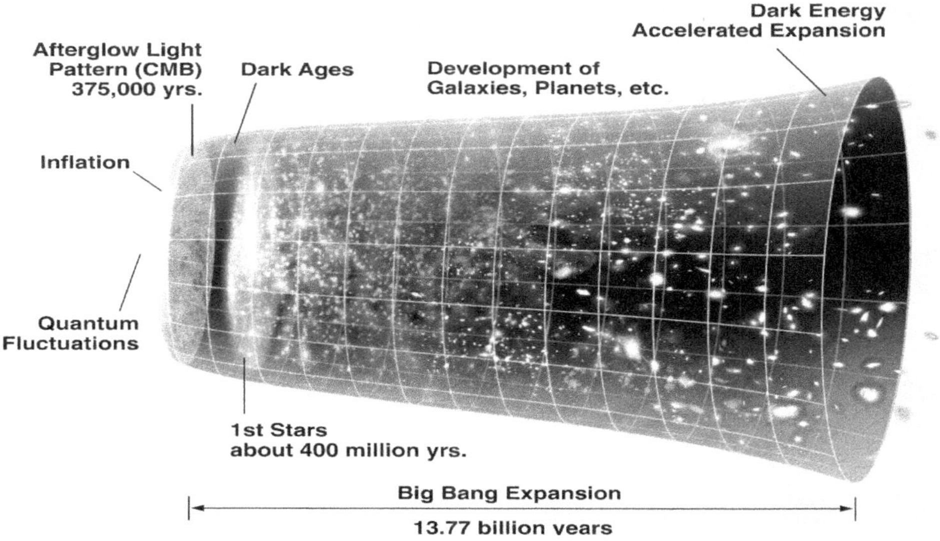

The Origin of the Universe (0 to 10^{-32} seconds)

The Big Bang theory describes the origin and expansion of the universe from a hot, dense state approximately 13.8 billion years ago. General Relativity establishes that gravity is the physical curvature of spacetime caused by the distribution of mass and energy. When physicists applied Einstein's field equations to the entire universe, they found that a static universe was unstable. Spacetime had to either expand or contract. Georges Lemaître proposed that the universe began as a single dense point that expanded outward, a concept that laid the operational groundwork for the modern Big Bang model.

The scientific community initially treated the expanding universe model with skepticism until Edwin Hubble's 1929 observations. Hubble measured the light shifts of distant galaxies and demonstrated that their recessional velocities are proportional to their distances from Earth. This discovery provided the first direct empirical evidence that space itself is expanding. In the mid-twentieth century, the model evolved as physicists calculated that an early, dense universe would synthesize light elements through nuclear fusion. This prediction materialized in 1964, when researchers discovered the Cosmic Microwave Background (CMB), a uniform thermal radiation left over from the time when the universe cooled down enough for atoms to form, roughly 380,000 years after the initial expansion.

To resolve structural puzzles like the horizon and flatness problems, Alan Guth proposed the theory of cosmic inflation in 1980. The horizon problem stems from measurements of the CMB. From opposite sides of the observable universe, the temperature is nearly identical, varying by only about one part in 100,000. According to standard Big Bang cosmology, these distant regions of space were separated by distances greater than the distance light could have traveled since the beginning of time. Because no signal can travel faster than the speed of light, these regions were causally disconnected. Standard physics requires a physical interaction to achieve thermal equilibrium, yet standard cosmic expansion prevents it. The horizon problem asks why the universe appears identical on large scales when those scales could never have communicated with one another.

The flatness problem concerns the geometric curvature of spacetime. In general relativity, the density parameter (Ω) represents the ratio of the actual density of the universe to the critical density required for a flat geometry. If Ω equals 1, the universe is flat. If Ω is greater than 1, it is closed; if less than 1, it is open. Current observational data indicates that Ω is remarkably close to 1 today. The equations governing cosmic expansion show that a flat geometry is an unstable equilibrium. If the early universe deviated from perfect flatness by even a minute fraction, that deviation would amplify rapidly over billions of years of expansion. For Ω to remain close to 1 today, it must have been equal to 1 within one part in 10^{60} during the Planck epoch. Standard cosmology provides no mechanism for explaining why the early universe had such an exact initial density.

Cosmic inflation solves these two issues by suggesting that the universe underwent a rapid, exponential expansion a fraction of a second after the initial state. This expansion smoothed out variations in density and geometry, creating the flat universe we observe today. Later space missions, including the Cosmic Background Explorer (COBE), the Wilkinson Microwave Anisotropy Probe (WMAP), and the Planck satellite, mapped the CMB temperature fluctuations with high precision, confirming the basic concept of inflation and establishing the era of precision cosmology.

While existing data strongly favor the Big Bang model, future experiments aim to probe the earliest fractions of a second to discover the precise mechanism of early expansion. The primary objective is to detect primordial gravitational waves generated during the inflationary epoch. These gravitational waves compress and stretch spacetime, imprinting a distinct swirl pattern known as B-mode polarization onto the Cosmic Microwave Background.

Several upcoming instruments are designed to detect this signal. Space missions such as the Japanese Aerospace Exploration Agency's LiteBIRD will map the full-sky CMB polarization with high sensitivity. On the ground, the Cosmic Microwave Background Stage 4 (CMB-S4) project will deploy telescopes in Antarctica and Chile to measure these polarization states. Discovering primordial B-modes would give you direct empirical confirmation of cosmic inflation and establish the exact energy scale of the early universe.

Future gravitational-wave observatories, such as the Laser Interferometer Space Antenna (LISA), will search for a background of gravitational waves generated directly during the Big Bang. These waves travel through the universe without interference, allowing them to carry clean data from periods much earlier than the 380,000-year limit of light-based observations. Radio telescopes like the Square Kilometer Array (SKA) will also trace the 21-centimeter line from neutral hydrogen to examine the transition periods before the first stars ignited, providing independent checks on the early expansion parameters of the universe.

The Hadron Epoch (10 microseconds)

The quark-hadron phase transition in the early universe occurred roughly 10^{-5} seconds (10 microseconds) after the Big Bang. Before this moment, the entire universe was an ultra-hot, dense soup called the quark-gluon plasma, where quarks and gluons roamed completely free. As the universe rapidly expanded and cooled, it reached a critical temperature, forcing these free particles to combine into composite particles called hadrons (such as protons and neutrons). This event is known as the Hadron Epoch.

It occurred at a critical temperature of roughly 155–165 MeV (about 1.8 trillion Kelvin). Unlike water freezing into ice (a sharp, first-order transition), early-universe physics confirms that this was a smooth, continuous crossover. Matter rapidly transformed from a plasma into a gas of hadrons without any abrupt shockwaves, boiling bubbles, or released latent heat. At this exact moment, quarks acquired their constituent masses. The intense vacuum energy of the strong nuclear force shifted, causing massless quarks to interact heavily with the gluon field, gaining the mass that now makes up 99% of the visible universe.

The Hot Phase (1 second)

The Cosmic Neutrino Background (CNB) is a uniform thermal bath of low-energy neutrinos that surrounds you and permeates the entire universe. It originated approximately one second after the Big Bang during the early, hot phase of the cosmos. At that stage, the early universe was close to thermal equilibrium, and all particles in the Standard Model were present in roughly equal abundance.

Neutrinos interacted constantly with electrons, positrons, and photons via weak nuclear forces. As the universe expanded, its temperature and density dropped. When the temperature fell to a few MeV, the interaction rate of the neutrinos became smaller than the expansion rate of the universe. At this point, neutrinos decoupled from other particles, ceased their regular interactions, and began to stream freely through space, as we observe today.

The Primordial Soup and Nucleosynthesis (3 minutes)

As temperatures dropped further, quarks bound together to form the first protons and neutrons. These particles then fused to create the nuclei of the lightest elements: primarily hydrogen, along with some helium and trace amounts of lithium.

Recombination and the Cosmic Microwave Background (380,000 years)

For the first 380,000 years, the universe was filled with a dense, opaque plasma. Light (photons) could not travel freely because it constantly collided with loose electrons. Eventually, the universe cooled enough for atomic nuclei to capture electrons, forming the first stable, neutral atoms.

With the electrons locked away, light was finally free to travel across space uninterrupted. This "first light" is still visible today as a faint glow of microwave radiation spanning the entire sky – the Cosmic Microwave Background (CMB). The CMB is the most critical piece of observational evidence supporting the Standard Model.

The Dark Ages and the First Stars (380,000 to ~ 200 million years)

Following the release of the CMB, the universe plunged into darkness. There were no stars or galaxies to produce light, only vast clouds of neutral hydrogen and cold dark matter. During this era, dark matter slowly clumped under gravity, drawing hydrogen gas into dense webs. Eventually, the gas became so compressed that nuclear fusion ignited, giving birth to the first stars.

Galaxy Formation & Accelerated Expansion (300 million years to present)

These first stars clustered into galaxies, and galaxies merged to form massive clusters. For the first half of the universe's life, dark matter's gravity slowed cosmic expansion. However, around 5 billion years ago, the density of matter dropped low enough for Dark Energy (λ) to become the dominant force. The universe transitioned from slowing down to accelerating outward, a phase it remains in today.

Astronomical observations suggest the earliest galaxies formed between 200 and 300 million years after the Big Bang. This period represents the first two percent of the 13.8-billion-year history of the cosmos. Before the operation of advanced instruments like the James Webb Space Telescope, earlier models estimated that regular galaxy assembly took 400–500 million years. However, new empirical data confirms that the collection of stars and gas into organized galactic structures occurred much faster than physicists previously assumed.

The λCDM (Lambda-Cold Dark Matter) model is accepted as the standard because it requires very few mathematical assumptions (only six fundamental parameters) yet successfully predicts a wide range of observations. It aligns perfectly with the detailed temperature ripples seen in the CMB, the observed distribution of galaxies across billions of light-years, and the accelerating expansion of space measured from distant supernovae. The model perfectly explains the temperature and density fluctuations observed in the CMB, which are the stretched-out seeds of all future structures. Dark matter clumps first, creating gravitational potential wells. Ordinary matter then falls into these wells, leading to the formation of stars, galaxies, and the vast cosmic web structure we observe today.

Einstein's Theory of Gravitation

The single most revolutionary idea of General Relativity is that gravity is not a force acting between two objects, as Isaac Newton proposed. Instead, Einstein's theory describes gravity as the consequence of spacetime being curved, or warped, by mass and energy. The central mantra, coined by physicist John Wheeler, is:

"Spacetime tells matter how to move; matter tells spacetime how to curve."

Imagine a trampoline stretched flat. This is "flat" spacetime. Place a heavy ball (like the Sun) in the center. The fabric of the trampoline stretches and curves. Now, roll a small marble (like the Earth) nearby. The marble will try to travel in a straight line, but because the fabric it's on is curved, its path is bent. It "falls" toward the bowling ball, "orbiting" it. The bowling ball isn't *pulling* the marble with an invisible "force." It's simply following the straightest possible path (called a geodesic) on a curved surface.

Einstein's journey began with a thought experiment he called his "happiest thought." Imagine you are in a windowless elevator in deep space, far from any gravity. A rocket on the bottom of the elevator accelerates you "upward" at 9.8 m/s^2. If you drop a ball, it will fall to the floor, just as it would on Earth. You'd feel "weight" pushing you to the floor. From the inside, this acceleration would be completely indistinguishable from being at rest in a gravitational field. This led to the Equivalence Principle, which states that the effects of gravity are equivalent to the effects of acceleration. This was a profound insight. If gravity and acceleration are indistinguishable, they are two sides of the same coin. This led him to realize that gravity isn't a force at all, but a property of spacetime itself.

Einstein's theory of Special Relativity united three-dimensions space and one dimension of time into a single, 4-dimensional block called spacetime. General Relativity states that this 4D spacetime fabric is not a static "stage" for events to play out on, but a dynamic, flexible *participant* that can be bent, twisted, and warped. Mass and energy (also momentum and pressure) are the sources that warp this spacetime fabric. The more mass (or energy), the more curvature. This curvature is what we perceive as gravity.

Crucially, Einstein discovered that this warping affects time as well as space. Because space and time are linked, the stronger the gravity (the deeper the fabric's curve), the slower time passes. This means that time actually ticks a tiny bit more slowly on Earth's surface than it does for a satellite high in space. General Relativity reveals a universe that is dynamic and flexible, where matter tells space how to curve, and space tells matter how to move.

How do things move in spacetime? A geodesic is the "straightest possible path" in a curved space, so in flat spacetime (no gravity), a geodesic is a straight line at a constant speed. In the curved spacetime around Earth, astronauts feel "weightless" not because there is no gravity, but because they are following their natural geodesic path. Weight is what you feel when a force *resists* your geodesic. The ground pushing up on your feet is a force that prevents you from free-falling to the center of the Earth.

Einstein's Field Equations (EFE) are at the mathematical heart of General Relativity, summarized in one (deceptively simple) tensor equation:

$$\mathrm{G}_{\mu\nu} + \Lambda g_{\mu\nu} = \frac{8\pi G}{c^4} T_{\mu\nu}$$

***$G_{\mu\nu}$* (The Einstein Tensor):** This represents the geometry of spacetime. It describes how much the "fabric" is warped, stretched, or curved at a specific point. If this value is zero, spacetime is flat.

***$T_{\mu\nu}$* (The Stress-Energy Tensor):** This represents the "stuff" in the universe. It accounts for mass, energy, momentum, and even pressure. In Einstein's view, all of these things–not just mass–contribute to the warping of spacetime.

***$\frac{8\pi G}{c^4}$* (The Coupling Constant):** This is essentially the "stiffness" of spacetime. Notice the c^4 in the denominator (c) is the speed of light. Because light travels so fast, c^4 is an unimaginably large number. This means you need a staggering amount of mass or energy ($\mathrm{T}_{\mu\nu}$) to create even a tiny bit of curvature ($\mathrm{G}_{\mu\nu}$). This is why we only notice gravity around massive objects like planets and stars.

***$\Lambda g_{\mu\nu}$* (The Cosmological Constant):** As seen in dark energy, this term represents the energy inherent to empty space itself. It acts as a repulsive force that pushes the universe apart, counteracting the "inward" curve caused by matter.

On the left is the "geometry" side – the Einstein tensor. It's a complex mathematical object (a tensor) that describes the curvature of four- dimensional spacetime. It essentially says, "Here is the shape of spacetime."

On the right is the Stress-Energy tensor. This is the "matter/energy" side. It describes the distribution of all mass, energy, momentum, and pressure in that spacetime. It says, "Here is the stuff in the universe." The equation is the grand statement: The "stuff" on the right tells the "geometry" on the left how to curve. Finding a solution to these equations means figuring out what a specific spacetime *looks like* (e.g, around a star or throughout the universe).

Rather than thinking of gravity as an invisible "pulling" force, Einstein realized that space and time are fused into a single four-dimensional fabric called spacetime. Large objects, like stars and planets, don't "pull" on things; they warp the fabric itself.

The Greek indices (μ, ν) are a mathematical shorthand. They indicate that this isn't just one equation, but a set of 10 independent equations working simultaneously. They range from 0 to 3, corresponding to the four dimensions of spacetime (one time dimension and three space dimensions) and track how space and time in one direction (like up-down) affect space and time in another (like left-right or past-future). This complexity is why the equation is so difficult to solve for anything other than perfectly symmetrical objects such as a non-rotating sphere.

General Relativity made several stunning predictions that have been verified to incredible precision. Astronomers had known for decades that Mercury's orbit "wobbles" (precesses) in a way that Newton's theory couldn't fully explain. When Einstein applied his equations, they predicted the *exact* amount of this wobble, accounting for the curvature of spacetime near the Sun.

I think it's worth explaining the Mercury wobble in more detail. In Newtonian mechanics, the gravitational force between two masses follows an exact inverse-square law: $F = G\frac{Mm}{r^2}$. In a pure two-body system governed strictly by a $\frac{1}{r^2}$ force, orbits are perfectly closed ellipses (Kepler's First Law). Mercury would return to the same closest point to the Sun (perihelion) on every orbit.

Newtonian physics *does* predict orbital precession when other planets (like Jupiter and Venus) pull on Mercury. However, these planetary perturbations account for only 532 arcseconds of shift per century. Astronomers measured 575 arcseconds per century, leaving an unexplained discrepancy of 43 arcseconds per century.

In General Relativity, gravity is not a Newtonian force pulling across space, but the geometry of curved spacetime. When solving the geodesic equation for a particle orbiting a static mass M in Schwarzschild coordinates, the effective radial potential energy equation acquires an extra non-Newtonian term:

$$V_{\text{eff}}(r) = -\frac{GM}{r} + \frac{L^2}{2r^2} - \frac{GML^2}{c^2r^3}$$

where L is the orbital angular momentum per unit mass and c is the speed of light. Comparing the classical and relativistic terms highlights the key difference:

- **Classical Effective Potential:**

  $$V(r) = -\frac{GM}{r} + \frac{L^2}{2r^2}$$

  The $-\frac{GM}{r}$ term represents standard Newtonian attraction, and the $+\frac{L^2}{2r^2}$ term is the centrifugal barrier that stops orbiting bodies from plunging straight into the center.

- **Relativistic Correction:**

$$\frac{GML^2}{c^2r^3}$$

This extra term drops off much faster with distance because of the $\frac{1}{r^3}$ dependence. Because it depends on $\frac{1}{r^3}$, its strength increases significantly when Mercury is near perihelion (its closest approach, around 46 million km from the Sun).

Because the true force law is slightly stronger at close distances than a pure $\frac{1}{r^2}$ force, Mercury experiences an extra "tug" near perihelion.

1. As Mercury plunges toward its closest point, the $\frac{1}{r^3}$ term pulls it slightly tighter than a Newtonian ellipse would.
2. Mercury overshoots the expected radial turning point and travels slightly farther around the Sun before reaching its new perihelion.
3. The ellipse fails to close on itself, rotating the entire orbit forward in a rosette pattern.

Evaluating this correction per orbit yields the relativistic shift per revolution ($\Delta\phi$):

$$\Delta\phi \approx \frac{6\pi GM}{a(1-e^2)c^2}$$

Plugging in the parameters for Mercury (semi-major axis a $\approx 57.9 \times 10^6$ km, eccentricity e ≈ 0.2056, and $M_\odot \approx 1.989 \times 10^{30}$ kg) gives 42.98 arcseconds per century, matching astronomical observation without relying on unseen planets.

If spacetime is curved, then the path of light should also be curved as it passes a massive object. In 1919, Sir Arthur Eddington led an expedition to observe a solar eclipse. He measured the position of stars near the eclipsed Sun and found that their light was indeed bent, "shifting" their apparent position in the sky exactly as predicted. This made Einstein a global celebrity overnight.

Here's an example of using Einstein's field equations to find the geometry of a universe that contains no matter or radiation but has a cosmological constant Λ. This scenario produces de Sitter space. The complete field equation includes the cosmological constant term:

$$G_{\mu\nu} + \Lambda g_{\mu\nu} = \tfrac{8\pi G}{c^4} T_{\mu\nu}$$

In this problem, we assume the universe is entirely empty of physical matter and radiation. Therefore, the stress-energy tensor $T_{\mu\nu}$ equals zero everywhere. The equation simplifies to:

$$G_{\mu\nu} + \Lambda g_{\mu\nu} = 0$$

The Einstein tensor $G_{\mu\nu}$ is defined by the Ricci curvature tensor $R_{\mu\nu}$, the scalar curvature R, and the metric tensor $g_{\mu\nu}$:

$$G_{\mu\nu} = R_{\mu\nu} - \tfrac{1}{2} R g_{\mu\nu}$$

Substituting this definition into the simplified field equation yields:

$$R_{\mu\nu} - \tfrac{1}{2} R g_{\mu\nu} + \Lambda g_{\mu\nu} = 0.$$

To solve this tensor equation, we must determine the value of the scalar curvature R. This is achieved by taking the trace of the entire equation, which means multiplying every term by the inverse metric tensor $g^{\mu\nu}$. In four-dimensional spacetime, contracting the metric tensor with its inverse yields the number of dimensions: $g_{\mu\nu} g^{\mu\nu} = 4$. Contracting the Ricci tensor yields the scalar curvature: $R_{\mu\nu} g^{\mu\nu} = R$. Multiplying the equation by $g^{\mu\nu}$ transforms the tensor terms into scalar numbers:

$$R - \tfrac{1}{2} R(4) + \Lambda(4) = 0$$

Simplifying the linear terms:

$$R - 2R + 4\Lambda = 0$$
$$-R + 4\Lambda = 0$$
$$R = 4\Lambda$$

These steps show that the total scalar curvature of this universe is a constant value directly proportional to the cosmological constant. Now substitute the value $R = 4\Lambda$ back into the original equation to find the relationship for the Ricci tensor:

$$R_{\mu\nu} - \tfrac{1}{2}(4\Lambda) g_{\mu\nu} + \Lambda g_{\mu\nu} = 0$$
$$R_{\mu\nu} - 2\Lambda g_{\mu\nu} + \Lambda g_{\mu\nu} = 0$$
$$R_{\mu\nu} = \Lambda g_{\mu\nu}$$

This relation indicates that the local curvature tensor is proportional to the metric tensor at every point in the coordinate space. To find the exact physical distances in this universe, you solve this relation for the line element ds^2. One standard solution uses flat spatial coordinates that expand over time. The resulting metric configuration produces the following spacetime interval:

$$ds^2 = -c^2dt^2 + e^{2Ht}(dx^2 + dy^2 + dz^2)$$

The constant expansion factor H relates directly to the cosmological constant by the formula $H = \sqrt{\frac{\Lambda}{3}}$. The math demonstrates that adding a positive cosmological constant to an empty universe forces the spatial components (dx, dy, dz) to grow exponentially over time (t).

If you ask a theoretical physicist about the two main principles of General Relativity, they will probably mention the Equivalence Principle and Invariance. Spacetime curvature and the warping of time are also important. We already talked about the first principle, so let's review it and then look at invariance.

Einstein's Equivalence Principle is the foundation of General Relativity – it states that the effects of gravity and acceleration are exactly the same. Picture yourself in a windowless elevator floating in deep space, feeling weightless. If a rocket engine pushes the elevator upward, your feet will hit the floor. If you let go of a tennis ball, the floor moves up to meet it, so it seems to fall just like it would on Earth.

Einstein realized that you cannot do any experiment inside that closed room to tell if you are sitting still on Earth under gravity or moving through space with constant acceleration.

General covariance means that the laws of physics must look the same in every coordinate system. Nature does not have a preferred grid, viewpoint, or universal clock. Whether someone is sitting still in a lab, speeding up in a rocket, or spinning in a whirlpool, the equations describing reality stay the same.

In classical mechanics and special relativity, physicists used fixed coordinate systems like Cartesian grids. These flat grids work well in simple, non-accelerating situations. But general relativity shows that mass and energy bend spacetime into curved shapes. In a curved universe, rigid grids cannot cover space without becoming distorted.

Coordinates are just labels we use to mark points in space and time. Imagine drawing a grid on a flat rubber sheet. If you stretch or twist the sheet, the grid lines bend, and the distances between points change a lot. Still, events on the sheet, like two marbles colliding, happen the same way no matter how the grid looks. General covariance says that the laws of physics must show this independence, so our grids never change the real events.

To write equations that keep their form in any coordinate system, Einstein used tensor calculus. A tensor is an object whose parts change predictably when you switch from one coordinate system to another. See the *Background Materials* for details.

If a physical law is expressed as a tensor equation – such as (A = B) – and it is true in one coordinate system, the rules of tensors make sure it stays true in every other coordinate system (A' = B'). Einstein's field equations use only tensors, so the link between matter and spacetime curvature is always the same and can be checked by any observer in the universe.

Einstein's Invariance and Kretschmann's Objection

In physics, an "invariant" is something that stays the same for every observer, no matter their speed or location. Before Einstein, classical mechanics said space and time were separate invariants. People thought everyone experienced the same clock and distances. Relativity changed this: speed stretches time and shrinks space. Einstein replaced the old constants with a new, unified four-dimensional invariant called the Spacetime Interval.

Imagine a broomstick. If you tilt it away from someone, it looks shorter to them, but its real length stays the same. You are just seeing it from a different angle. The Spacetime Interval is similar. Two people moving at different speeds will not agree on the time (Δt) passed or the distance traveled (Δx). But when they use the interval formula, their differences cancel out.

$$s^2 = (c\Delta t)^2 - (\Delta x)^2$$

Every observer in the universe calculates the same value for s^2. This mathematical invariance ensures that even though perspective is relative, the fundamental laws of nature and causal reality remain absolute and identical for everyone.

However, the physicist Erich Kretschmann stated that any physical theory can be rewritten in a generally covariant language using sophisticated mathematics, so general covariance by itself possesses no physical content.[43] It is really about how flexible our math language is, not about the universe itself. Einstein agreed with this. General covariance is not a strict physical rule, but a helpful tool that led him to the right equations.

If general covariance is just about language and the Equivalence Principle is about acceleration, what is the real physical principle that makes General Relativity unique? Most physicists today point to *Background Independence*: the idea that, unlike other theories that depend on a fixed, absolute stage of space and time, General

Relativity defines the geometry of spacetime as a dynamical entity determined by the distribution of matter and energy. This means that spacetime itself is not a static backdrop but evolves according to the Einstein field equations, so the gravitational field (the metric) is not prescribed in advance, but is fully determined by the physical content of the universe.

In almost every other physical theory, from Newton's mechanics to Quantum Field Theory and the Standard Model, there is an absolute, unmoving background. This background is like a stage in a theater. The actors – particles, fields, and forces – move and interact, but the stage never changes.

In Special Relativity, the background is flat Minkowski spacetime. Space and time can stretch and contract based on an observer's speed, but the overall structure of spacetime stays fixed. It is an absolute setting that controls how things move, but nothing can change it in return.

General Relativity removes the stage completely. Background Independence means there is no fixed arena. Spacetime is not just a backdrop; it is a changing, active field (the metric tensor) that takes part in what happens. As John Wheeler put it, "*Spacetime tells matter how to move; matter tells spacetime how to curve.*"

If you rewrite Newton's laws in a covariant way using Newton–Cartan theory (created by Élie Cartan in 1923, which treats gravity as spacetime curvature), the equations look like those in General Relativity. But the physics is still "background dependent." The time and space structures in Newton–Cartan theory are fixed and do not change when matter moves. You can move planets, explode stars, or shift galaxies, but Newtonian time slices stay the same. The background stage is still present, just hidden by complex math.

In General Relativity, you cannot separate the shape of the universe from the matter inside it. If there is no matter or energy, spacetime has one shape. If you add even one electron, the whole geometry changes everywhere.

This is why Background Independence is the most important principle of General Relativity. It also explains why it is so difficult to combine General Relativity with Quantum Mechanics. Quantum mechanics depends on a fixed, predictable background to define things like wavefunctions and particle probabilities. General Relativity says there is no such fixed stage, so physicists are looking for a quantum theory of gravity that can create its own background.

If you want to learn more about General Covariance and Erich Kretschmann, you can read J.D. Norton's paper below. It is only 16 pages, so it is not a long read.

Norton, J. D. (1993). General covariance, gauge theories, and the Kretschmann objection. *In History of Philosophy of Science Part B: Studies in History and Philosophy of Modern Physics*, *24*(3), 115–159. https://sites.pitt.edu/~jdnorton/papers/NortonGCGTKO_final.pdf

Gravitational Redshift and Time Dilation

Time itself runs more slowly in stronger gravity. A clock at sea level (deeper in Earth's gravity well) ticks *more slowly* than a clock on a mountain. This has been verified with ultra-precise atomic clocks. If a massive object accelerates, it should create "ripples" in the fabric of spacetime, much like throwing a rock in a pond. These ripples are gravitational waves. Einstein predicted them in 1916 and assumed they would be too weak to ever be found. For 100 years, they were indeed too faint to find. In 2015, the LIGO observatory made the first-ever direct detection of gravitational waves from two colliding black holes, opening a brand new field of astronomy and providing the most stunning confirmation of General Relativity.

EFE predicts that if enough mass is concentrated in a small enough space, spacetime curves so extremely that it creates a region from which not even light can escape, a black hole. When applied to the entire universe, the EFE predicts that the universe cannot be static; it must be expanding or contracting, leading to the Big Bang theory, which is the dominant model of our universe's origin and evolution.

What is Spacetime?

Stop thinking of space as an empty "room" and time as a ticking "clock". Instead, imagine they are woven together into a single, flexible fabric that fills the entire universe. We move in three dimensions of space: length, width, and height. If you want to meet a friend, you give them a location (the corner of 5th and Main) and a floor number (the 10th floor). However, those coordinates are useless without a fourth dimension: Time. If you don't agree on a time to meet, you will never find each other. Einstein realized that space and time aren't separate things; they are inextricably linked. You cannot move through space without also moving through time. Together, they form a 4D "map" called the spacetime continuum.

This is usually explained using the "Rubber Sheet" analogy. In this view, gravity is not a "pull" coming from the Sun. Instead, the Sun has changed the room's shape. The Earth is just following the "curved floor" of the universe.

This is the part that is hardest to grasp: when the fabric of space stretches, time stretches with it. Imagine that the "threads" of the spacetime fabric are actually small clocks. When a massive object (like a Black Hole) stretches the fabric, those threads get pulled thin. In deep space, the fabric is flat, and the clocks tick at a normal speed.

Near a massive object, the fabric is stretched, and the clocks actually tick more slowly. Time is not a universal constant. It flows at different speeds depending on how much "weight" is sitting on the fabric of spacetime near you. Spacetime tells us that the universe is like a living geometry. It can ripple (gravitational waves), it can stretch (the expanding universe), and it can even be punctured (black holes).

Time is Relative and Can Slow Down

Clocks run more slowly when they are closer to a massive object, where gravity is stronger. Gravitational Time Dilation isn't just a theoretical effect; it's a real phenomenon that must be accounted for in modern technologies like GPS, since satellites orbit at high velocity. Constant corrections for this effect are required to keep GPS accurate.

Near a black hole's event horizon, time can seem to stop. An observer far away, watching an object fall in, would see it slow down, freeze in time at the event horizon, and fade away forever. However, the object falling in would experience time normally and cross the event horizon in a finite amount of its own time.

Spacetime Can Be Warped and Shaken (*not stirred...*)

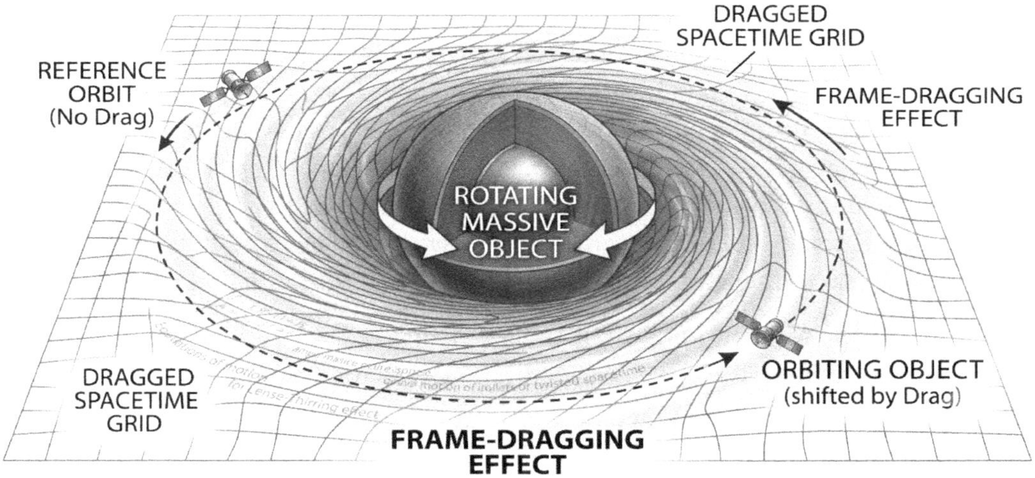

When massive objects accelerate rapidly, such as when two black holes collide, they create ripples in the fabric of spacetime called gravitational waves. The LIGO observatory first directly detected these waves in 2015, confirming a major prediction of Einstein's theory. The effects of these waves are tiny; the spacetime ripples from a black hole merger may stretch a detector by less than one ten-thousandth the width of a proton.

Pulsar timing array (PTA) collaborations worldwide attempting to find compelling evidence of an isotropic, low-frequency (nanohertz) stochastic gravitational wave background. When gravitational waves pass through the Milky Way, they stretch and squeeze space-time, introducing tiny variations (timing residuals) in the arrival times of pulses from millisecond pulsars. In General Relativity, a truly stochastic background creates a characteristic quadrupolar angular correlation pattern across pairs of pulsars – known as the Hellings–Downs curve.

The major global PTA groups – including NANOGrav (North America), EPTA (Europe), PPTA (Australia), CPTA (China), and InPTA (India) – have published datasets measuring this distinct Hellings-Downs spatial correlation at statistical significance ranging between 3 to 4.5 standard deviations (σ) depending on the array.

The current focus centers on interpreting the spectral index (γ) and amplitude (A) of the timing residual power spectrum:

- ***Supermassive Black Hole Binaries (SMBHBs):*** The leading astrophysical interpretation is an unresolved cosmic background generated by inspiraling pairs of supermassive black holes (10^8 to10^{10} solar masses) orbiting at center-of-galaxy mergers. For circular, purely gravitational-wave–driven orbits, theory predicts a power-law strain spectral index of $\alpha = -2/3$ (or a timing residual spectral index $\gamma = 13/3 \approx 4.33$).
- ***Environmental Coupling:*** Observed spectra show a slightly flatter slope than the idealized $13/3$ value. Detailed noise modeling and astrophysics suggest this flattening reflects environmental coupling – the binary black holes interacting with surrounding gas disks and stellar populations during their final parsec inspiral.
- ***Cosmological Hypotheses:*** While SMBHBs remain the standard model, researchers continue testing alternative early-universe sources, such as cosmic string networks, first-order cosmological phase transitions, or primordial inflationary perturbations.

Individual PTAs are combining their long-baseline datasets under the International Pulsar Timing Array (IPTA) Data Release 3 (DR3) framework. By pooling over a hundred millisecond pulsars timed across 15 to 25+ years, this combined analysis is expected to cross the definitive 5σ discovery threshold and begin searching for individual, continuous-wave supermassive binary systems standing out above the background hum.

A massive, rotating object literally drags the spacetime around it like a spoon stirring honey. The frame-dragging effect was predicted by general relativity and later confirmed by the Gravity Probe B mission.

One effect of frame-dragging around a super massive black hole, such as Sagittarius A* at the center of the Milky Way, is to force material on one side of the black hole to rotate toward us, while the material on the opposite side moves away. The material moving toward you emits light that undergoes Doppler boosting, making it appear much brighter. The material moving away fades. This creates a bright crescent shape rather than a uniform ring. If the incoming matter rotates at an angle to the black hole spin, frame-dragging forces the inner region of the disk to align with the equatorial plane of the black hole. This creates a physical twist in the disk structure, known as the Bardeen-Petterson effect.

To understand what it means to "warp" spacetime, you have to stop thinking of space as an empty "nothingness" and start thinking of it as a physical material – like a fabric or a medium that fills the entire universe. Spacetime isn't just a background where things happen; it is a flexible, dynamic participant in the universe. When we say spacetime is warped, we mean its geometry has changed.

In a "flat" universe with no mass, light travels in perfectly straight lines, and the shortest distance between two points is a straight line. However, when you add mass or energy (like a star or a planet), it pushes against that fabric. The mass creates a "dip" or a curve in the dimensions of space and time. Objects moving near that mass aren't being "pulled" by a mysterious force; they are simply following the new "straight" path along the curved surface.

Earlier, we learned that Sir Arthur Eddington measured the position of stars near the eclipsed Sun and found that their light was indeed bent, "shifting" their apparent position in the sky exactly as Einstein had predicted.

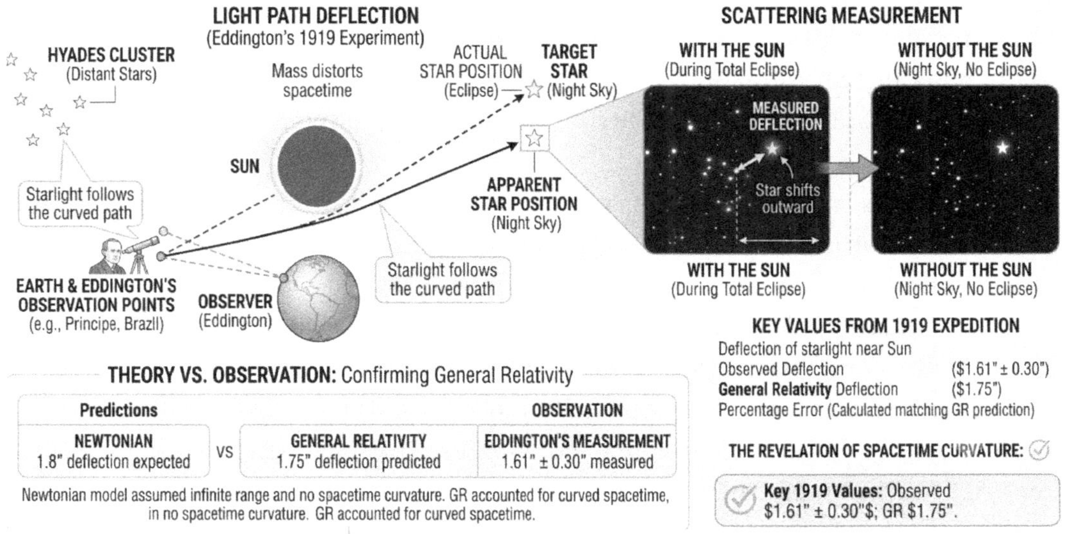

The term is *spacetime* because space and time are inextricably linked. When space warps, time warps with it. When space warps, it causes light to bend and planets to

orbit. When time warps (time dilation), the "stretching" of the fabric also stretches the intervals between seconds. Near a massive object where the warp is intense, time actually moves more slowly compared to a region where spacetime is flat. Think of a marathon runner. In flat spacetime, the track is straight and the runner's watch ticks normally. In warped spacetime, the track is hilly (space is curved), and the runner's legs move through "thick air" that slows their movement (time is dilated).
According to Einstein's Field Equations, energy and momentum are the "sources" of this warping. The more mass or energy you pack into a small area, the more violent the warp. Earth creates a gentle warp (enough to keep the Moon in orbit). The Sun creates a deeper warp (enough to keep all planets in orbit). A Black Hole creates a warp so steep and deep that it becomes a "bottomless pit" from which even light cannot climb out.

Black Holes

The history of black holes is a journey from mathematical curiosities to the universe's most extreme physical realities. Once considered a "scandal" of general relativity, they are now central to our understanding of gravity, time, and the evolution of galaxies.

John Michell (1783) and Pierre-Simon Laplace (1796) proposed the concept of a "dark star" first in the late 18th century. They reasoned that if a star were massive and dense enough, its escape velocity would exceed the speed of light, rendering it invisible to the outside world. However, the modern era of black holes began with Einstein's theory of General Relativity. The physicist Karl Schwarzschild found a solution to Einstein's equations describing the gravitational field around a point mass. It revealed a mathematical boundary – the Schwarzschild radius – where spacetime becomes so warped that nothing can escape.

For decades, many physicists, including Einstein himself, believed these were mathematical artifacts rather than physical objects. This changed in 1939, when Robert Oppenheimer and Hartland Snyder published a paper on "continued gravitational contraction," showing that a sufficiently massive star would inevitably collapse into a singularity once its nuclear fuel was exhausted.

The term "black hole" was popularized in the 1960s by John Wheeler, marking a "Golden Age" of theoretical research led by figures like Stephen Hawking and Roger Penrose. Penrose mathematically proved that singularities were a strong prediction of general relativity, a feat for which he received the Nobel Prize in 2020.[44] The search for physical evidence began in the 1960s with the rise of X-ray astronomy. In 1964, a powerful X-ray source, Cygnus X-1, was discovered. By 1971, observations showed it to be a massive invisible object orbiting a blue supergiant star. The invisible object was too massive to be a neutron star, making it the first widely accepted black hole candidate.

Evidence for black holes is categorized into several "layers" of certainty:

- ***X-ray Binaries***: Observing gas being pulled from a companion star into an invisible compact object, heating up and emitting X-rays as it spirals in.
- ***Stellar Dynamics***: Astronomers have tracked the orbits of stars at the center of our galaxy. The star S2 was observed rapidly orbiting a tiny, invisible point. This point, Sagittarius A*, was confirmed to be a supermassive black hole with 4 million times the mass of the Sun.
    - Newer studies propose it could be an ultra-dense concentration of dark matter. This would bend light and exert enough gravity to mimic the high-speed orbits of the stars (such as S2). The new model helps explain features that traditional theories struggle to explain.
    - Sgr A* is unusually quiet and lacks the aggressive, and expected, "tidal disruption events". While the EHT captured an image of the glowing gas ring around Sgr A*, it is still a challenge confirming whether a physical event horizon exists or if it is a massive dark object.
- ***Active Galactic Nuclei (AGN)***: Quasars and Blazars–the brightest objects in the universe–are powered by supermassive black holes at the centers of distant galaxies, consuming vast amounts of matter.

In 2015, the LIGO and Virgo detectors made history by detecting gravitational waves, the "ripples" in spacetime, caused by the collision of two black holes 1.3 billion light-years away.[45]

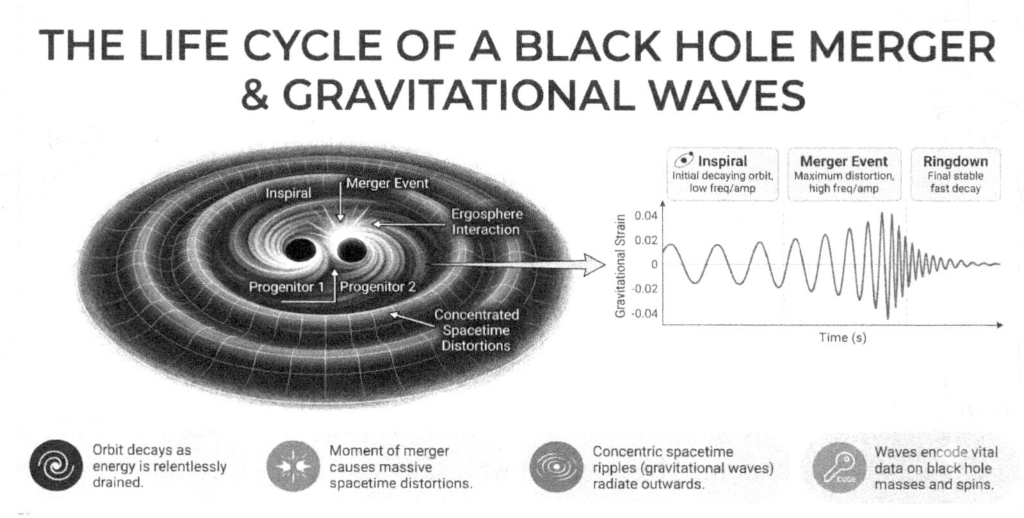

In 2019, the Event Horizon Telescope (EHT), a global network of radio telescopes, released the first-ever image of a black hole's "shadow" in the galaxy M87. The image showed a dark central region (the event horizon) surrounded by a bright ring of glowing gas. In 2022, the EHT followed up by releasing the image of the supermassive black hole (Sagittarius A*) at the center of our own Milky Way, confirming that Einstein's equations hold even in the most extreme gravitational environments.

Today's observations focus on the magnetic fields and the "spin" of black holes. In 2021, the EHT released a polarized-light image of M87*, revealing strong, twisted magnetic fields that help launch massive jets of matter across space. Later discoveries also include "intermediate-mass" black holes – the missing link between stellar-mass and supermassive black holes–and Gaia BH1, the closest known black hole to Earth, located just 1,560 light-years away. Physicists are now exploring the Information Paradox and whether black holes act as "mirrors" to other sectors of the universe, suggesting that our understanding of these objects is still very much in its infancy.

The Black Hole Information Paradox

The black hole information paradox is one of the most significant unresolved problems in theoretical physics. It represents a fundamental conflict between two of the most successful theories in science: General Relativity, which describes the universe at the largest scales (gravity and spacetime), and Quantum Mechanics, which describes the universe at the smallest scales (particles and fields).

The paradox is how these two theories treat information. A core principle of quantum theory is "unitarity." It dictates that the state of a system at any given time must contain the information needed to determine its past and future states. In short, information cannot be created or destroyed; it can only be scrambled. According to general relativity, black holes are incredibly simple objects. The "No-Hair Theorem" states that once a black hole is formed, it is characterized solely by its mass, electric charge, and angular momentum (spin). All other information about the matter that fell into the black hole – its chemical composition, the arrangement of its atoms, its temperature – is effectively lost to the outside universe.

The problem became a "paradox" in 1974, when Stephen Hawking mathematically demonstrated that black holes are not completely black.[46] Due to quantum effects near the event horizon, black holes emit radiation, now known as Hawking radiation. Hawking's original calculation suggested that this radiation is purely thermal and random. This means that as a black hole slowly evaporates and eventually disappears, it leaves behind only this random radiation. If the radiation carries no information about what fell in, then once the black hole is gone, the information that was once inside is destroyed. This directly violates the quantum principle of unitarity. Saying that Hawking radiation is caused by particle-antiparticle pairs at the horizon is a useful metaphor; a more accurate description is that it involves the "warping" of vacuum fluctuations by gravity, leading to a thermal spectrum of radiation.

Proposed by Leonard Susskind and others in 1993, the Black Hole Complementarity theory suggests that information is both reflected at the event horizon (from the perspective of an outside observer) and passes through it (from the perspective of an infalling observer).[47] According to this view, no single observer can ever witness a violation of physics, so the paradox is avoided through a kind of "observer-dependent" reality.

The Holographic principle suggests that all the information contained within a volume of space can be described as a "hologram" on the boundary of that space. In the context of a black hole, interior information would be stored on the surface area of the event horizon. This idea was reinforced by string theory, which maps a gravitational system onto a lower-dimensional quantum field theory.

In 2012, physicists later realized that "complementarity" might lead to a new problem – the firewall paradox. If information is preserved, the laws of quantum entanglement suggest that the event horizon should be a high-energy barrier, or a "firewall," that would incinerate anything falling in. This, however, violates Einstein's Equivalence Principle, which states that an observer falling into a black hole should feel nothing special at the horizon.

In 2019 and 2020, researchers used new mathematical techniques to calculate the Page Curve – a graph showing how the entropy of Hawking radiation changes over time.[48] They discovered that "islands" of the black hole interior effectively become part of the radiation itself as the black hole evaporates. This allows the information to "escape" through the radiation, preserving unitarity without requiring a firewall.

Research has shifted toward CPT-symmetric models and Direct-Sum Quantum Theory (DQFT). These theories propose that black holes act as "mirrors" or bridges to another sector of the universe where time runs in reverse. In these models, the information paradox is resolved by viewing the black hole not as a "drain" where information disappears, but as a "handshake" between two halves of a larger, symmetric system. If the universe exists as a direct-sum of states across these mirror sectors, information is never lost; it transitions into a parity-reversed state that remains mathematically accessible to the total system.

The Black Mirror Hypothesis

The "Black Mirror" hypothesis was proposed by physicists Latham Boyle and Neil Turok in late 2024.[49] Their paper, *Black Mirrors: CPT-Symmetric Alternatives to Black Holes*, suggests a radical departure from the standard view of black holes by applying CPT symmetry to the event horizon itself.

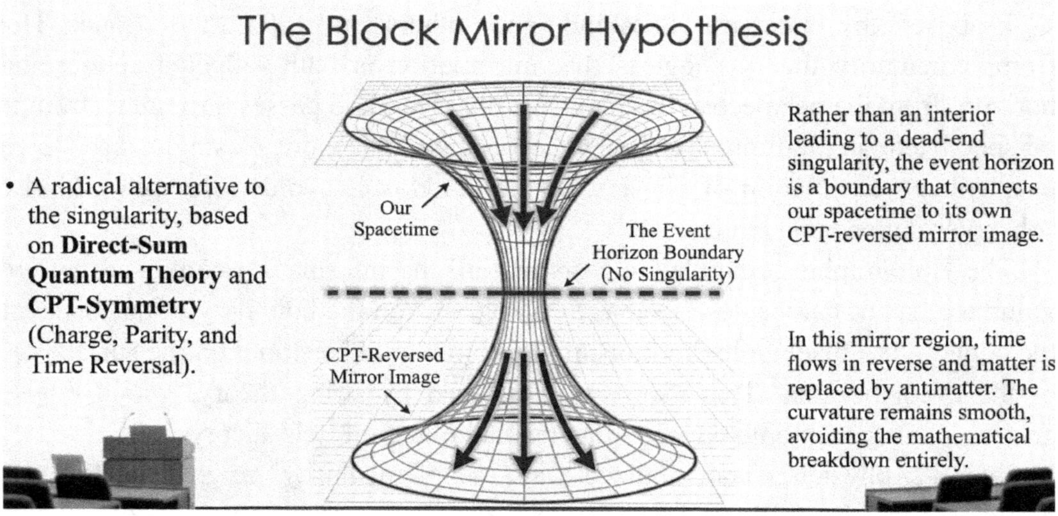

In standard general relativity, a black hole is thought to contain a singularity–a point of infinite density where the laws of physics break down, and information is effectively destroyed. The "Black Mirror" model proposes that, rather than an interior leading to a singularity, the event horizon is a boundary connecting our region of spacetime to its CPT-reversed mirror image. In this mirror region, time flows in reverse – the temporal direction is the opposite of ours. Matter is replaced by antimatter – charge and parity are inverted. The curvature of spacetime remains smooth and bounded, removing the mathematical "dead end" of a singularity.

This model addresses the black hole information paradox – the conflict between general relativity (which says information is lost) and quantum mechanics (which says it must be preserved) – by ensuring that the system remains unitary. Because there is no singularity to "crush" information out of existence, every bit of data that enters the black hole is preserved within the CPT-symmetric structure. The information effectively "reflects" or exists in the mirror counterpart. This allows the universe to maintain a consistent history, as the "mirror" region ensures that the system's state at one time can always determine its state at another, satisfying the requirements of quantum theory.

This theory is a direct extension of the authors' earlier work on the CPT-Symmetric Universe, which suggests that the Big Bang served as a similar mirror point between our universe and an "anti-universe" expanding backward in time.

Direct-Sum Quantum Theory (DQFT): The Mirror Universe Patch

In a paper by Gaztañaga, Kumar, and Marto (2026), the authors offer a compelling idea for one of the most persistent problems in our universal blueprints: the Black Hole Information Paradox.[50] By introducing the framework of Direct-Sum Quantum Theory, the authors suggest that Einstein-Rosen bridges are not physical tunnels through space, but rather the mathematical "handshake" between two distinct sectors of the universe.

In our standard view of the universe, we encounter a fatal error at the event horizon of a black hole: General Relativity suggests information is destroyed, whereas Quantum Mechanics demands it be conserved (Unitarity). Direct-Sum Quantum Theory (DQFT) resolves this by redefining the architecture of a quantum state as a "direct-sum" of two components existing in different regions of spacetime. Instead of treating the universe as a single, continuous "operating system," DQFT proposes that the Hilbert space (the mathematical space of all possible quantum states) is partitioned into discrete Superselection Sectors. These sectors represent regions of spacetime related by CPT symmetry (Charge, Parity, and Time reversal). In our universe (sector A), we are moving forward in time with a standard orientation of space. The mirror universe (sector B) is a parity-reversed region where the arrow of time runs in the opposite direction.

In this framework, the Einstein-Rosen (ER) bridge is reclassified. It is no longer a physical wormhole that matter "travels" through; instead, it is the mathematical connection – the "bridge" – that allows a quantum state to exist simultaneously in both sectors. Because every piece of information entering a black hole in Sector A has a corresponding "mirror" state in Sector B, the information is never truly lost. An observer bounded by a horizon sees what appears to be "lost" information, but the total system remains unitary because the information is perfectly preserved across the direct-sum of both sectors.

Standard inflationary models struggle to explain why certain large-scale patterns in the CMB appear to exhibit "handedness" or parity asymmetry. The authors' Direct-Sum Inflation (DSI) model predicts this exact asymmetry. Statistical analysis of Planck satellite data shows that this "Mirror Universe" architecture is 650 times more likely to be the correct description of reality than the Standard Model.

DQFT suggests that the universe maintains internal consistency by running two synchronized "backups" of itself, each with an opposite temporal arrow. This ensures that even when the local "spacetime metric" fails (at a singularity), the global "system state" remains pure and reversible.

Kerr Black Holes

A Kerr black hole does away with the central point-like singularity. It's a rotating, uncharged black hole first described mathematically by Roy Kerr in 1963,[51] it represents the most realistic model of astrophysical black holes, as all massive stars spin. It is characterized only by two properties: its mass and its angular momentum.

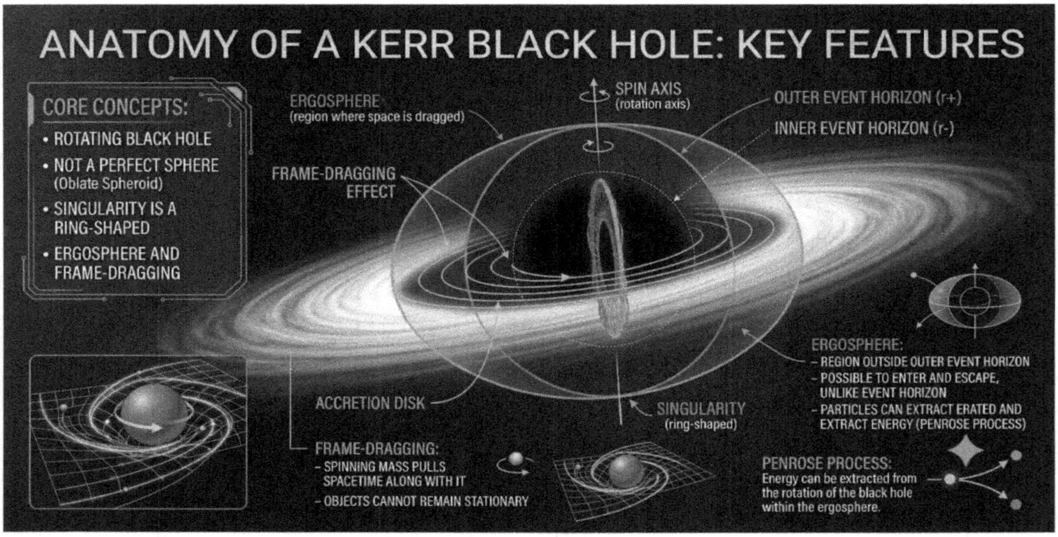

Because the black hole spins, it drags the surrounding spacetime and everything within it along its rotational path (frame-dragging). This swirling effect is called the Lense–Thirring precession.

Just outside the event horizon is a flattened region known as the ergosphere. Within this zone, spacetime is twisted so intensely that it is physically impossible for any object or light to remain stationary. Everything is forced to co-rotate with the black hole. Because of the ergosphere, it is theoretically possible to extract energy from a Kerr black hole via the Penrose process. Matter entering the ergosphere can split, allowing one piece to fall into the black hole while the other accelerates outward, carrying away some of the black hole's rotational energy.

Unlike non-rotating black holes, where the singularity is infinitely dense at a single point, a Kerr black hole's immense rotation flattens the singularity into an infinitely thin, spinning ring.

The Kerr metric describes a spacetime that is axisymmetric (completely symmetrical around a central axis) and stationary. Geometry yields concentric horizons. The outer horizon, the event horizon, marks the point of no return for light and matter. The inner horizon, also known as the Cauchy horizon, encloses the ring singularity. In the idealized mathematical solution, passing through the inner horizon might theoretically take an object into a different region of spacetime, potentially acting as a wormhole or leading to alternate universes.

In real-world astrophysics, matter falling into black holes naturally forms accretion disks that have angular momentum. This causes the black hole to continuously spin, meaning the Kerr solution accurately describes essentially all observed black holes in the universe.

Planck Star

In classical general relativity, when a massive star collapses under its own gravity at the end of its life cycle, nothing can stop the contraction. The mathematics dictates that matter collapses to a point of zero volume and infinite density. This point is a singularity.

A singularity represents a breakdown in the mathematics of general relativity. Infinities in physics usually indicate that a theory is being applied outside its scope of validity. General relativity describes gravity on large scales but ignores quantum mechanics, which governs the subatomic scale. To resolve the infinity of a singularity, you must incorporate quantum effects into the description of gravity.

The Planck star theory, developed by physicists Carlo Rovelli and Francesca Vidotto, uses loop quantum gravity to eliminate the singularity.[52] Loop quantum gravity proposes that spacetime is not a smooth, continuous fabric. Instead, it consists of discrete, indivisible packets or "atoms" of space. Because space cannot be infinitely divided, matter cannot be crushed into a point of zero volume.

As a dying star collapses, its volume shrinks and its density increases. When the density of the collapsing matter approaches the Planck density, the behavior of spacetime changes. The Planck density is approximately 5.1×10^{96} kilograms per cubic meter, a fundamental upper limit determined by the constants of nature:

$$\rho_P = \frac{c^5}{\hbar G^2}$$

At this scale, the quantum discrete units of space can no longer be compressed. This creates a powerful, repulsive quantum pressure. The force of gravity effectively reverses, changing from attractive to repulsive. The repulsive quantum pressure halts the gravitational collapse before a singularity can form. Instead of collapsing to zero volume, the matter reaches a minimum size and undergoes a "bounce." The resulting object is a Planck star.

A Planck star is not an infinitely small point. For a stellar-mass black hole, the star would have a physical size of roughly 10^{-10} centimeters. While microscopic, this volume is still approximately 10^{23} times larger than the fundamental Planck length (1.6 $\times 10^{-35}$ meters), meaning it avoids infinite density entirely.

The main reason we don't see Planck stars bouncing open in the night sky is extreme gravitational time dilation. According to general relativity, gravity slows the passage of time. Because a Planck star is incredibly dense, the gravitational field is immense. From the perspective of the matter inside the star, the collapse and subsequent bounce happen almost instantaneously, lasting only a fraction of a second. However, for an outside observer far from the black hole, the fraction of a second inside the black hole translates to billions or trillions of years on the outside. To an observer in the universe today, the object appears as a stable black hole.

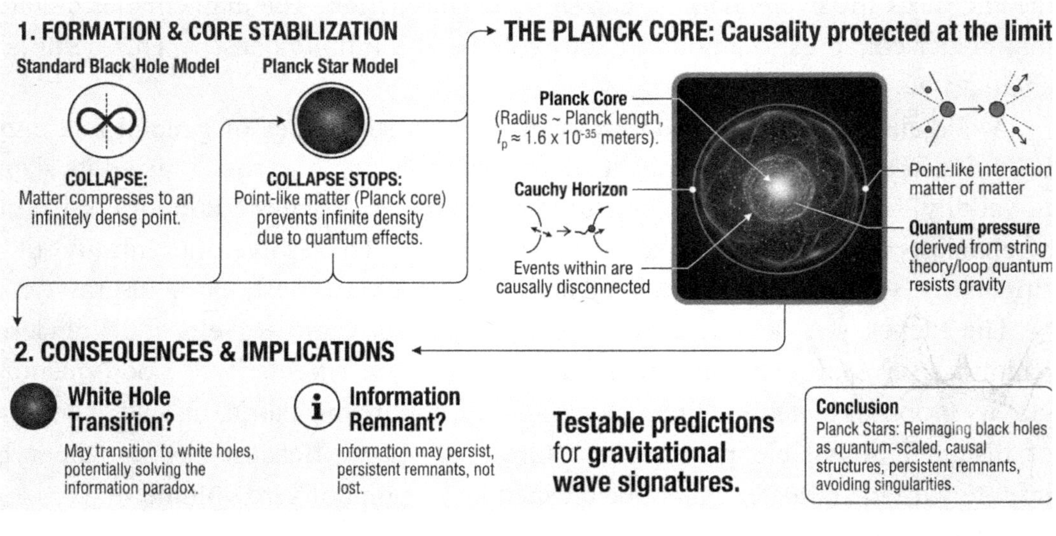

By avoiding a singularity, the Planck star theory offers a solution to the black hole information paradox. Quantum mechanics states that physical information about a system's state cannot be destroyed.

In the Planck star model, the black hole eventually loses mass through Hawking radiation over trillions of years. As mass decreases, gravitational time dilation weakens. Eventually, the internal quantum bounce reaches the outside observer's timeline. The Planck star explodes, releasing all the trapped matter and energy back into the universe, thereby preserving the information.

Consequences of Special and General Relativity

Special relativity fundamentally restructured the foundations of physics by asserting that the laws of nature are the same for all observers moving at constant speed. This theory is built on the radical premise that the speed of light in a vacuum is an absolute, universal limit that never changes, regardless of how fast an observer or a light source moves.

By treating light as a cosmic constant, Einstein demonstrated that space and time are not separate but are intertwined into a single, dynamic continuum known as spacetime. Because the speed of light is fixed, other physical measurements must vary to compensate for high speeds. This results in time dilation, in which time passes more slowly for an object moving near the speed of light than for one at rest. Similarly, length contraction occurs, causing fast-moving objects to appear shorter in the direction of their motion to a stationary observer.

The theory most famously shows that mass and energy are two forms of the same thing, as expressed by the equation $E = mc^2$. This principle shows that even a minuscule amount of matter contains a staggering amount of concentrated energy. By linking mass and energy, special relativity provided the essential framework for modern particle physics. (I should note that $E=mc^2$ is strictly true only for an object at rest. m represents the "rest mass" (also called invariant mass), which is the mass of an object measured by an observer who is moving along with it. When an object is moving, it possesses kinetic energy in addition to its rest energy. To calculate the total energy of a moving object, you must use Albert Einstein's complete relativistic energy equation: $E^2 = (mc^2)^2 + (pc)^2$. In this complete formula: m is the rest mass, p is the momentum of the object, and c is the speed of light.)

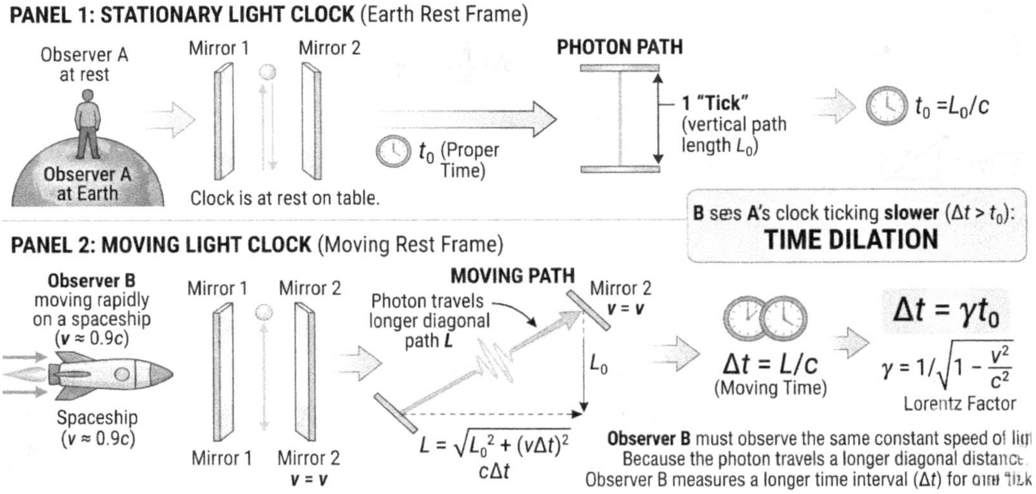

Special relativity invites us to discard our common-sense notions of how the universe behaves. The theory rests upon a single, startling contradiction to everyday experience: the speed of light is the same for everyone, regardless of how fast they are moving or where the light comes from. This constancy is the "special" rule that forces space and time themselves to bend and stretch to compensate for the math.

To visualize how time can slow down, we often use the analogy of a light clock. Imagine a clock made of two mirrors with a single photon bouncing up and down between them. Each bounce is one "tick" of the clock. If this clock is stationary on a table, the photon travels in a straight, vertical path. But if that clock is placed on a rocket ship zooming past at nearly the speed of light, an observer on the ground sees the photon traveling a much longer, zigzag path to keep up with the moving mirrors. Because the photon cannot speed up to cover that extra distance – remember, its speed is fixed – it takes longer to complete each bounce. As a result, the person on the ground sees the moving clock tick more slowly. This is not a mechanical error; time itself is physically dilating.

Special relativity also shatters the idea that "now" happens everywhere at once. Einstein illustrated this with a thought experiment involving a train and two lightning bolts. Suppose a passenger is sitting in the exact middle of a train car moving at a high velocity. Two lightning bolts strike the front and back of the train at the same moment. To a person standing on the tracks, the flashes reach them simultaneously, so the events occur at the same time. However, the passenger on the train is rushing toward the light from the front bolt and away from the light at the back. They see the front flash first, followed by the rear flash. Both observers are technically correct

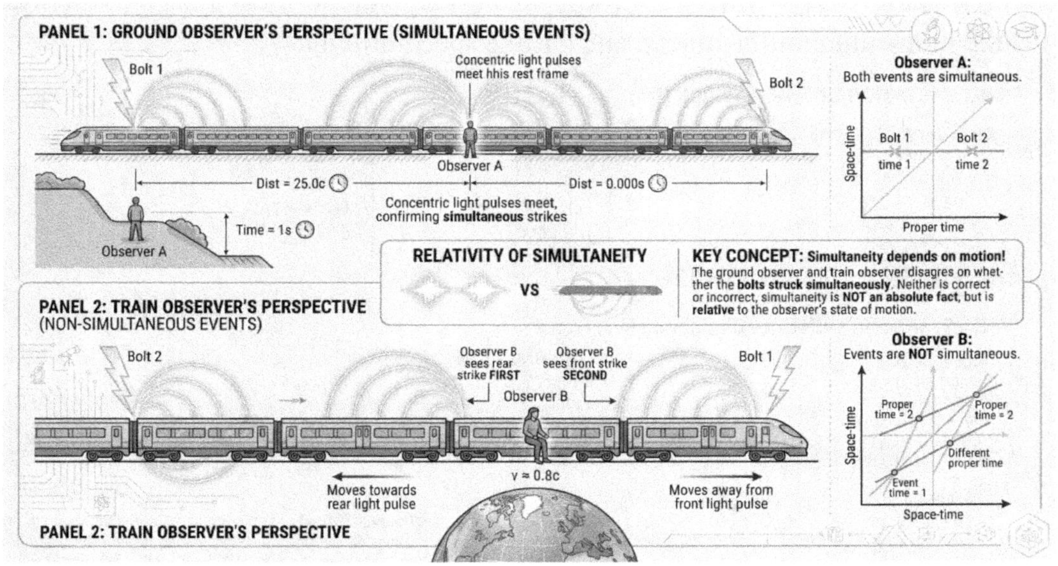

within their own frames of reference, proving that simultaneity depends entirely on your motion.

If time stretches, then space must shrink to maintain the balance of the universe. This phenomenon, known as length contraction, means that objects moving at relativistic speeds appear shorter in the direction of their travel. This is best evidenced by subatomic particles called muons. Created in the upper atmosphere, muons should decay long before they hit the ground. However, because they move so fast, their internal clocks slow down from our perspective, and the distance they have to travel shrinks from their perspective. These two relativistic effects are the only reasons we can detect them at sea level.

In special relativity and Newtonian mechanics, kinetic energy explains why no object with mass can ever reach the speed of light, as relativistic kinetic energy diverges as the speed of light is approached. As an object is pushed faster and faster, the energy used to accelerate it doesn't just increase its velocity; it also increases its "effective mass." The heavier the object, the more energy is required to push it farther. To reach the speed of light, an object would require an infinite amount of energy and would become infinitely heavy.

In this unified view, mass is nothing more than highly condensed energy, held together by the very geometry of space and time. It's not intuitively obvious why two fast objects approaching each other don't simply add their speeds ($v_1 + v_2$). The relativistic velocity addition formula explains how two velocities combine when an observer measures objects moving at extreme fractions of the speed of light, $v = \frac{v_1 + v_2}{1 + \frac{v_1 v_2}{c^2}}$. In classical mechanics, you add the velocities together. If a spaceship travels at 70% the speed of light ($0.7c$) and fires a missile forward at 60% the speed of light (0.6c) relative to the ship, a stationary observer expects to see the missile traveling at 130% the speed of light (1.3c). This violates the cosmic speed limit. Einstein's formula resolves this paradox by adjusting the denominator to account for how space and time warp at high speeds.

The mathematical proof that nothing can exceed c becomes clear when you substitute the speed of light into the equation. Imagine the spaceship switches on a laser beam, firing photons forward at exactly c ($v_2 = c$).

$$v = \frac{0.7c + c}{1 + \frac{0.7c \cdot c}{c^2}}$$

$$v = \frac{1.7c}{1 + 0.7} = \frac{1.7c}{1.7} = c$$

No matter how fast the spaceship moves, dividing the numerator by the boosted denominator forces the final observed speed of the light beam to equal exactly *c*. Time dilation stretches the clock ticking inside the ship, and length contraction shortens the meters it travels, perfectly balancing the measurements so that the speed of light remains an absolute constant for all observers.

Time dilation describes the fact that time passes differently for observers in relative motion or in different gravitational fields. Special-Relativistic time dilation occurs due to relative motion. If two observers, A and B, are moving relative to one another, each will observe the other's clock running slower than their own. The faster the relative speed, the greater the time difference. The saying sums it up: "Moving clocks run slow."

Gravitational Time Dilation arises from differences in gravitational potential. Clocks run slower in a stronger gravitational field (closer to a massive object) and faster when they are in a weaker field (farther away). For example, a clock at the top of a mountain runs slightly faster than a clock at sea level.

Gravitational waves are ripples or perturbations in the curvature of spacetime that propagate outward from their source at the speed of light. They are generated by the acceleration of large masses, particularly those that are not spherically symmetric (like two massive objects orbiting each other). The strongest sources include black hole mergers, neutron star collisions, and core-collapse supernovae. Gravitational waves squeeze and stretch spacetime. When a wave passes Earth, it momentarily changes the distance between objects. This change is minuscule – less than the width of an atom's nucleus. The LIGO and Virgo observatories detect these changes using massive laser interferometers.

General Relativity has extreme consequences, such as black holes, in which spacetime curvature becomes infinite (a singularity). The event horizon is a boundary around the (presumed) singularity where the gravitational pull is so strong that the escape velocity exceeds the speed of light. Once an object crosses the event horizon, it cannot escape. At the center of the black hole, matter is compressed to an infinitely dense point called the singularity. Physics breaks down at this point, requiring a theory of quantum gravity to accurately describe it.

A geodesic is the straightest possible line in a curved spacetime. In General Relativity, all objects subject only to gravity (i.e., in free fall) follow geodesics. For example, the path of a baseball thrown on Earth is a parabola, which is the geodesic it takes in the curved spacetime around the Earth. An astronaut in orbit is not fighting gravity; they are simply following the natural geodesic path–they are in continuous free fall.

Originally, Einstein added the cosmological constant (λ) term to his equations to force the universe to be static (as was believed in 1917). After Edwin Hubble discovered the expanding universe, Einstein called λ his "biggest blunder." The cosmological constant was revived to explain dark energy, the mysterious force driving the *universe's accelerated expansion*. In modern cosmology, λ is now accepted as the term for the energy of space (vacuum energy), which exerts a repulsive gravitational force.

The Lorentz Length contraction is a purely Special Relativistic effect describing the shortening of an object along the direction of its motion relative to an observer. If an observer is at rest relative to a moving object, the observer will measure the object's length to be *shorter* than if the object were at rest. The effect is negligible at everyday speeds but becomes significant as the object approaches the speed of light c. An object traveling at c would be measured as having zero length in its direction of motion.

Note: Length contraction and time dilation arise from the same mathematical transformation (the Lorentz transformation) required to ensure that the speed of light is the same for all inertial observers. Though derived during Special Relativity, $E = mc^2$ is fundamental to General Relativity, as it states that mass and energy are interchangeable, the mass-energy equivalence. This is crucial because all forms of energy, not just mass, cause spacetime curvature. Pressure, momentum, and stress also contribute to the gravitational field (as described by the Stress-Energy Tensor in Einstein's Field Equations).

Alternative and Extended Theories of Gravity

Alternatives to general relativity modify or extend Einstein's equations to address its limitations, such as dark energy and quantum-gravitational effects. Examples include:

- Starobinsky ($R+R^2$) gravity was an early extension of general relativity that adds a quadratic term to the Einstein-Hilbert action.
- The Gauss–Bonnet theory includes terms quadratic in the curvature tensor.
- f(R) gravity is a class of theories where the Einstein-Hilbert action is replaced by a more general function of the Ricci scalar (R).
- The Lovelock theory is a generalization of Einstein's theory of gravity.
- Scalar-Tensor Theories (e.g., Brans-Dicke) introduce a scalar field that couples to gravity, modifying its strength.
- Modified Newtonian Dynamics (MOND) proposes that gravity behaves differently at low accelerations, explaining galaxy rotation without dark matter.

- Teleparallelism & Symmetric Teleparallelism reinterpret gravity using torsion (Teleparallelism) or non-metricity (Symmetric Teleparallelism) instead of curvature, offering alternatives to the metric framework.
- Biometric Theories involve two metrics to describe gravity, separating it from matter.

Timescape Cosmology

Timescape cosmology is a fascinating, non-mainstream cosmological model that attempts to explain the apparent accelerated expansion of the universe without invoking dark energy. It's rooted in General Relativity but challenges some fundamental assumptions of the Standard Model of Cosmology (λCDM). The central premise of Timescape cosmology is that the universe is not as smooth and homogeneous on large scales as the Standard Model assumes. Instead, it proposes that the universe's structure is fundamentally inhomogeneous, consisting of voids, vast regions of space that are significantly underdense, containing very little matter. These are expanding faster than average. David L. Wiltshire proposed the timescape cosmology model. He introduced the framework in 2007 as a radical alternative to standard homogeneous cosmologies that require dark energy.[53] Wiltshire argues that cosmic acceleration is an apparent effect rather than a real physical expansion driven by a cosmological constant. His model accounts for the universe's inhomogeneous structure, recognizing that gravity affects clocks differently in vast, empty cosmic voids than in dense galaxy filaments where observers reside.

Dense regions form walls, filaments, and clusters of galaxies, where matter is concentrated. These regions are expanding more slowly than average and are gravitationally bound. Timescape cosmology argues that the apparent accelerated expansion we observe is not due to a mysterious dark energy, but rather an artifact of how we average observations in this inhomogeneous universe. It's an "apparent" acceleration, not an absolute, uniform acceleration of spacetime everywhere. Here's how Timescape cosmology proposes the apparent acceleration emerges:

Because voids are underdense, they experience weaker gravity and expand faster. Walls/Filaments (denser regions) experience stronger gravity, slowing down their expansion. Gravitational Time Dilation (Differential Aging) is a crucial element. According to General Relativity, time is affected by gravity. In the underdense voids, where gravity is weaker, clocks tick slightly faster relative to observers in denser regions. In the denser walls, where gravity is stronger, clocks tick slightly slower relative to observers in voids.

We, as observers in galaxies, are in denser regions. Our clocks run slower than hypothetical observers in a vacuum. When we observe distant supernovae (which are used to measure cosmic expansion), we are essentially looking out into a mix of voids and walls. However, we interpret these observations under the assumption of a homogeneous universe with a single cosmic time, thereby misinterpreting redshifts. Because of differential expansion and time dilation, when we *average* over these inhomogeneous regions and interpret the data through the lens of a homogeneous model, we get the *illusion* of accelerated expansion. It's like comparing apples and oranges – we're not accounting for the different "aging rates" (time dilation) across regions, leading to a misinterpretation of the expansion.

A common analogy is raisins in rising bread. The voids are like dough that expands faster than average. The walls and filaments are like raisins; they are denser, closer together, and move apart more slowly than the expanding dough around them. We observe the universe from within the "raisins," so from the perspective of an observer on a raisin, it might seem like the overall "bread" is expanding in an accelerating way, because the regions between raisins (dough/voids) are stretching out faster. However, this "acceleration" is not uniform; it results from averaging different expansion rates.

The most striking feature is that it aims to explain the apparent acceleration without needing dark energy or a cosmological constant. It operates within the standard framework of General Relativity, while emphasizing the importance of inhomogeneity and averaging effects. It also predicts that the universe is dominated by voids, which occupy a substantial fraction of the total volume. Estimates indicate a void volume fraction of around 60-80%. Thus, the distribution of matter is often described as fractal-like, with structure on many scales (voids within voids, walls around voids, etc). It predicts subtle but potentially observable differences in cosmological parameters and relationships compared to λCDM.

Although it can fit supernova data, the best-fit parameters and interpretation differ from those of the λCDM model. Predictions for the CMB are still being refined, but subtle differences in the angular power spectrum may exist. Baryon Acoustic Oscillations (BAO) are being investigated and may predict a slightly different local Hubble expansion rate compared to the global average, potentially explaining some local Hubble tension observations.

Timescape cosmology addresses the "dark energy problem" by suggesting that it may not be needed and that it fits naturally with the observed large-scale structure of the universe, which *is* known to be inhomogeneous. It addresses the challenge of averaging Einstein's equations in an inhomogeneous universe, a known issue in cosmology. And finally, it makes predictions that, in principle, are testable against observations (though distinguishing them from λCDM can be challenging).

However, there are some problems and criticisms. Dealing with inhomogeneous cosmologies in General Relativity is mathematically more complex than homogeneous models. Timescape cosmology is not yet widely accepted by the cosmology community. λCDM is currently favored for its simplicity and excellent fit to a broad range of data (though it also has problems). Distinguishing between the subtle predictions of Timescape cosmology and those of λCDM requires precise observations, which can be challenging.

Gödel's Universe and Einstein's Field Equations

I was reading an article when something about the rotation of our universe came up, which caught my attention – rotating in relation to what? It's not like you can stand outside of our universe and see the rotation. What I discovered was this bizarre universe described by Kurt Gödel. I felt it was appropriate to put it here.

In 1949, the mathematician Kurt Gödel discovered an exact solution to Einstein's field equations of general relativity. Einstein's equation dictates how the geometry of spacetime interacts with mass and energy. Gödel proved that these equations allow for universes with bizarre temporal properties. His model describes a homogeneous, static universe filled with a uniform distribution of pressureless matter, often called *dust*. To prevent this swirling matter from collapsing under gravity or flying apart due to rotational forces, Gödel introduced a negative cosmological constant (Λ). The attractive gravitational effect of this negative constant perfectly counters the centrifugal force of the rotating matter, keeping the universe stable and non-expanding.

Gödel's universe directly contradicts Mach's principle, which heavily influenced Einstein in developing general relativity. Mach posited that local inertial forces are entirely determined by the distribution of matter throughout the cosmos. According to this idea, a local inertial frame should not rotate relative to the average distribution of distant stars and galaxies.

In Gödel's universe, if you set up a gyroscope to establish a local inertial compass, you will observe that the entire matter distribution of the universe rotates relative to your gyroscope. The rotation is an absolute property of the spacetime geometry itself, rather than an effect generated by the matter. This proved that Einstein's field equations do not automatically enforce Mach's principle.

The global rotation of matter in Gödel's solution alters the causal structure of spacetime through a particularly severe manifestation of frame dragging. As you travel radially outward from any chosen origin point, the intensity of this rotational dragging increases. This forces the local light cones – which dictate the boundaries of all possible future paths for matter and radiation – to tilt progressively forward.

Key Differences to Our Universe

- ***Global Rotation versus Cosmic Expansion:*** Gödel's universe features a uniform global rotation and remains static in size. Our universe exhibits no measurable global rotation and expands continuously from a hot big bang.
- ***The Sign of the Cosmological Constant:*** Gödel's model requires a negative cosmological constant to balance the centrifugal forces of rotation. Observations confirm that our universe contains a positive cosmological constant, causing the expansion of space to accelerate over time.
- ***The Rules of Causality:*** Gödel's universe contains closed timelike curves that break the linear progression of time, allowing causes to loop back on their own effects. Our universe maintains strict chronological protection; you cannot find paths that loop into the past, and causality remains absolute.

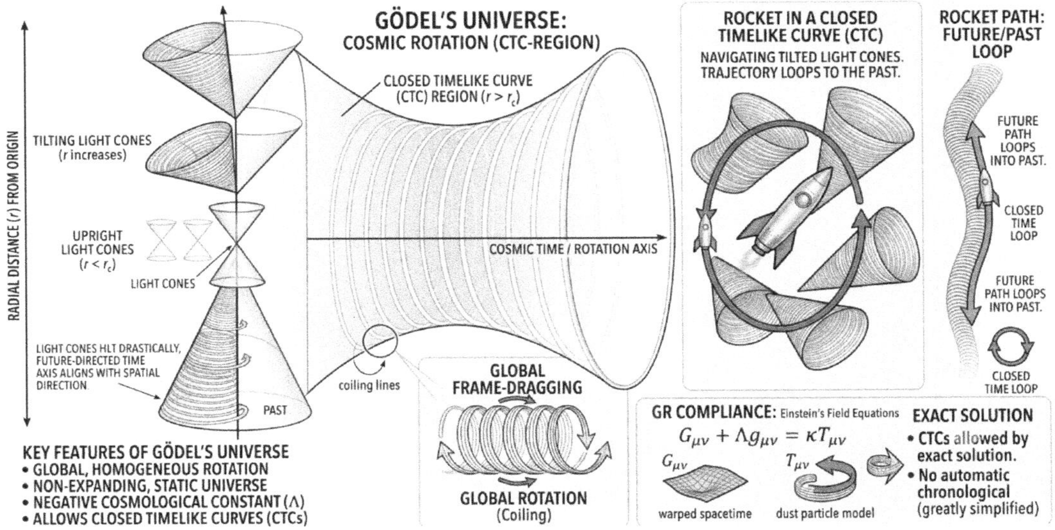

So, if you thought it was only quantum theory that is weird…

Summary

Transitioning to the macroscopic cosmic scale, Chapter Five introduces Albert Einstein's masterpiece: General Relativity. It starts with the Lambda-CDM model, the standard framework of modern cosmology, detailing how the universe is composed of roughly 68% Dark Energy, 27% Cold Dark Matter, and only 5% ordinary baryonic matter. It then moves on to Einstein's core concept: gravity is not a traditional pulling force but the geometric curvature of the 4D fabric of spacetime caused by mass and energy.

The chapter reviews the Equivalence Principle, Background Independence (Covariance), the warping of time (gravitational time dilation), the history and paradoxes of black holes (including Stephen Hawking's theory of radiation and the information paradox), and non-mainstream alternatives such as Timescape Cosmology. It also shows that the equations of General Relativity are complex, and can be used to describe otherworldly universes.

Chapter Six: Quantum Gravity Theories

Ψ Tμν

Quantum Gravity History

In the 1930s, physicists started exploring the concept of "quantum gravity" by attempting to quantize the gravitational field using perturbative methods, but faced many challenges.

The Arnowitt-Deser-Misner (ADM) formalism is a Hamiltonian formulation of Albert Einstein's general relativity. Developed in the early 1960s by Richard Arnowitt, Stanley Deser, and Charles W. Misner, it translates complex spacetime geometry into standard dynamical variables, serving as the foundational framework for numerical relativity and quantum gravity.[54]

Instead of viewing space and time as an inseparable 4D block, the ADM formalism analyzes gravity as an evolving 3D "Cauchy problem". It decomposes spacetime into a stack of 3D spatial slices (hypersurfaces) that evolve over time, much like flipping through the pages of a book. In astroparticle physics, kinetic Boltzmann equations are used to compute the relic abundance of dark matter (frequently abbreviated as xDM for extended Dark Matter models). The Cauchy problem in this setting refers to integrating the coupled differential equations for particle density from the early universe forward to the present day.

In the 1970s and 1980s, string theory was a potential candidate for quantum gravity. It proposes that fundamental particles are tiny vibrating strings in higher-dimensional space. In the late 1980s and 1990s, Carlo Rovelli and Lee Smolin developed loop quantum gravity (LQG), which describes spacetime as a network of interconnected "spin networks" at the quantum level.[55]

Other theories include Causal Set Theory, Asymptotically Safe Gravity, Gravity from Qubits, Noncommutative Geometry, Dynamical Triangulation, and Spinfoam Formalism. There is no single, universally accepted "latest theory" of quantum gravity. Think of it more as a collection of the latest directions or the most promising avenues of investigation.

Loop Quantum Gravity

Loop Quantum Gravity (LQG) attempts to quantize spacetime itself, directly quantizing the geometry of space and time using techniques from canonical quantization. It predicts that spacetime is fundamentally discrete, composed of discrete "quanta" of volume and area. It has evolved into spin foam models, which provide a path-integral formulation of the theory. The theory has also been applied to cosmology (Loop Quantum Cosmology, or LQC) to address the Big Bang singularity and explore early-universe scenarios. A big challenge is to show how LQG connects to the familiar world of low-energy physics and general relativity in a smooth limit.

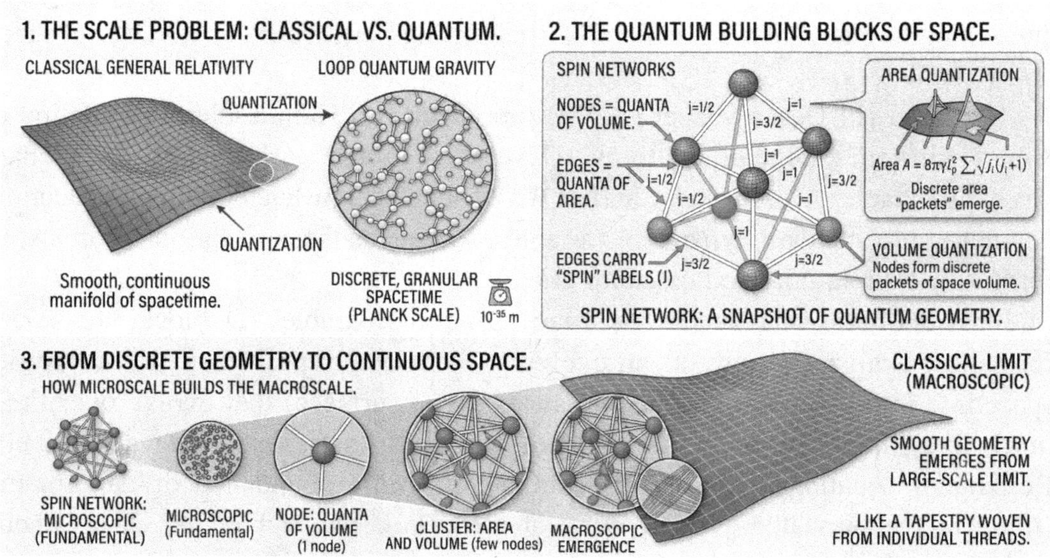

Group Field Theory (GFT) is a related approach that aims to define a "second quantization" of LQG spin networks and foams, potentially offering a more complete and reliable framework. Tensor networks are being explored as tools for describing and simulating LQG and GFT states, with connections to condensed matter physics and quantum information.

Modern physics relies on two separate mathematical frameworks that contradict each other. General relativity explains gravity through the smooth, continuous geometry of space and time. Quantum mechanics describes the subatomic world in terms of individual, distinct packets of energy. Loop Quantum Gravity is a theoretical framework designed to resolve this mathematical conflict by applying quantum principles directly to the structure of space.

In classical physics, you can divide a distance or a volume into infinitely smaller pieces. Loop Quantum Gravity changes this assumption. It states that space possesses a fundamental, indivisible unit of size. The smallest possible physical length is the Planck length – approximately 1.6×10^{-35} meters. Consequently, space has a minimum possible volume, measured at roughly 10^{-105} cubic meters. If you attempt to calculate anything smaller than these values, the mathematics of the theory shows that the state cannot exist.

To track the geometry of this discrete space, physicists use mathematical structures called spin networks.

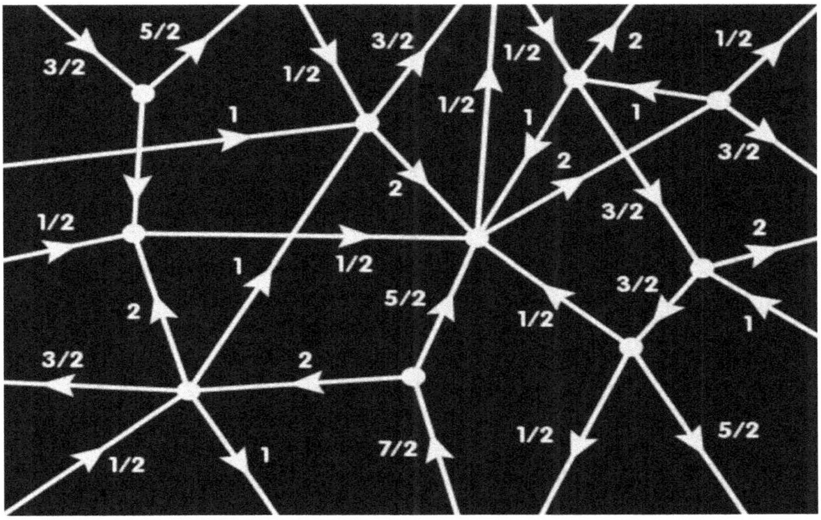

Nodes + Lines + Arrows + Labels = Spin Network

The diagram illustrates the structural components of a spin network:

- **Nodes:** The intersection points represent individual units of volume.
- **Lines:** The paths linking the points represent the intersecting surface areas where those volumes meet.
- **Labels:** The fractional and whole numbers, such as 1/2, 1, and 2, denote quantum spin values. These specific numbers dictate the exact amount of physical area and volume present in that section of geometry.

In this framework, matter and energy do not move through an existing background container of space. Instead, the continuous reconfiguration of these nodes and lines gives rise to the physical manifestations of space, time, and gravity.

There is an interesting consequence of Loop Quantum Gravity regarding Black Holes and its singularity.[56] In general relativity, when a massive star collapses under its own gravity, nothing stops the contraction. The mathematics dictates that all matter forces itself into an infinitely small point of infinite density at the center of a black hole – a singularity. At a singularity, the laws of physics break down completely because mathematical equations cannot handle dividing by zero.

Loop Quantum Gravity alters this scenario by imposing a fundamental limit on how small space can become. Space consists of indivisible units of volume. Because these units possess a minimum physical size, you cannot compress matter or energy into an infinitely small point.

When a dying star collapses, the matter reaches a critical threshold known as the Planck density – approximately 10^{96} kilograms per cubic meter. At this extreme density, the quantum structure of space resists further compression. Instead of creating a singularity, the extreme density triggers a powerful repulsive force due to quantum geometry effects. This repulsive behavior halts the gravitational collapse and reverses the inward motion of the matter. Physicists describe this reversal as a quantum bounce.

The collapsing black hole undergoes a transition, shifting into an expanding structure known as a white hole. A white hole behaves exactly opposite to a black hole, expelling matter and energy outward rather than pulling it inward.

You do not see black holes actively exploding into white holes throughout the universe because gravity warps the passage of time. Near the immense mass of a black hole, time passes much more slowly relative to regions far away.

While the bounce occurs almost instantly for the collapsing matter, severe time dilation makes the process take billions of years when viewed by an observer outside the black hole. Massive black holes appear completely stable and long-lived to modern astronomers, even though the internal singularity never forms.

String Theory and M-Theory

As discussed earlier, String theory postulates that fundamental particles are not point-like but rather tiny vibrating strings. Different vibration modes correspond to other particles, including gravitons (the hypothetical quantum of gravity). String theory naturally includes gravity and aims to unify all fundamental forces and particles. M-theory is considered a more fundamental, though less well-defined, theory that encompasses different string theories.

String theory gives rise to a vast number of possible universes, but most are thought to be inconsistent with low-energy effective field theories. The "Swampland" program aims to identify which effective field theories *cannot* be consistently embedded in string theory and, conversely, what "consistent" universes look like.

The holographic principle, particularly embodied in the AdS/CFT correspondence, remains a central and powerful concept. It provides a deep connection between gravity in Anti-de Sitter (AdS) spacetime and quantum field theories living on its boundary. Research continues to explore applications in black holes, condensed matter physics, and even flat-spacetime holography.

Researchers continue to explore specific string theory compactifications (reducing the number of extra dimensions) to derive models that could resemble the Standard Model of Particle Physics and Cosmology. However, connecting string theory to testable experimental data remains a major challenge.

Conformal Gravity Theories

Philip Mannheim's work on Conformal Gravity is an alternative approach to understanding gravity that attempts to fix two of the biggest problems in modern physics: the mysterious presence of dark matter and the mathematical breakdown of Einstein's General Relativity at subatomic scales.[57] His theory is based on the Weyl Conformal Gravity theory.

In standard physics, Einstein's theory treats gravity as the warping of space and time by mass. While General Relativity works beautifully for our solar system, it runs into severe complications when applied to the scales of entire galaxies or the subatomic quantum world. Mannheim's theory alters the core math of gravity to resolve these issues without inventing undetected forms of matter.

To understand Mannheim's approach, it helps to understand the difference between a standard ruler and a conformal relationship. In Einstein's relativity, the absolute scale or distance matters. If you change the scale of space, the gravitational equations change their behavior.

Mannheim introduces a strict mathematical symmetry called conformal invariance. This symmetry states that the fundamental laws of physics should only care about angles and relative shapes, not absolute size scales. Imagine a map of a city. If you zoom in or out, the absolute distances change, but the angles of the intersections and the relative shapes of the blocks stay identical. Conformal gravity demands that the universe operates on this scale-free principle at its most fundamental level.

Because absolute scales are removed from the baseline equations, mass cannot be an inherent, hardcoded property of the geometry itself. Instead, particles must acquire mass dynamically through a mechanism similar to the Higgs field, where a smooth background field breaks the scale symmetry and gives objects weight.

The primary observational success of Mannheim's theory is how it handles galaxy rotation curves.[58] In the 1970s, astronomers noticed that stars at the outer edges of galaxies rotate just as fast as stars near the center. According to Newton and Einstein, gravity should weaken with distance, meaning outer stars ought to travel much slower, just as Neptune orbits the Sun slower than Mercury. To make Einstein's equations match observations, standard physics requires packing galaxies with a massive, invisible halo of dark matter.

Mannheim's Conformal Gravity alters the gravitational potential formula. In standard gravity, the potential drops off strictly as a fraction of distance:

$$V(r) = -\frac{MG}{r}$$

When Mannheim solves the equations for Conformal Gravity, the math naturally yields an extra, linear distance term:

$$V(r) = -\frac{MG}{r} + \gamma r$$

Where γ (gamma) is a tiny cosmic constant.[59]

At solar system distances, the 1/r term dominates completely, matching our local measurements perfectly. However, across galactic distances – tens of thousands of light-years – the 1/r term fades away, and the linear γr term takes over. This extra term provides a gentle, structural pull that holds the outer stars in their fast orbits. Mannheim has used this formula to plot the rotation curves of over a hundred galaxies without invoking a single particle of dark matter.[60]

The second major achievement of Mannheim's work addresses the mathematical conflict between gravity and quantum mechanics. When physicists try to calculate the gravitational force between subatomic particles using Einstein's equations, the math yields nonsense answers of absolute infinity. This occurs because Einstein's equations allow gravity to become infinitely strong as the distance between two point-like particles shrinks to zero. This is known as a renormalization problem.

Conformal Gravity changes the underlying geometry to include fourth-order derivatives (tracking how rapidly the acceleration of curvature changes) rather than Einstein's second-order derivatives. This higher-order math alters the behavior of gravity at subatomic distances. Instead of spiking to infinity when two particles get close, the gravitational potential levels out and remains finite. This mathematical smoothing makes Conformal Gravity a "renormalizable" theory, meaning it can be calculated cleanly at the quantum scale without collapsing into infinite values.

While Conformal Gravity explains individual galaxy rotation curves well, standard General Relativity (combined with dark matter) is highly successful at explaining the Cosmic Microwave Background and how massive clusters of galaxies bend light (gravitational lensing). Replicating all of these large-scale cosmological achievements using only conformal equations remains an ongoing challenge.

Mannheim's position is that standard General Relativity appears successful at large scales only because it uses dark matter as a "flexible fitting parameter" to compensate for using second-order equations in regions where global cosmological effects dominate. In contrast, Conformal Gravity attempts to solve galaxy curves, cluster lensing, and CMB fluctuations through a single, unified mechanism: the gravitational influence of the rest of the visible universe.

Theories with fourth-order derivatives suffer from a quantum mechanical issue known as *ghosts*. These are mathematical states that possess negative energy or negative probability, which would cause the universe to become fundamentally unstable. Mannheim and his collaborators have published proofs arguing that these ghosts do not actually cause instabilities when calculated correctly. (Mannheim also said, tongue-in-cheek perhaps, *"But I said, never mind. Let's see if nature likes the theory. Because if it does, nature will know how to solve the ghost problem even if I don't know how to solve it"*).

Mannheim, alongside collaborator Carl Bender argues that ghosts only appear because physicists traditionally force quantum Hamiltonians to be *Dirac-Hermitian.* He demonstrates that fourth-order derivative theories (like the Pais-Uhlenbeck oscillator used to model higher-derivative gravity) are not Hermitian in the standard sense. Instead, they possess PT (parity-time) symmetry or broader antilinear symmetries.

Because the Hamiltonian is PT-symmetric rather than Hermitian, the standard quantum inner product (using Hermitian conjugate bras and kets) yields negative norms (ghosts). Mannheim and Bender showed that using a corrected inner product – built using the antilinear symmetry operator – results in a positive-definite probability metric. In certain limits where the Hamiltonian becomes a non-diagonalizable Jordan-block operator, the dangerous ghosts morph into zero-norm states that safely preserve probability without destabilizing the vacuum.

Despite these rigorous mathematical proofs published across decades, Mannheim's solution remains outside the mainstream consensus. Most mainstream quantum field theorists and string theorists remain skeptical. The skepticism stems from deep concerns about whether non-Hermitian or PT-symmetric quantum systems can maintain physical consistency, causality, and stability when coupled to interacting quantum fields in four-dimensional curved spacetime.

True to his tongue-in-cheek quote, Mannheim continues to treat conformal gravity as a viable path forward, betting that if nature genuinely relies on fourth-order derivatives to solve the cosmological constant problem, the mathematics of quantum mechanics will accommodate it regardless of conventional resistance, but many theorists remain unconvinced.[61]

Frankly, I find his attitude refreshing – he's not getting upset or angry, just optimistic that it's going to work out. Because the requirement of scale-independence is mathematically powerful, several other physicists have built distinct conformal frameworks to address quantum gravity, dark energy, and the Big Bang singularity.

Hoyle–Narlikar Gravity (Conformal Machian Gravity): Developed by Fred Hoyle and Jayant Narlikar in the 1960s, this theory blends conformal invariance with Mach's Principle (the idea that the inertia of a local particle is determined by the distribution of all the mass in the rest of the universe).[62]

Instead of treating the gravitational constant G as a fixed number, Hoyle–Narlikar gravity makes mass a variable field that depends on the positions of all other particles. By applying a conformal transformation, the math can be rewritten so that particle masses stay constant while the spacetime geometry stretches. This framework was historically used to attempt to explain cosmic expansion without a traditional Big Bang singularity.

Shaposhnikov & Zenhausern – Conformal Standard Model Gravity:[63] This approach attempts to unify gravity with particle physics by matching the conformal symmetry already found in the equations of the Standard Model.

In this theory, the fundamental equations start completely massless and scale-invariant. The theory introduces two scalar fields: the familiar Higgs field and a new gravity-related field called the "dilaton" – a hypothetical particle or scalar field that dictates the strength of quantum interactions and determines how extra dimensions change in size. When these fields interact in the early universe, they spontaneously break the scale symmetry. This simultaneous event generates the masses of elementary particles and sets the fixed value of the Planck mass (the fundamental scale of gravity).

Loop Quantum Gravity's Conformal Formulation: Within Loop Quantum Gravity (LQG), some theorists use a conformal framework to handle the mathematics of spacetime at the smallest possible scale.[64]

In standard LQG, space is made of discrete, quantized loops of area and volume. By separating the physical geometry into a "conformal class" (which dictates angles and shapes) and a separate scale factor (which dictates absolute size), physicists can isolate the quantum fluctuations of geometry more easily. This helps smooth out the equations when calculating how quantum space transitions from one state to another.

Roger Penrose – Twistor Theory: While not a traditional alternative to Einstein, Sir Roger Penrose's Twistor Theory relies entirely on conformal geometry.[65] Penrose argued that at the fundamental quantum level, the universe is not made of space and time points, but of light rays.

Because light has no mass, it never experiences the passage of time or absolute distance; it only experiences angles. Therefore, the foundational geometry of Twistor Theory is completely conformally invariant. Physical spacetime, along with mass and gravity, only emerges as a secondary property when these conformal light-ray relationships intersect.

Non-Local Conformal Gravity:[66] To avoid the mathematical problem of "ghosts" (unstable negative energy states) that can plague Mannheim's higher-derivative theory, some modern theorists study non-local conformal gravity.

Instead of adding traditional higher derivatives to the equations, this theory introduces mathematical operators that contain an infinite series of derivatives (non-local terms). This infinite arrangement preserves the scale-free conformal symmetry at high energies, tames the infinities of quantum gravity, and mathematically prevents the dangerous ghost states from forming.

Transactional Interpretation (TI)

Transactional Interpretation (TI) offers a unique framework for addressing quantum gravity by shifting the focus from a one-way evolution of spacetime to a symmetrical, time-symmetric interaction. TI provides a conceptual connection between the quantum and the cosmic realms through the following mechanisms.

Standard quantum field theory usually requires a fixed "background" of spacetime to function. Quantum gravity is difficult because spacetime itself must be quantized. In TI, spacetime is not just a stage on which events occur. Instead, spacetime intervals are defined by the "handshakes" (transactions) between emitters and absorbers. This suggests that the geometry of the universe could emerge from the network of these completed transactions. If there are no handshakes, there is no defined spacetime interval.

General relativity treats time as a dimension that is woven into the fabric of space (x, y, z, t). Quantum mechanics treats time as an external parameter that just "ticks" along. Because TI uses both retarded waves (moving forward in time) and advanced waves (moving backward in time), it satisfies the time-symmetry condition of Einstein's field equations. This bidirectional flow helps reconcile the quantum world's "arrow of time" with the cosmic world's four-dimensional "block universe" where past, present, and future are equally real.

Quantum gravity must account for non-locality–the fact that entangled particles remain connected over vast distances, like the million light-year scenario we discussed. Classical gravity is local (it travels at the speed of light). In TI, the handshake is a "standing wave" that exists across a four-dimensional interval. The connection between particles is not "traveling" through space; it is a completed circuit in 4D spacetime. This provides a physical mechanism for how gravity might coordinate quantum states across large distances without violating relativity.

In general relativity, mass curves spacetime. In TI, mass and energy act as emitters and absorbers. Some physicists explore the idea that the "force" of gravity is actually a manifestation of the massive number of transactions occurring between all matter in the universe. The backward-moving "confirmation waves" from distant matter (the absorbers) might exert a physical influence on the present, contributing to what we perceive as the gravitational field or inertia.

While TI provides a more compatible "language" for quantum gravity than the Copenhagen interpretation, it is not yet a complete mathematical solution. Calculating the total gravitational potential of the universe by summing over every possible emitter-absorber handshake is mathematically overwhelming. We still lack a single equation that perfectly links the Guiding Equation or the Transactional handshake to the curvature of the spacetime manifold. TI treats the universe as a self-constructing web of handshakes. Gravity, in this view, is the collective result of these transactions "pulling" the fabric of spacetime together.

Asymptotically Safe Gravity

Asymptotic Safety suggests that quantum gravity might be a consistent theory not by being perturbatively renormalizable (like the Standard Model), but by being "non-perturbatively renormalizable" at a high-energy fixed point. This means that the theory remains well-defined at arbitrarily high energies due to a subtle balance of quantum corrections. It's the idea that gravity can be safely described using standard quantum physics at extreme energies because the strength of gravity stops changing and locks into a constant, stable value instead of blowing up to infinity.

If you calculate what happens to gravity at extreme subatomic distances, the equations output infinity. It means the theory cannot make reliable predictions at the smallest scales, such as the exact center of a black hole.

Asymptotically Safe Gravity offers a way to fix the math without inventing entirely new entities like extra dimensions or strings. Steven Weinberg proposed this idea in 1976.[67]To understand how it works, look at how forces change with distance.

In the quantum world, the strength of a force depends on how close you get to the object. For example, the electric charge of an electron changes depending on the energy you use to probe it.

Standard calculations suggest that as you zoom closer to a particle, the strength of gravity increases without limit. Asymptotic safety proposes that this increase stops. As energy levels rise toward the Planck scale – an extreme energy level of 10^{19} gigaelectronvolts, or distances around 10^{-35} meters – gravity reaches a maximum strength. Physicists call this limit an ultraviolet fixed point. Once gravity reaches this point, its strength locks into place and stops growing.

Imagine driving a car up a mountain road that gets steeper the higher you climb. Standard quantum gravity theories predict that the road eventually becomes a vertical wall. Your car crashes, and you cannot go any further.

Asymptotic safety changes the shape of the mountain. It suggests that the road gets steeper for a while, but as you approach the very peak, it flattens out into a high, level plateau. You can drive onto the plateau safely. The slope drops to zero, the physics remains predictable, and the math no longer spits out infinities.

Because the Planck scale requires energies far greater than what modern particle colliders can produce, scientists cannot test this idea directly. The Large Hadron Collider operates at roughly 1.3×10^4 gigaelectronvolts, which is far too low to see the fixed point. Instead, researchers use computer simulations and advanced mathematical modeling to track how gravitational force flows at high energies. If the mathematics continues to show that gravity hits a reliable plateau, the theory remains a viable candidate for explaining the quantum fabric of space and time.

Emergent Gravity and Quantum Information

These approaches suggest that gravity might not be a fundamental force but instead emerges as a collective phenomenon from more basic, non-gravitational degrees of freedom, much like thermodynamics emerges from the statistical mechanics of atoms. Quantum information concepts, such as entanglement, are often suggested as playing a role in this emergence.

Ideas linking entanglement in quantum field theories to spacetime geometry and gravity (following the work of Ryu-Takayanagi and others in holography) are being actively pursued. This includes exploring whether and how spacetime geometry and dynamics can be derived from the entanglement properties of an underlying quantum system. Building on the idea that gravity is related to thermodynamics (Hawking radiation), researchers are exploring whether gravity can be derived as an emergent thermodynamic phenomenon.

These approaches explore spacetime discretizations using matrices or random geometries, aiming to find quantum gravity through non-perturbative methods and by connecting to statistical mechanics and condensed matter physics. There's growing interest in using quantum information theory as a foundational framework for understanding spacetime and gravity. Concepts like quantum error correction, quantum complexity, and tensor networks are being explored in this context.

Many approaches suggest that quantum gravity may involve non-locality, challenging our classical notions of spacetime. LQG and some other approaches predict a fundamental discreteness of spacetime at the Planck scale. Quantum gravity might require modifications or generalizations of spacetime symmetries, such as Lorentz symmetry. The holographic principle remains a guiding principle, suggesting deep connections between gravity and non-gravitational theories. Quantum information concepts are increasingly seen as essential tools and fundamental constituents of spacetime and gravity.

While direct experimental tests of quantum gravity are incredibly challenging at the Planck scale, there's growing interest in exploring the potential for subtle phenomenological consequences that might be testable in cosmology, black hole physics, or future high-energy experiments.

The "latest theory of quantum gravity" isn't a single entity but rather a dynamic evolution of theoretical frameworks and research directions. There is no consensus on which approach is "correct," and the ultimate theory may incorporate aspects from multiple frameworks or even be something entirely new.

Let's look at some other quantum gravity theories.

Noncommutative Geometry

Noncommutative Geometry (NCG), a branch of mathematics where multiplication is not always commutative, is being explored as a potential framework for understanding quantum gravity, as it offers a way to describe spacetime at the Planck scale where the very notion of a smooth manifold might break down, potentially allowing for a unified description of gravity and other fundamental forces; this approach is fascinating because it could provide a way to incorporate non-commutativity naturally into the geometry of spacetime, which is thought to be necessary for a quantum theory of gravity.

In classical geometry, the algebra of functions on a space is commutative; NCG explores "spaces" in which this is no longer true, effectively replacing points with algebraic structures such as operator algebras. The foundational framework in NCG consists of an algebra, a Hilbert space, and a self-adjoint operator, often a Dirac operator. This "triple" encodes the metric and differential structure of a noncommutative space without needing the concept of a "point". NCG primarily uses *C*-algebras* and von Neumann algebras to describe noncommutative spaces. These act as the "noncommutative coordinates" of the system.

Some key applications include providing a natural language for quantum mechanics, in which physical observables (such as position and momentum) do not commute. It is also used in research on the Quantum Hall Effect and String Theory. The Noncommutative Standard Model (Spectral Standard Model) attempts to unify gravity with the other fundamental forces by treating spacetime as a product of a 4D manifold and a discrete noncommutative space.

Causal Set Theory

This approach is built on the hypothesis that the fundamental structure of spacetime is not a continuous manifold but a discrete collection of points, or "spacetime atoms," linked by causality. Causal Set Theory (CST) is an approach to quantum gravity that posits that spacetime is a discrete set of "spacetime atoms" ordered by causal relationships. This partial order, which represents a "before and after" relationship among spacetime events, is the theory's fundamental structure. Rafael Sorkin, one of the initiators of the causal set program, famously summarized the theory's core idea with the slogan: "Order + Number = Geometry. "

CST assumes that spacetime is not the smooth, continuous manifold described by General Relativity but is fundamentally a discrete, microscopic structure. The "elements" of a causal set are the indivisible building blocks, or "atoms," of spacetime. A causal set is a partially ordered set (poset) where the ordering relation is the physical cause-and-effect relationship between discrete spacetime events.

Unlike General Relativity, CST treats causality as the most fundamental feature of reality. The ordering is *partial* because some events are "spacelike separated" and have no causal relation to one another. For a causal set to approximate a smooth, continuous spacetime (a "manifold-like" causal set), its elements are "sprinkled" into the manifold. The number of elements within a given region of the causal set is, on average, proportional to the volume of that region in the continuum spacetime.

The geometry of spacetime is thus derived from the causal ordering and the number of elements (determining the volume). Although spacetime is discrete, CST is designed to maintain local Lorentz invariance, a key feature of relativity that states the laws of physics are the same for all inertial observers. The random nature of the sprinkling process avoids the problem of a preferred reference frame that arises with a regular, crystal-like lattice.

Work on the dynamics of CST is ongoing, and some approaches, such as sequential growth models, propose that the causal set "grows" by adding new elements one by one according to specific probabilistic rules. Another is the sum over causal sets, similar to the path-integral formulation in quantum mechanics. This approach computes quantum effects by summing over all possible causal sets, each with a specific quantum weight.

Despite being an incomplete theory, CST has produced intriguing results: early simulations based on it predicted a tiny, fluctuating cosmological constant. This idea predated its discovery as the driving force behind the universe's accelerated expansion. CST provides a way to explain black hole entropy, suggesting that it arises from counting discrete spacetime elements on the black hole's event horizon.

Gravity from Qubits

The idea that gravity might emerge from qubits in the quantum foam is a highly theoretical concept in quantum gravity research. It uses quantum information theory, specifically the behavior of qubits, to understand and simulate the fundamental nature of gravity. This research explores whether spacetime itself emerges from quantum entanglement.

A central theme in modern physics, often associated with the "It from Qubit" collaboration, is the idea that spacetime geometry is not fundamental, but an emergent property of quantum information. Researchers use the AdS/CFT correspondence to show how a "bulk" gravitational theory can be described by a quantum theory on its "boundary". Physicists have used quantum processors to simulate "traversable wormholes," sending information between entangled qubits in a way that mimics gravitational dynamics.

While gravity is often ignored in quantum computing, studies have shown that it acts as a "universal dephasing channel," limiting qubit stability. Differences in vertical height (even by 1 cm) cause qubits to experience time at different rates. This leads to subtle phase shifts that contribute to decoherence in large-scale quantum systems. For vertically aligned qubit chips, Earth's gravity induces measurable "dephasing," in which quantum states lose synchronization.

The extreme sensitivity of qubits to gravity is being repurposed to create next-generation quantum sensors. By measuring phase shifts in entangled qubits, scientists can detect minute variations in local gravitational acceleration with high precision (relative sensitivity around 10^{-7}). Experiments such as the QGEM (Quantum Gravity-Induced Entanglement of Masses) protocol aim to detect whether gravity can entangle two masses, providing experimental evidence for the quantization of gravity.

Unlike classical bits that can be either 0 or 1, a qubit can exist in a superposition of both states simultaneously. Qubits can also be entangled, meaning their states are linked so that they share the same fate, regardless of the distance separating them. One intriguing idea is that the fundamental building blocks of spacetime at the Planck scale are not points or strings, but instead these units of quantum information – qubits. In this view, the seemingly continuous spacetime we experience at macroscopic scales emerges from the collective behavior and interactions of these underlying qubits.

The crucial link between these qubits and gravity lies in entanglement. Several theoretical frameworks suggest that the geometry of spacetime, and so gravity, might arise from the entanglement structure of these fundamental qubits. The ER=EPR Conjecture (Maldacena & Susskind) proposes a deep connection between quantum entanglement and wormholes. It suggests tiny wormholes connect entangled particles in spacetime. This indicates a connection between quantum information and spacetime.

I think the above "indication" depends upon your view of quantum entanglement. I concede that these particles are correlated, but that's different from being linked or connected in some physical reality – like wormholes. Indeed, a 2026 high-precision, subatomic experiment tested the ER=EPR conjecture by analyzing the quantum-mechanical entanglement of the proton and electron spins in a hydrogen atom. By measuring the electric field behavior when this entanglement is intentionally broken, researchers found that the data matches standard quantum mechanics, contradicting the existence of a physical wormhole bridging the particles.

I think a brief discussion of entanglement vs. correlation is needed: By measuring the electric-field behavior when this entanglement is intentionally broken, researchers found that the data matches standard quantum mechanics, contradicting the existence of a physical wormhole bridging the particles. I think a brief discussion of entanglement vs correlation is needed:

Based on my understanding, Entanglement does not require a physical tether or any communication channel. That is, doing something to one entangled particle doesn't get "communicated" to the other particle. *"But how does the second particle know what to 'be'?"* The simple answer is that entanglement is a fundamental conservation law written into a *single*, shared math equation (a wave function). Entangled quantum particles can be thought of as one "entity".

Imagine you have two sheets of paper, one red and one blue. You place them into two identical envelopes and mix them. One envelope goes to Los Angeles and the other to New York. If you open the envelope in Los Angeles and find the blue paper, you instantaneously know the envelope in New York contains the red paper. Did the Los Angeles envelope send a faster-than-light signal to New York? Did a microscopic wormhole tunnel between them? Unlikely. In this non-quantum example, the outcomes are perfectly correlated simply because they share a common origin.

Quantum entanglement works on the same premise of common origin, but with a quantum twist: superposition. If two subatomic particles are created from the same event – say, a zero-spin parent particle that decays into two daughter particles – physics mandates that their spins must balance out to zero. One must spin UP, and the other must spin DOWN. Unlike the paper in the envelope (which was always blue or red), quantum particles do not have a fixed identity while traveling. They exist in a hazy state of probabilities where both particles are simultaneously UP and DOWN. When an experimenter measures Particle A in Los Angeles, the haze clears. Particle A "chooses" to be UP. Because total spin must equal zero, Particle B in New York must instantly be DOWN.

This isn't "action at a distance" because nothing (no information) was sent. The person in Los Angeles cannot force Particle A to be UP or DOWN; the result is completely random – so they cannot use it to send a message. As mentioned earlier, the two particles are governed by a single mathematical description – a wave function. When you look at one part of the equation, the rest of the equation naturally updates.

Trying to explain entanglement using FTL particles or wormholes is an attempt to force our everyday, macroscopic intuition onto the quantum world. We assume that for two things to agree, they must talk to each other. Quantum mechanics tells us that two things can agree simply because they are two parts of the *same* system. They don't need a physical connection because, mathematically, they never truly became separate entities until the moment they interacted with the outside world. Yeah, it's still weird, but not as complicated as inventing FTL or wormholes to make it work.

The Holographic Principle, arising from black hole thermodynamics and string theory, suggests that the description of a volume of space can be encoded on a lower-dimensional boundary to the region, like a hologram. This implies that the information content (potentially in the form of qubits) on the boundary might be fundamental, and that the higher-dimensional spacetime emerges from it.

Some theoretical models use tensor networks, mathematical structures that represent the entanglement patterns of multiple qubits, to describe how emergent geometries may arise. Different patterns of entanglement in the network can correspond to distinct spacetime curvatures and topologies. Building on these ideas, some researchers, notably Erik Verlinde, have proposed that gravity might be an entropic force. This means that gravity is not a fundamental force mediated by a particle (like the graviton), but rather an emergent phenomenon arising from the system's tendency to increase its entropy (or maximize its information content).

The distribution and entanglement of qubits in the quantum foam might define an "information density" of spacetime. The gravitational force experienced by objects may result from the system's attempt to maximize its entropy or information content. When matter is present, it might create gradients in this information density, and the "entropic force" would manifest as gravity. Objects might be driven to move in ways that increase the overall entanglement of the underlying qubit system, and this movement is what we perceive as gravitational attraction.

Quantum Gravity offers an entirely different approach to unifying quantum mechanics and gravity, starting from quantum information rather than quantizing classical gravity. It suggests that spacetime is not a fundamental entity but rather an emergent phenomenon arising from the underlying quantum information structure.

The exact mechanism by which the entanglement of qubits in the quantum foam gives rise to the specific laws of gravity (such as Einstein's Field Equations) is not fully understood. It is also incredibly challenging to devise experiments to probe the Planck scale and test these ideas directly.

Dynamical Triangulations

A "dynamical triangulation" is a method in theoretical physics, particularly in the field of quantum gravity, where spacetime is constructed by dividing it into tiny, triangular-like building blocks called "simplices," essentially creating a "lattice" that allows for studying the quantum properties of spacetime by simulating its behavior at the most fundamental level; this approach is often referred to as "Causal Dynamical Triangulation" (CDT) and is considered a way to study spacetime without relying on a pre-existing background geometry, meaning the fabric of spacetime itself is considered dynamic and can evolve. CDT is the most successful modern evolution of dynamical triangulation, introduced by Jan Ambjørn, Jerzy Jurkiewicz, and Renate Loll.[68] It adds a critical constraint: *causality*.

Spacetime is built from four-dimensional simplices (pentachorons) glued together in a way that preserves a clear distinction between space and time. CDT is the first approach to demonstrate that a classical, four-dimensional universe can emerge dynamically from random quantum fluctuations. At extremely small scales (the Planck scale), spacetime appears to transition from 4 dimensions to approximately two dimensions.

Unlike string theory, CDT does not assume a preexisting space; the triangulated fabric itself evolves to create space. Curvature is measured by calculating "deficit angles" around the edges where these triangles meet, similar to how a geodesic on a sphere differs from one on a flat plane. Researchers use these to sample millions of geometries on computers to calculate the most probable structure of the universe.

The basic unit is a "simplex," which can be thought of as a tiny triangle in higher dimensions. Connecting these simplices creates a network that represents spacetime, similar to a lattice. In CDT, the arrangement of simplices is designed to maintain a consistent time direction, thereby preserving causality. This method enables researchers to investigate the quantum properties of spacetime (quantum gravity) by simulating the interactions and fluctuations of its fundamental building blocks. Imagine trying to understand the surface of a mountain by covering it with tiny, irregularly shaped tiles. Each tile represents a simplex, and how they fit together represents the curvature and structure of the mountain's surface.

Spin-foam Formalism

A "spin-foam formalism" is a mathematical framework in theoretical physics used to quantize gravity by representing spacetime as a network of tiny, discrete building blocks called "spins". These spins are connected in a foam-like structure, essentially describing the geometry of spacetime at the quantum level, where the fabric of space and time is not smooth but granular; it's a way to calculate quantum gravitational effects by summing over all possible "spin foam" configurations, similar to how Feynman diagrams are used in quantum field theory. The spin-foam formalism is the path-integral approach to Loop Quantum Gravity.

Key models include the Barrett-Crane and EPRL models, which use spin foams to describe the evolution of spacetime at the Planck scale. The formalism completes Loop Quantum Gravity by providing a dynamic, covariant framework that offers insights into the Planck scale by showing discrete spacetime, connecting to classical gravity in the low-energy limit, and enabling quantum-gravity calculations, such as black-hole transitions.

While spin networks describe the instantaneous, three-dimensional "atoms of space," spin foams represent the four-dimensional "history" or evolution of these networks. A spin foam is mathematically defined as a 2-complex – a collection of vertices, edges, and faces.

- ***Faces*** are labeled with group representations that represent quantized areas.
- ***Edges*** are labeled by intertwiners, which represent quantized volumes.
- ***Vertices*** represent points where the topology or geometry of space changes.

In somewhat simpler terms, Loop Quantum Gravity (LQG) scientists use simplified mathematical frameworks called spin-foam models to calculate how tiny, quantum building blocks of space and time evolve over time.

The EPRL Model (Engle–Pereira–Rovelli–Livine) is currently the most popular and actively studied.[69] It provides a more complete picture of four-dimensional quantum gravity by successfully bridging two major ideas in physics:

1. ***Path Integrals:*** A method that sums up all possible historical pathways a system can take from start to finish.
2. ***Spin Networks:*** The fundamental, net-like structures that represent the quantum geometry of space at a single frozen instant.

The Barrett-Crane Model was an earlier pioneering model. It relied heavily on Regge calculus – a mathematical technique that breaks down smooth, curved spacetime into small, flat geometric building blocks (much like building a curved dome out of flat triangular panels). While foundational, it lacked some key quantum properties that the newer EPRL model successfully includes.

Cosmological Applications are used in spin-foam cosmology to describe quantum fluctuations and, potentially, resolve the Big Bang singularity with a "bounce". The primary goal is to compute the probability that space evolves from one spin network state to another. The formalism is "background independent," meaning it does not assume a pre-existing spacetime stage; the foam generates spacetime itself. This framework is used to study the early universe, with an explicit focus on quantum fluctuations at the "Big Bounce," where a quantum transition replaces the initial singularity.

The Graviton

Let's talk about gravity's elusive quantum particle, the Graviton. Whether gravity is quantized remains open, with considerable debate and speculation. The graviton is the hypothetical elementary particle that stands as the ultimate bridge between the smooth, warped geometry of General Relativity and the grainy, probabilistic world of Quantum Mechanics.

In the Standard Model, forces are not just "influences" that act at a distance; they are interactions mediated by the exchange of specific particles. Just as the photon carries the electromagnetic force and gluons bind the nuclei of atoms, the graviton is predicted to be the "packet" of energy that carries the force of gravity.

From a theoretical perspective, the graviton must possess a very specific set of characteristics to align with our observations of the universe. Because gravity has an infinite range and appears to travel at the speed of light, the graviton must be massless and stable. Its most defining feature, however, is its spin-2 nature. In quantum field theory, the spin of a particle dictates how it couples to matter. While a spin-1 particle like the photon can both attract and repel, a spin-2 particle is uniquely required to produce a universally attractive force that couples directly to the energy and momentum of everything in existence.

While gravitons are presumed to be massless, they would still carry energy, as does any other quantum particle, but they would be electrically neutral. Massless particles also carry photon energy and gluon energy. Alternatively, if gravitons are massive at all, the analysis of gravitational waves yields a new upper bound on the mass of gravitons. The graviton's Compton wavelength is at least 1.6×10^{16}m, or about 1.6 light-years, corresponding to a graviton mass of no more than $7.7 \times 10^{-23} eV/c^2$. This relation between wavelength and mass-energy is calculated with the Planck–Einstein relation, the same formula that relates electromagnetic wavelength to photon energy.

At the Planck scale, gravity becomes as strong as the other fundamental forces, requiring a unified theory of quantum gravity. Here, spacetime is believed to cease being smooth and to exhibit violent quantum fluctuations, or "spacetime foam," where general relativity breaks down, and microscopic black holes may form.

The Planck scale is reached when a particle's Schwarzschild radius equals its Compton wavelength. Here, gravitational, quantum, and relativistic effects (combined by constants $G, \hbar, c$) are all equally important. Instead of a smooth, classical geometry, spacetime at this scale is hypothesized to be a turbulent "foam" of constant, tiny fluctuations in black holes.

Our current understanding of physics fails at this scale – the energy required to probe this distance would be sufficient to create a black hole. String Theory proposes that particles are not points but tiny vibrating strings, thereby removing singularities. Loop Quantum Gravity suggests that space itself is quantized into discrete loops or chunks.

Theories abound, but so far there has been little optimism about the prospects for an experiment to detect them. The last time I looked, the idea was that detecting individual gravitons requires energies approaching the Planck scale, far beyond current capabilities. An accelerator capable of this would likely need to be as large as the Solar System or even the Milky Way galaxy. Detecting a single graviton is often cited as the hardest experimental challenge in the history of science. Because gravity is orders of magnitude weaker than the other fundamental forces, the "cross-section" - or the likelihood of a graviton interacting with a particle of matter – is vanishingly small.

Traditional calculations suggest that a detector the size of a planet orbiting a dense star might only catch one graviton every thousand years. This led many to believe that, while we could measure gravitational waves – trillions of gravitons moving in a classical wave – we might never prove the existence of the individual particles. Due to gravity's extreme weakness, direct detection is considered practically impossible with current technology. Estimates range from an accelerator that spans the Solar System to one as large as the Milky Way. Graviton interaction cross-sections are so low that they are $\sim 10^{39}$ orders of magnitude weaker than electromagnetic forces at current accessible scales.

Some researchers suggest using extremely high-frequency, resonant detectors near existing, less powerful accelerators to detect synthetic gravitational waves, rather than building a dedicated particle accelerator. Detecting a single graviton is so difficult that it may be nearly impossible, as gravity is effectively a cumulative, macroscopic force rather than an easily isolated individual particle interaction. While gravitons might be produced at lower energy scales, creating enough of them to be detected and separating them from the immense background noise requires a collider of unprecedented, likely impossible, dimensions.

Graviton Counting via Quantum Acoustodynamics (QAD): Igor Pikovski, a theoretical physicist at Stevens Institute of Technology and Stockholm University, primarily leads this experiment, which aims to provide the first empirical proof of quantum gravity by detecting individual "quanta" of the gravitational field – gravitons. The experiment uses a massive acoustic resonator, such as a 0.5-meter sapphire cylinder or specialized aluminum bar, supercooled to its quantum ground state (near 10^{-6} Kelvin).

At this temperature, thermal "noise" is almost eliminated, allowing the cylinder to vibrate only when it absorbs discrete energy packets. When a gravitational wave passes, it interacts with the cylinder. If gravity is quantized, the cylinder will absorb energy in discrete "quantum leaps" as it transitions between vibrational levels. This process converts a single graviton into a phonon (a quantum of sound/vibration) that ultra-precise quantum sensors can measure. To ensure the signal isn't an experimental error, or local interference, the experiment is synchronized with the LIGO – which acts as the "validation partner," detecting the massive classical wave while the sapphire cylinder "counts" the individual gravitons within that same wave. A simultaneous signal across both systems effectively confirms that gravity is a quantum field.

The Bose-Amaretto-Vedral (BMV) Experiment: The BMV experiment is a promising test for quantum gravity. The logic is elegant: if a system (gravity) can act as a mediator to entangle two quantum objects, then that mediator itself must be quantum. In our systems-architect view, a classical "cable" cannot transmit a quantum "signal" without destroying its coherence.

Two tiny, identical masses – such as micro-diamonds containing a single nitrogen-vacancy center – are placed into a "superposition of locations" using an interferometer. Each mass is effectively in two places at once. These two masses are positioned close enough to interact primarily through their mutual gravitational pull, while being shielded from all other electromagnetic interference. If gravity is a quantum field, the two masses will become "entangled" – their states will become inextricably linked. If gravity is strictly classical, no entanglement will occur.

While this experiment is theoretically straightforward, it is technologically daunting since gravity is exceptionally weak. It requires using masses of about a nanogram (10^{-12} kilograms), which is massively larger than anything previously put into a quantum superposition. The interferometers holding these masses must be placed extremely close to one another – roughly a micron to a few hundred microns apart – and the superposed state must be maintained for about a microsecond. During this time, the system is highly susceptible to microscopic noise and stray vibrations that act on timescales much shorter than a microsecond, which could easily destroy the entanglement before it can be measured.

Macroscopic Quantum Resonators (MAQRO): While the BMV experiment looks for entanglement, proposals like MAQRO focus on the point of "Decoherence." In a systems context, decoherence is the process by which the quantum "blur" resolves into classical "certainty." This proposal involves a space-based platform to conduct experiments in a high-vacuum, microgravity environment. Scientists would observe large molecules or nanoparticles to see at what point gravity causes their quantum wave function to collapse. If gravity has a "quantum noise" signature, it will affect these particles in a very specific, measurable way. This experiment seeks to determine if gravity itself is the "observer" that forces the universe to choose a single state.

Tabletop Optomechanics and Levitated Crystals: Breakthroughs in levitated optomechanics allow us to "trap" tiny glass spheres or crystals using lasers. By cooling these objects to their "ground state"–the absolute minimum energy level–we can treat them as large-scale quantum systems.

By placing two of these levitated spheres near each other, researchers can measure the infinitesimal gravitational pull one exerts on the other. The goal is to observe "quantum jumps" in the motion of one sphere caused by the gravity of the other. If these jumps are discrete, it proves that gravity is not a smooth, continuous field but rather operates in "packets" or quanta.

Atomic Clouds: Researchers from Stockholm University have calculated that gravitational waves could subtly alter how atoms emit light. For this, the authors consider a cloud of 10-100 million atoms cooled to near absolute zero. They assume that the atoms are in an "excited" state, meaning they will release energy in the form of photons when slightly disturbed. The authors then calculate that when a gravitational wave passes through the cloud, it very slightly shifts the frequencies of the emitted photons in certain directions. While these frequency shifts are tiny, the researchers say the directional distortion might be measurable if enough photons are collected. The atoms emit light like a music player that keeps a steady tone, but a gravitational wave changes how the note sounds in different directions.

The atom cloud gravitational wave detector would measure gravitational waves with wavelengths on the order of millions of kilometers, considerably longer than those measured by LIGO.

Summary

Chapter Six tackles the primary conflict of "The Great Divide" – the fundamental incompatibility between the smooth geometry of General Relativity and the discrete, probabilistic quantum realm. It reviews the leading mathematical candidates trying to unify these frameworks into a singular "Theory of Everything" (TOE).

The primary theories explored include Loop Quantum Gravity (LQG), String Theory and M-Theory, applications of the Transactional Interpretation, Asymptotically Safe Gravity, Emergent Gravity, Noncommutative Geometry, Causal Set Theory, Dynamical Triangulations (specifically Causal Dynamical Triangulation), and the concept of deriving gravity directly from quantum bits (qubits).

Chapter Seven: Other Theories and Ideas

Tμν

"Why did the cosmologist break up with their partner? There was just too much tension, and they couldn't agree on how fast they were moving apart."

Hubble Tension

The "Hubble tension" refers to a discrepancy between the measured expansion rate of the universe from nearby observations and the expansion rate inferred from the early universe's afterglow (the Cosmic Microwave Background), suggesting that the universe appears to be expanding faster now than expected from its early conditions. This created a mismatch between the two measurements – direct observation of nearby galaxies (the "Late Universe") and the CMB data extrapolation based on the Standard Model of Cosmology (The "Early Universe").

The two values disagree by about 8-10%, which is a highly significant difference in physics. The Hubble Constant is the current rate at which the universe is expanding. It relates the velocity (v) at which a distant galaxy moves away from us to its distance $v = Dv = D(d)$. The Hubble Constant is usually measured in kilometers per second per megaparsec (km/s/Mpc). The core of the tension is the highly precise disagreement between two distinct "rungs" of the cosmic distance ladder:

Direct Measurement: Higher Value $\approx 73 \pm 1$ km/s/Mpc. Uses a cosmic "distance ladder," relying on local, direct measurements of celestial objects like Cepheid variable stars and Type Ia supernovae (Standard Candles) to measure distances to nearby galaxies.

Model-Dependent Measurement: Lower Value $\approx 67.4 \pm 0.5$ km/s/Mpc. Based on analyzing the CMB data (primarily from the Planck satellite), and then using the (λCDM) model to extrapolate that early expansion rate forward to the present day.

The persistence of this gap suggests two possible scenarios, both of which have deep meaning for our understanding of the universe. If the tension is resolved by identifying an error in the measurements, it would likely indicate an unknown systematic error in calibrating the "standard candles" (Cepheids and supernovae). For example, dust might be affecting their brightness in unaccounted-for ways, or their intrinsic properties may vary in ways we don't understand. Or, while less likely, given the rigor of the Planck data, there could be a subtle error in the analysis of the early-universe data.

If both measurements are accurate, the tension implies that the Standard Model of Cosmology (λCDM) is incomplete and needs modification. This exciting "New Physics" would have to change the rate of expansion sometime between the early universe (CMB epoch) and the late universe (today).

Leading new physics solutions include a proposed form of dark energy (EDE) that briefly dominated the universe immediately after the Big Bang but then quickly decayed. This brief energy boost would increase the early expansion rate, leading to the higher H_0 value we observe today, thus bridging the gap between the CMB prediction and the local measurement.

Early Dark Energy (EDE) is typically modeled as an ultra-light, axion-like scalar field. A secondary class of models involves late-time dynamical fields, such as quintessence and phantom energy. EDE introduces a scalar field ϕ that temporarily contributes 5% to 10% of the universe's total energy density near matter-radiation equality ($z \sim 3000$–5000) before rapidly decaying.

- ***Potential Function:*** The field is often described by an axion-like periodic potential: $V(\phi) = \Lambda^4 \left[1 - \cos\left(\frac{\phi}{f}\right)\right]^n$
  where f is an axion decay constant, Λ sets the energy scale, and $\mathrm{n} \geq 2$ (frequently $\mathrm{n} = 3$).
- ***Freezing Phase:*** Early in the radiation era, Hubble friction (3H) is high relative to the field mass ($\mathrm{m}_\phi \approx V''(\phi)^{1/2}$). The field remains stuck on its potential slope, acting like a temporary cosmological constant with the equation of state $\mathrm{w} \approx -1$.
- ***Decay Phase:*** Once the Hubble rate drops below the field's effective mass ($\mathrm{H} \lesssim m_\phi$), the field rolls into the minimum and oscillates. For an $\mathrm{n} = 3$ potential, the cycle-averaged equation of state during oscillation is:

$$w_\phi = \frac{n-1}{n+1} = \frac{1}{2}$$

  Because $\mathrm{w}_\phi > 1/3$, the field's energy density $\rho_\phi \propto a^{-3(1+w_\phi)} = a^{-4.5}$ dilutes faster than radiation ($\rho_r \propto a^{-4}$), disappearing before it can disrupt standard Big Bang nucleosynthesis or the late-time matter era.

Dynamical dark energy models active in the late universe ($z < 2$) treat the current acceleration as a rolling scalar field rather than a static cosmological constant (Λ):

- ***Standard Quintessence*** **($\mathrm{w} \geq -1$):** A canonical scalar field with a slowly varying potential $\mathrm{V}(\phi)$. Standard quintessence generally worsens the Hubble tension; if dark energy was less dense in the past ($w > -1$), matching the CMB distance scale requires an even lower H_0 today.

- ***Phantom Fields*** **($w < -1$):** A scalar field with a negative kinetic term. Phantom fields increase dark energy density toward the present, which can accommodate a higher H_0, but they introduce theoretical instabilities and fine-tuning challenges.
- ***Interacting Dark Energy (IDE):*** The scalar field directly couples to cold dark matter through an interaction term Q. Energy transfer between the dark sector components alters the background expansion history to relieve tensions between local and high-redshift data.

While an EDE scalar field raises H_0 to $\sim$ 72–73 km/s/Mpc, it alters the growth rate of cosmological structure. Increasing pre-recombination expansion requires a corresponding increase in the cold dark matter density (ω_c) to fit the CMB damping tail. This higher matter density can elevate the amplitude of matter fluctuations, exacerbating the secondary S_8 tension (the clustering discrepancy observed between CMB predictions and weak gravitational lensing surveys).

The properties of dark matter or dark energy might change over time, or they might interact in unexpected ways, accelerating the expansion in the recent past. Modified Gravity (e.g., modified λCDM) that changes the laws of gravity itself at very large scales, affecting how structure forms and how the universe expands.

The Hubble Tension is currently the biggest outstanding challenge to the Standard Model of Cosmology, and resolving it is a top priority in astrophysics.

Modified Newtonian Dynamics (MOND)

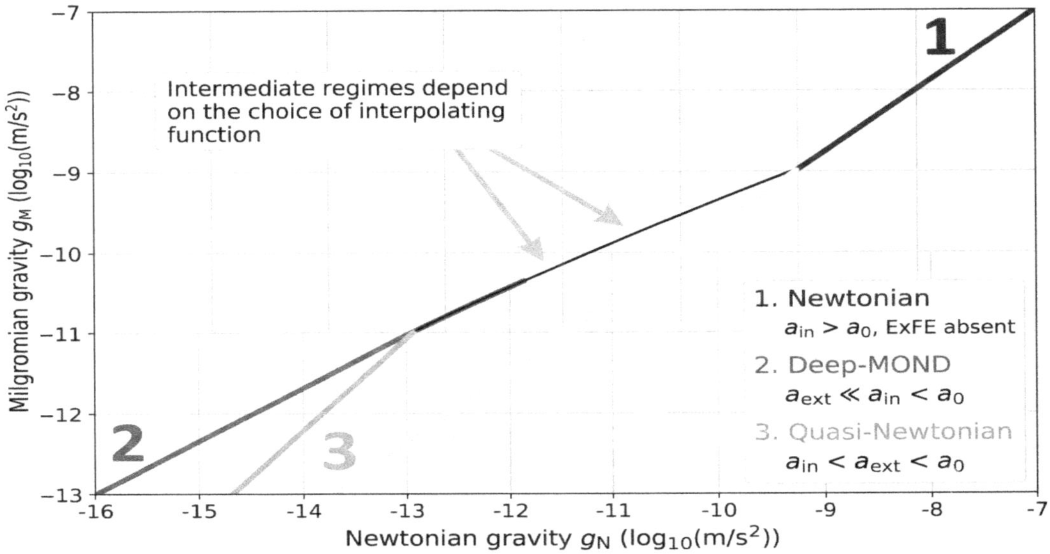

The four different limits of MOND. The black line connecting the first and second regimes is the simple interpolation function. The quasi-Newtonian regime can lie along any line parallel to the green lines between the Deep-MOND and forced-Newtonian limits, depending on the strength of the external field.

MOND is a theory proposed by physicist Mordehai Milgrom in the early 1980s as an alternative to the prevailing dark-matter model, which explains the observed motions of stars within galaxies.[70] A primary piece of evidence for dark matter comes from the observation that stars in the outer regions of galaxies orbit the center at speeds much higher than predicted by Newton's law of gravity based solely on visible matter (stars, gas, dust).

According to Newtonian gravity, the orbital speed should decrease with distance from the galactic center. However, observations show that these rotation curves tend to "flatten out" at large distances. The standard explanation for this discrepancy is the existence of a large amount of unseen, non-luminous matter called dark matter, which provides the extra gravitational pull needed to keep the outer stars moving at such high speeds.

MOND offers a different approach. Instead of postulating the existence of unseen matter, it proposes that Newton's law of gravity is not entirely accurate at very low accelerations, such as those experienced by stars in the outer reaches of galaxies. It introduces a new fundamental constant with the dimensions of acceleration, denoted as α_0. Its value is approximately $1.2 \times 10^{10} m/s^2$. This acceleration scale is incredibly small – much smaller than the acceleration we experience on Earth.

MOND proposes a modified gravitational force law that alters the relationship between the gravitational force (F), the masses (M and m), and the distance (r) when the acceleration due to gravity ($g = \frac{F}{m}$) falls below a critical value, a_0. In regions of high gravitational acceleration, such as within the solar system or in the inner parts of galaxies, MOND's predictions closely match those of Newtonian gravity:

$$F \approx G \times M \times \frac{m}{r^2}$$

In regions of very low gravitational acceleration such as the outer parts of galaxies, MOND proposes a modification in which the force is proportional to the square root of the mass instead of directly proportional: $F = m\sqrt{\frac{GM_{a0}}{r^2}}$. This modification at low accelerations predicts flat rotation curves. In the outer parts of galaxies where the acceleration is very small, the orbital speed (v) becomes approximately constant. It is related to the enclosed mass (M) by the formula: $v \approx (GMa_0)^{\frac{1}{4}}$. This naturally explains why the observed rotation speeds remain roughly constant even at large distances from the galactic center.

MOND has been remarkably successful at predicting the observed rotation curves of a wide variety of galaxies based solely on their visible matter content. It often fits the data without requiring adjustable dark matter halos. MOND also predicts a specific relationship between the observed mass discrepancy (the difference between the mass inferred from gravity and the visible mass) and the gravitational acceleration. This relationship has been observed to hold for a large sample of galaxies. MOND has successfully explained other correlations in galaxy properties that are difficult to account for within standard dark matter models.

However, MOND faces some serious issues and limitations. Newtonian gravity is a non-relativistic theory. To be a complete theory of gravity, MOND needed a relativistic extension applicable to cosmology and other phenomena where gravity is strong, or speeds are high. Jacob Bekenstein's Tensor-Vector-Scalar (TeVeS) theory, a relativistic theory that reduces to MOND in appropriate limits, partially addressed this. MOND struggles to account for galaxy cluster dynamics solely in terms of visible matter. Observations of galaxy clusters often require the presence of some form of non-baryonic dark matter.

The Standard Model of Cosmology, incorporating dark matter and dark energy, provides a highly successful explanation for the anisotropies in the CMB and the formation of large-scale structure in the universe. MOND, in its original non-relativistic form, doesn't offer a natural explanation for these observations and usually requires some form of additional non-standard matter or modifications to the standard cosmological framework.

MOND is primarily a phenomenological modification of gravity. It describes *how* gravity might behave differently at low accelerations, but doesn't provide a fundamental explanation for *why* this modification occurs. Dark matter, while its nature is unknown, is at least a proposal for a new type of matter within the existing framework of physics. While MOND has successes at the galactic scale, some newer observations, particularly involving colliding galaxy clusters (like the Bullet Cluster), are often cited as strong evidence for dark matter, as they seem to show a separation between the gravitational potential and the visible matter in a way that is difficult for MOND to explain without additional components.

Analysis of galaxy clusters argues that much of the "missing" mass may not be exotic dark matter, but ordinary matter in the form of faint stellar remnants. Galaxy clusters have long required about five times as much mass as is visible to explain the observed galaxy motions, an observation usually attributed to dark matter. But the authors combined gravitational lensing, X-ray, and optical data with models of star formation – they found that the total mass in stars and their remnants could be almost twice as high as previously estimated. This matters because MOND does not seem to work well for galaxy clusters.

The dominant paradigm is that galaxies are embedded in halos of non-baryonic dark matter, which interacts gravitationally but not electromagnetically, such as through light. This explains the flat rotation curves and other phenomena. The challenge for dark matter is to detect its particles directly and to fully reconcile certain observations at small scales (such as those from dwarf galaxies).

MOND has strong predictive power at the galactic scale but faces challenges at larger cosmological scales and with certain specific observations. While it has achieved successes, it also faces challenges in explaining observations at larger cosmological scales and lacks an entirely accepted fundamental theoretical basis.

In April 2026, Patricio A. Gallardo and a team of over 40 researchers analyzed data to verify whether the inverse-square law of gravity holds at a cosmological scale, examining how gravity behaves over immense distances in the universe.[71] To measure gravity at this scale, the team evaluated the movement of massive galaxy groups, known as halos. The motion of these halos relative to one another reveals the strength and nature of the gravitational forces pulling them together. The researchers measured this movement by analyzing the kinematic Sunyaev-Zeldovich (kSZ) effect. This effect occurs when low-energy photons from the cosmic microwave background scatter off the hot, ionized gas within galaxy clusters, shifting their energy.

The team combined cosmic microwave background maps from the Atacama Cosmology Telescope with galaxy catalogs from the Sloan Digital Sky Survey. By analyzing these datasets, they estimated the mean pairwise velocity of the halos over spans ranging from 30 to 230 megaparsecs – meaning the study tested gravity across distances up to 750 million light-years.

The researchers modeled the gravitational acceleration between the halos using the formula $g \propto 1/r^n$, where r represents the separation distance and n determines how quickly the force weakens over distance. In standard Newtonian gravity and General Relativity within an expanding universe, n equals exactly 2. The study concluded that $n = 2.1 \pm 0.3$. This numerical result shows that gravity continues to follow the inverse-square law even across vast cosmic voids.

The measurement matches the predictions of the standard Lambda cold dark matter (ΛCDM) cosmological model. It seems to rule out several alternative gravity models that predict different rates of decay over long distances. The team noted that upcoming cosmological surveys will have enough precision to rule out alternative models, such as $n = 1$, with a high statistical significance of 10 sigma.

Double Special Relativity (DSR)

Double Special Relativity is an extension of Einstein's theory that addresses one of the most significant hurdles in modern physics: the conflict between the smooth geometry of relativity and the "pixelated" nature of the quantum world.

In standard Special Relativity, there is only one universal constant that all observers agree on: the speed of light (c). No matter how fast you are moving, light always passes you at the same speed. Double Special Relativity introduces a second constant that is also observer-independent: a minimum length scale (the Planck length) or a maximum energy scale (the Planck energy). This means that just as everyone agrees on the speed of light, everyone also agrees on the size of the "pixels" of spacetime, regardless of their relative velocity.

According to Einstein's original equations, if you move fast enough toward an object, that object will appear shorter due to length contraction. If you move fast enough, a distance could theoretically be contracted down to zero. However, quantum mechanics suggests there is a "bottom" to the universe – the Planck length (10^{-35} meters). If length contraction were absolute, a fast-moving observer might see a distance as being smaller than the Planck length, which would break the rules of quantum physics. DSR modifies the math so that the Planck length remains a "floor" that can never be contracted further, no matter how fast you move.

DSR is often used as a framework for Loop Quantum Gravity and other theories that treat spacetime as a discrete "fabric." By ensuring that the "grains" of space appear the same to everyone, DSR allows a version of relativity to coexist with a quantized universe. It suggests that at extreme energies – such as those found at the moment of the Big Bang – the laws of physics might "bend" to preserve the integrity of the Planck scale. This framework is a key candidate for solving the "Great Divide" because it provides a mathematical language that respects both the speed limit of the cosmos and the size limit of the quantum realm.

Theoretical physicist Sabine Hossenfelder has raised several critical arguments against DSR and related theories of "quantized" spacetime. Her critiques generally focus on mathematical consistency and the lack of empirical evidence. One of Hossenfelder's primary technical refutations of DSR is known as the soccer ball problem. This is a consistency issue regarding how the rules of DSR scale from individual subatomic particles to macroscopic objects. In DSR, the laws of physics are modified at the Planck scale to keep the Planck length constant. However, if you combine billions of particles into a larger object, the mathematical modifications that work for a single particle can lead to absurd results for the whole object. Hossenfelder argues that many versions of DSR fail to provide a "natural" way for these quantum effects to disappear at the macroscopic level, potentially leading to violations of basic physics that we should have already observed.

Hossenfelder points out that despite decades of theoretical work, there is zero experimental evidence that the Planck length is a universal constant or that Lorentz symmetry (the basis of standard relativity) is violated. She criticizes the field for "moving the goalposts." When experiments (such as observations of high-energy gamma rays from distant bursts) fail to find the timing delays predicted by some DSR models, proponents often tweak the theory to make the effects even smaller and harder to detect. Another concern involves locality. Some versions of DSR lead to a "relative locality," in which the position of an event depends on the observer's energy. Hossenfelder has argued that this can lead to non-physical consequences, such as particles appearing to be in two places at once, in ways that contradict our basic understanding of how objects move through space.

Another strong counterargument comes from the experimental side, led by physicists like Ted Jacobson, Stefano Liberati, and David Mattingly. They specialize in quantum gravity phenomenology – the study of how to actually test these theories. DSR predicts that the speed of light should change slightly at extremely high energies (a "Modified Dispersion Relation"). However, by analyzing high-energy light from distant gamma-ray bursts and cosmic-ray behavior, this group has set incredibly tight constraints on these effects.

Their work effectively shows that if these relativistic "deformations" exist, they must be orders of magnitude smaller than the Planck scale would suggest. For many, the lack of evidence for time delays in high-energy photons traveling across the universe serves as a practical refutation of the most common versions of DSR.

Initially, DSR gained traction because some experiments seemed to show that cosmic rays exceeded the Greisen-Zatsepin-Kuzmin "GZK" limit – a theoretical energy limit for particles traveling through space. DSR offered a way for these particles to "break" the limit. However, newer, more precise observations have since confirmed that the GZK limit likely exists and is not being violated.

When the "GZK Paradox" disappeared, DSR's primary experimental motivation evaporated. This has shifted the consensus back toward standard Lorentz Invariance, leaving DSR as a fascinating mathematical exercise with no clear role in the actual architecture of the cosmos.

Tachyons

"Engineering to Bridge. We're detecting a tachyon field emanating from the surface of the planet!"

Since a tachyon moves faster than the speed of light, we can not see it approaching. After a tachyon passes nearby, we would see two images of it, one appearing and the other departing, moving in opposite directions.

Tachyons represent a fascinating, though hypothetical, deviation from the standard relativistic interaction protocols we have discussed in previous chapters. Within the framework of cosmic architecture, Einstein's Special Relativity established the speed of light as a universal ceiling for all objects with real mass. . However, the mathematical scaffolding of relativity actually allows particles to be born on the other side of that speed barrier. These are the tachyons – particles that never slow down to the speed of light and instead inhabit a world where the standard rules of kinetic energy and momentum appear to be inverted.

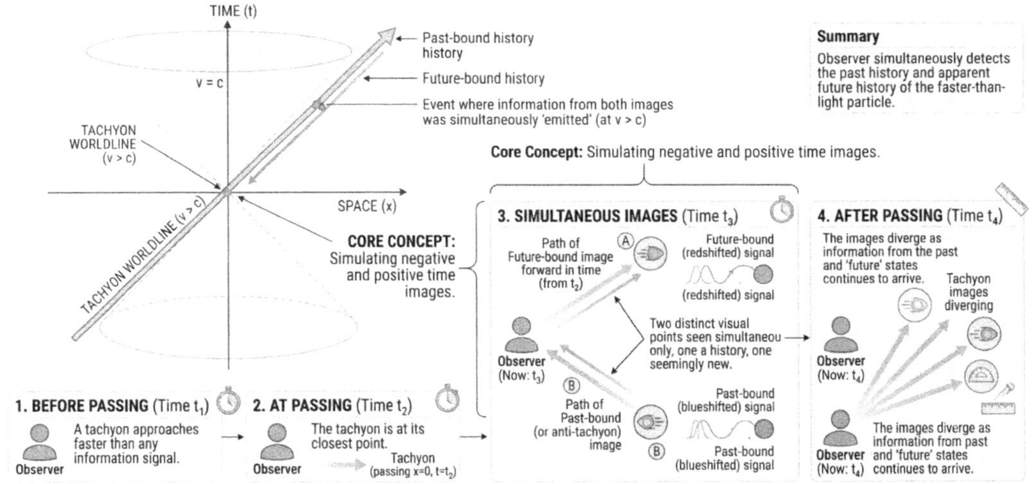

From a systems perspective, the most distinctive feature of a tachyon is its imaginary mass. While the particles in our "Particle Zoo" have a real, positive mass that increases with speed, a tachyon is modeled as having negative mass squared. This creates a strange functional dynamic: as a tachyon loses energy, it actually speeds up, accelerating toward infinite velocity. Conversely, as it gains energy, it slows down toward the speed of light but can never cross that threshold from above, just as we can never cross it from below. This suggests an architectural divide in the universe, where two different classes of matter exist in separate "spacelike" and "timelike" regions, separated by an impenetrable wall of light.

Detecting such an elusive component of the universal system requires looking for "leaks" or specific signaling artifacts. The most promising experimental approach is to detect tachyonic Cherenkov radiation. In our classical world, when a particle moves through a medium, such as water, faster than light can travel in that medium, it creates a "sonic boom" of light known as Cherenkov radiation. If tachyons exist, they would be traveling faster than light in a vacuum. Theoretically, they would spontaneously emit radiation as they move, creating a signature blue glow even in the void of space. By monitoring the vacuum of deep space for these specific photonic signatures, detectors could potentially capture evidence of a particle moving through the "prohibited" speed zones.

Beyond the blue glow, researchers have considered looking for "pre-arrival" signals in high-energy particle collisions. In a high-speed "system audit" of particle impacts, a tachyon might appear to leave a detector before the interaction that supposedly created it ever occurred. This highlights the major challenge for the tachyon theory: causality. Because tachyons move faster than light, they could, in theory, be used to construct a "tachyon anti-telephone," allowing a signal to be sent into the past. This creates a logical loop that challenges our fundamental understanding of cause and effect.

The tachyon remains a mathematical possibility awaiting a definitive experimental handshake. Whether they are actual building blocks of the cosmos or merely a theoretical ghost in the machine, they remind us that the blueprint of reality may still have rooms we have yet to enter. For an amateur astronomer watching the horizon at Oak Island, it is a reminder that the quiet dark between the stars might be more restless than it appears, filled with messengers traveling in reverse through the very fabric of time.

Come To The *Dark* Side

I am (almost) certain that the scientific process of theory, observation, and experiment will eventually answer the unknowns surrounding these crucial (and fundamental) “dark” issues that are key to understanding the universe we live in. I am also (almost) certain that it is not due to some dark and sinister wizardry – but you never know!

Dark Energy

Dark energy is a mysterious force that is believed to be accelerating the universe's expansion, pushing space itself apart. It is akin to negative pressure pushing the universe outward. The late 1990s marked a "tectonic shift" in our understanding of the cosmos. Before this period, the prevailing question wasn't whether the universe's expansion was slowing, but how fast it was decelerating.

Physicists were obsessed with a value called Ω (Omega)–the ratio of the actual density of the universe to the "critical density" required to stop its expansion. In the mid-90s, the scientific community was split into camps regarding the fate of the universe based on the "Omega mass" (specifically Ω_m, the density of matter):

- ***If*** $\mathbf{\Omega_m > 1}$: The universe has enough mass to eventually halt expansion and collapse in a "Big Crunch."
- ***If*** $\mathbf{\Omega_m < 1}$: The universe would expand forever, but at a steadily decreasing rate as gravity put on the "brakes."
- ***If*** $\mathbf{\Omega_m = 1}$: The "Critical" or "Flat" universe, where expansion slows but never quite stops.

Two rival teams – the Supernova Cosmology Project (led by Saul Perlmutter) and the High-Z Supernova Search Team (led by Brian Schmidt and Adam Riess) – set out to measure this deceleration. To measure the change in expansion, these teams used Type Ia Supernovae as "standard candles." Because these stellar explosions always occur at nearly the same mass threshold, they have a predictable intrinsic brightness. By comparing how bright they appeared with how far away they should be given their redshift, physicists could map the history of the universe's expansion.

In 1998, both teams reached a shocking conclusion: the distant supernovae were dimmer than predicted. For them to be that dim, they had to be much further away than a decelerating (or even a constant) expansion would allow. The result showed that the expansion of the universe isn't slowing down under the weight of "Omega mass"; it is accelerating.

This discovery required a new "repulsive" force to counteract gravity. Physicists have resurrected Albert Einstein's *Cosmological Constant* (Λ), which he originally added to his equations to keep the universe static and later discarded as his "biggest blunder." This unknown force was dubbed Dark Energy. This led to the creation of the ΛCDM model (Lambda Cold Dark Matter), which is now the "Standard Model of Cosmology – also known as the "concordance" model.

- $\mathbf{\Lambda}$ ***(Lambda):*** Represents Dark Energy, making up roughly 68% of the universe. It acts as a constant energy density inherent to empty space, pushing galaxies apart.
- ***CDM (Cold Dark Matter):*** Represents the invisible matter ($\approx 27\%$) that provides the gravitational "glue" to form galaxies.
- ***Normal Matter:*** The stuff we can see (stars, planets) makes up a mere 5%.

While ΛCDM was wildly successful, a massive discrepancy has emerged in the last decade regarding the Hubble Constant (H_0)–the value that represents the current rate of expansion. This is known as the Hubble Tension. The problem arises because we have two different ways of measuring H_0, and they no longer agree:

1. The "Late-Time" Observation (The Standard Ladder): Using the same "local" methods that discovered acceleration – Type Ia Supernovae and Cepheid variables – astronomers (like the SH0ES team) measure how fast nearby galaxies are receding. The result was $H_0 \approx 73$ to 74 km/s/Mpc.

2. The "Early-Time" Observation (The CMB): Using the Cosmic Microwave Background (CMB) – the afterglow of the Big Bang – satellites like Planck measure the "sound horizon" of the early universe. By feeding these "early" conditions into the ΛCDM model, they predict what the expansion rate *should* be today. The result was $H_0 \approx 67$ to 68 km/s/Mpc. (km/s/Mpc means that for every megaparsec (3.26 million light-years) distant, it recedes an additional x kilometers per second due to expansion. So a galaxy 326 million light-years from us would be receding an additional 100,000 km/s.)

The gap between 67 and 73 might seem small, but the statistical significance is now so high that it's unlikely to be a simple measurement error. This "tension" suggests that our ΛCDM model – the very one born from the "Omega" searches of the 90s – might be missing a crucial piece of physics. It could mean Dark Energy isn't constant, or that our understanding of gravity itself needs revision.

We can't directly observe Dark Energy; we infer its existence from the observed accelerating expansion of the universe. Scientists estimate that it represents about 70% of the universe's total energy content. We don't understand what dark energy is or how it works, making it a major area of research in cosmology.

The Vacuum Energy Discrepancy

Often called the "worst prediction in the history of physics," the vacuum energy discrepancy is a massive disagreement between quantum field theory and cosmology regarding the energy of "empty" space.

Quantum Field Theory suggests that the vacuum is teeming with virtual particles and zero-point energy ("the quantum foam"). If you add up all this energy, the density is gargantuan. Based on the observed expansion of the universe, the actual energy density of the vacuum is incredibly small. Prediction and observation differ by approximately 120 orders of magnitude. We do not know why the vacuum energy isn't as strong as quantum theory predicts.

"*Something is wrong with gravity*" encompasses a broad range of theories suggesting that a mysterious force does not drive the accelerating expansion, but rather that Einstein's general theory of relativity needs to be modified or is incomplete on cosmic scales. Finally, some alternative studies suggest that the observed acceleration is an illusion caused by the universe's clumpiness and by how we measure cosmic distances, without invoking dark energy at all. While the two are currently considered separate, some hypotheses suggest that dark energy could be a combination of the two. However, this is not the mainstream view.

The Finsler-Friedmann Equation

Finsler geometry is an alternative that suggests accelerating expansion could be a natural property of spacetime's geometry rather than requiring a separate dark energy component. Finsler geometry is a generalization of Riemannian geometry, the framework that underpins Albert Einstein's General Relativity.

In standard Riemannian geometry, if you have a "ruler" to measure spacetime, that ruler stays the same length regardless of which way you point it. The geometry is isotropic, meaning it looks the same in every direction. Finsler geometry removes this restriction. In a Finslerian manifold, the "length" of a vector depends on the direction it is pointing. making it a much more flexible tool for describing systems in which the environment is not uniform, such as light traveling through a complex crystal or a universe that might expand at different rates in different directions.

The Friedmann (Alexander Friedmann) equations are a specific set of equations in cosmology that describe how the universe expands or contracts over time.[72] They are usually derived using the Friedmann-Lemaître-Robertson-Walker (FLRW) metric, which assumes the universe is perfectly homogeneous and isotropic (the Cosmological Principle). The Finsler-Friedmann equation is the result of applying the Friedmann derivation to a Finslerian spacetime rather than a Riemannian one.

Because Finsler geometry allows for direction-dependency, the resulting expansion equation often contains additional terms. These terms account for potential anisotropies – directions in space that might be expanding at different rates. Physicists use the Finsler-Friedmann framework to investigate cosmological puzzles that the Standard Model struggles to explain, such as the "Axis of Evil", an apparent alignment in the Cosmic Microwave Background, or the Hubble Tension.

In some Finslerian models, the extra geometric terms in the Finsler-Friedmann equation behave mathematically like Dark Energy. This suggests that the accelerated expansion of the universe might be a natural consequence of a more complex geometry rather than an unknown fluid or constant.

Finsler geometry is a primary candidate for theories involving Lorentz Violation. Standard relativity relies on Lorentz symmetry (the idea that the laws of physics remain the same for all observers moving at constant velocity). If space itself has a direction-dependent structure at the smallest scales – as some theories of quantum gravity suggest – then Finsler geometry provides the necessary mathematical language to join the smooth large-scale universe and a potentially "grainy" or oriented subatomic reality.

The Finsler-Friedmann equation is a generalization of the standard Friedmann equation, derived from Finsler gravity. It describes the evolution of the cosmic scale factor by coupling the full one-particle distribution function (1PDF) of a kinetic gas directly to the geometry of spacetime.

In its most general form, it relates the expansion rate and its acceleration to the moments of the kinetic gas. A simplified version for the evolution of the scale factor in a vacuum or with specific matter distributions is given as:

$$a\frac{d^2a}{dt^2} = c_1\left(\frac{da}{dt}\right)^2 + c_2$$

Where:

- $a(t)$: The cosmic scale factor.
- c_1, c_2 : Constants derived from the underlying Finsler geometry and kinetic gas vacuum properties.

Unlike standard general relativity, which requires a cosmological constant (λ) for acceleration, the Finsler-Friedmann equation naturally admits an exponentially expanding vacuum solution $(a(t) = d_2e^{d_1t})$ when $c_1 = 1$. This suggests that the accelerated expansion of the universe could be a geometric property of spacetime itself rather than the result of "dark energy".

Dark Energy from Black Holes

Are black holes the actual source of dark energy? Championed by researchers like Duncan Farrah and Kevin Croker, and updated in 2024, this hypothesis seeks to address two major issues in our cosmic model at once: the mysterious nature of dark energy and the mathematical impossibility of the black hole singularity.[73]

In the Standard Model of General Relativity, a black hole is treated as an isolated object sitting *inside* the universe. Its center is presumed to be a singularity – a point of infinite density where the equations of spacetime lose all predictive power. This new theory proposes an elegant structural patch called Cosmological Coupling. It suggests two fundamental changes to how we view the cosmic scaffolding: First, instead of crushing matter into an infinitely dense point, the extreme gravitational collapse of a massive star traps pure "vacuum energy" at its core. Secondly, black holes are not isolated entities; they are inextricably linked (coupled) to the fabric of spacetime itself. As the universe expands, the vacuum energy inside the black hole scales proportionally. Because energy and mass are interchangeable ($E = mc^2$) the black hole gains mass simply because the universe is stretching.

Observational data supporting this hypothesis have unfolded in two main phases:

1. **Dormant Galaxies (2023):** Researchers looked at ancient, giant elliptical galaxies spanning 9 billion years of cosmic history. These galaxies are "dead" – they have no remaining gas, dust, or star formation, meaning their central supermassive black holes have absolutely nothing left to consume. Without a fuel source, their mass should be static. However, the data showed that modern black holes are 7 to 20 times more massive than their ancient counterparts. Their mass grew exactly in sync with the universe's expansion rate, a behavior perfectly predicted by the cosmological coupling model.
2. **The DESI Data (Late 2024):** To verify this mechanism on a universal scale, the team used the Dark Energy Spectroscopic Instrument (DESI) to track the exact moments when dark energy began to dominate the cosmos. Compared to the cosmic star-formation rate, which dictates how many massive stars are dying and collapsing into stellar-mass black holes, the data aligned beautifully. The total amount of dark energy in the universe increased at the same rate as new black holes were being minted.

If this theory survives ongoing scientific scrutiny, it fundamentally rewrites the relationship between the macroscopic universe and the quantum realm. We would not need to invent a new, mysterious force or an undiscovered subatomic particle to explain dark energy. The source is already built into Einstein's gravity; it is just hiding behind event horizons.

Replacing the infinite singularity with vacuum energy removes the catastrophic mathematical divide at the center of black holes. It suggests that when gravity becomes overwhelmingly strong, matter undergoes a phase transition – almost like a Big Bang played in reverse – turning back into the quantum vacuum energy that pushes the surrounding space apart. It proposes a beautifully cyclical system: the deaths of massive stars create localized nodes of vacuum energy, and the collective outward pressure of billions of these microscopic nodes drives the accelerating expansion of the entire macroscopic cosmos. If the Cosmological Coupling theory – where black holes act as nodes of vacuum energy – proves correct, it doesn't just fix dark energy and singularities. It potentially resolves three other critical issues in the λCDM operating system.

In the standard λCDM model, there is a glaring, unexplained coincidence: why are we living at the exact moment in cosmic history where the density of dark matter and dark energy are roughly comparable? For the first 9 billion years, matter dominated, slowing the universe's expansion. Then, roughly 5 billion years ago, dark energy took over, accelerating it. Classical theory has no explanation for why this transition happened *when* it did; it is treated as a random, hard-coded parameter.

If dark energy comes from the interior of black holes, its growth is tied directly to the life cycle of stars. The universe experienced a massive boom in star formation – often called "Cosmic Noon" – around 10 billion years ago. As those massive stars died and collapsed, they minted millions of new black holes. So dark energy *had* to become the dominant force in the universe shortly after this period, once a critical threshold of black holes had formed. It turns a random coincidence into a logical, sequential dependency.

The Standard Model assumes that the Cosmological Constant (λ) is exactly that – constant. It assumes a smooth, steady outward pressure from the beginning of time. However, Cosmological Coupling *dynamically* injects dark energy. As the rate of black hole formation peaks and valleys, the injection of dark energy into the system also fluctuates. This variable energy curve changes the mathematical bridge between the early universe and today, potentially perfectly aligning the early and late measurements and resolving the tension.

Moving from the cosmic scale down to the subatomic, standard black holes present a massive problem for quantum mechanics: the Information Paradox. Quantum theory strictly dictates that information (the exact physical state of particles) cannot be destroyed. But if a particle falls into a black hole and is crushed into an infinite singularity, and that black hole eventually evaporates via Hawking radiation, the information appears to be permanently deleted from the universe.

By replacing the infinite singularity with a stable core of quantum vacuum energy, the mathematical "crash point" is eliminated. A finite core implies that the quantum information of the collapsed matter forming the black hole is not destroyed. It is preserved, rearranged, or encoded within that vacuum structure, keeping the laws of quantum mechanics intact. The theory is compelling because it provides exactly the kind of elegant, unified architecture that a physical system demands. It demonstrates how the subatomic (vacuum energy) can directly govern the macroscopic (cosmic expansion).

Dark Matter

Dark matter is an invisible, non-baryonic substance that accounts for approximately 85% of the universe's total mass. It's a form of matter that we cannot see, feel, or directly detect because it does not interact with the electromagnetic spectrum. Its gravitational influence is essential for explaining the structure and motion of galaxies. Dark matter is believed to serve as gravitational scaffolding for cosmic structures.

After the Big Bang, dark matter clumped into blobs along narrow filaments, with superclusters of galaxies forming a cosmic web at scales on which entire galaxies appear like tiny particles.

In the standard λCDM model of cosmology, the mass–energy content of the universe is 5% ordinary matter, 26.8% dark matter, and 68.2% a form of energy known as dark energy. Thus, dark matter constitutes 85% of the total mass, while dark energy and dark matter constitute 95% of the total mass–energy content.

While the density of dark matter is significant in the halo around a galaxy, its local density in the Solar System is much less than that of normal matter. The total of dark matter in Neptune's orbit would be about 10^{17} kg, the same as a large asteroid. The term "dark" simply means that it does not interact with light. It does not emit light (like stars do), absorb light, or reflect light. Dark matter interacts only through one known fundamental force: gravity. For this reason, it cannot be made of ordinary (baryonic) matter like atoms, protons, or neutrons.

Extensive searches for large reservoirs of non-luminous normal matter (like rogue planets or non-glowing hydrogen clouds) have ruled them out as the primary component.

The existence of dark matter is inferred entirely from its gravitational effects on visible matter, providing overwhelming indirect evidence. Stars and gas clouds orbiting the center of a spiral galaxy are expected to slow down the farther they are from the center. American astronomer Vera Rubin showed that stars far from a galaxy's center orbit just as quickly as those near the center, resulting in a flat rotation curve. This flat curve can only be maintained if the visible matter is embedded within a massive, invisible halo of dark matter that extends far beyond the luminous edges of the galaxy.

Large amounts of mass – whether visible or invisible – bend the fabric of spacetime, distorting, magnifying, and warping the light from distant background galaxies. This effect is called gravitational lensing. By measuring the precise distortion of background light, astronomers can create a map of the total mass distribution in a galaxy cluster. These maps consistently show that mass is concentrated well beyond visible stars and gas.

One of the most compelling pieces of direct evidence is the "bullet cluster", the result of two galaxy clusters colliding. Scientists tracked two components of mass. The first was visible hot gas, composed of ordinary matter, from the two colliding clusters, slowed by electromagnetic drag and clumping in the center (as seen via X-rays). The second was the mass map from gravitational lensing, which showed that most of the mass passed straight through the collision without slowing, resulting in a separation between the visible gas and the gravitational center.

Since dark matter is not ordinary matter, many physicists hypothesize that it is composed of new, undiscovered elementary particles. Theoretical physics proposes several alternative dark matter candidates across a vast mass spectrum:

Sterile Neutrinos: Right-handed neutrinos with masses in the $1 \text{ to } 50 \text{ keV}$ range that interact almost purely through gravity, along with weak mixing into standard active neutrinos. Because their mass places them in the warm dark matter regime, they damp small-scale structure formation compared to cold dark matter. Space telescopes like XMM-Newton search for an unaccounted 3.5 keV X-ray emission line in galaxy clusters, hypothesized to stem from sterile neutrino decays.

WIMPzillas: Non-thermal particles with masses reaching 10^8 GeV to 10^{15} GeV (10^{-16} to 10^{-9} grams). Created by rapid gravitational expansion at the end of cosmic inflation, these ultra-heavy states avoid thermal equilibrium constraints. Detectors such as the Pierre Auger Observatory monitor the upper atmosphere for ultra-high-energy cosmic rays produced if WIMPzillas undergo rare decays.

Dark Photons: Hypothetical vector gauge bosons belonging to a hidden sector. Masses span from sub-eV to GeV scales, coupling to ordinary matter through kinetic mixing with standard photons. Particle accelerator experiments, including NA64 at CERN and DarkLight at Jefferson Lab, search for dark photons converting into electron-positron pairs in beam-dump targets.

Asymmetric Dark Matter: Particles with masses around $1 \text{ to } 10 \text{ GeV}$ set by a dark-baryon asymmetry similar to standard matter. This framework accounts for why the universe contains roughly 5 times more dark matter mass than visible matter mass ($\Omega_{DM} \approx 5\Omega_b$), rather than differing by orders of magnitude. In astrophysics, dark matter accumulating inside dense neutron stars could reach a critical mass and trigger collapse into micro black holes.

Gravitinos: The spin-3/2 superpartner of the graviton predicted by supergravity models, with masses spanning eV to TeV scales. Gravitinos interact weakly and form non-thermally via the decays of heavier particles in the early universe. The Large Hadron Collider searches for long-lived superpartners (like staus or neutralinos) that decay into gravitinos outside collision points.

Macro Dark Matter: Composite bound states of dark quarks, strange quark nuggets, or dark atoms with masses ranging from grams to 10^{24} grams (asteroid masses). Instead of single fundamental particles, these macroscopic states form during early universe phase transitions. Researchers search ancient mica deposits for physical track damage or scan Earth's atmosphere for acoustic shock waves created by high-speed macro impacts.

Modified Gravity Theories: Frameworks like Modified Newtonian Dynamics (MOND) alter gravitational equations at low accelerations ($a < 1.2 \times 10^{-10}\ \mathrm{m/s^2}$) rather than introducing unseen mass. MOND matches observed galaxy rotation curves without dark matter particles, though it requires extra neutrino mass to account for gravitational lensing in galaxy cluster collisions like the Bullet Cluster. (See the section on MOND.)

Axions: A hypothetical particle much lighter than an electron. They are theorized to transform into a detectable photon in the presence of strong magnetic fields, a process that is a focus of current laboratory searches.

Primordial Black Holes: These would be small black holes formed in the early universe. While most stellar-mass black holes are ruled out as the primary component, asteroid-sized or smaller primordial black holes remain a possible, albeit less favored, candidate.

Self-Interacting Dark Matter (SIDM): This theory proposes that dark matter particles can interact with each other through a "dark force," which could explain high-density dark matter halos observed in galaxies, according to the University of California, Riverside.

Fuzzy Dark Matter

This theory suggests that dark matter particles are extremely lightweight and behave more like waves than particles, according to Scientific American.

Strangelets: Strange matter, in the form of *strangelets*, would be extremely dense, and could potentially explain dark matter. Normal matter (like you, me, and the stars) is made up of up and down quarks. However, there is a third, heavier type of quark called the strange quark. While strange quarks usually decay instantly in our world, the Strange Matter Hypothesis suggests that if you clump enough up, down, and strange quarks together in roughly equal numbers, they become absolutely stable.

A strangelet is essentially a "nugget" of this strange matter. They can range in size from the subatomic level to the size of a mountain. Because the strong nuclear force holds them together but contains three types of quarks instead of two, they can be packed much tighter than normal matter. One of the most famous (and slightly terrifying) theories about strangelets is that they are infectious. If a stable, negatively charged strangelet touched normal matter, it could theoretically convert that matter into more strange matter, creating a chain reaction. If they exist, they are likely only stable at extreme pressures, such as those found in the cores of neutron stars. However, strangelets are a compelling candidate for dark matter because they are invisible, massive, and provide the heavy gravitational "tug" needed to explain galaxy rotation. Because they are so small and dense, they could pass through normal matter like a "ghost" unless they hit an atomic nucleus directly.

The idea is that if strangelets are the answer to dark matter, they would have formed during the Quark-Hadron Phase Transition, just microseconds after the Big Bang. Early in the universe, everything was a hot "soup" of free-floating quarks. As the universe cooled, most quarks grouped into protons and neutrons. According to some models, large "lumps" of strange quark matter could have formed and survived the cooling process. These "primordial strangelets" would have remained as relics of the Big Bang, scattered throughout the universe. Today, they would drift through space as cold, heavy, invisible nuggets – exactly matching the behavior we observe in dark matter. Analyses of data from the Fermi Gamma-ray Space Telescope detected a gamma-ray halo around the Milky Way that matches the predicted signature of WIMP annihilation, potentially providing the first direct observational evidence of dark matter.

Two recent theories suggest that dark matter might have originated from a hidden "mirror world" or from quantum radiation at the edge of the observable cosmos shortly after the Big Bang. The other new theory proposes that dark matter might not be entirely invisible, but could subtly change the color of light passing through it, creating a detectable "fingerprint."

The Cosmic Neutrino Background (CNB) radiation, if detected, would take us back to one second after the Big Bang. For comparison, the Cosmic Microwave Background radiation takes us back to 380,000 years after the Big Bang. It may help identify the neutrino's mass and might lead to new theories about dark matter.

Are Primordial Black Holes Dark Matter?

The hypothesis that Primordial Black Holes (PBHs) make up 100% of dark matter faces tight constraints from observational data, though specific mass windows remain possible. Observational bounds (such as Hawking radiation limits, gravitational microlensing surveys, CMB distortion measurements, and fast radio burst lensing) seem to rule out PBHs as making up all of the dark matter .
However, there is a key exception:

- ***The Asteroid-Mass Window:*** (10^{17} g to 10^{22} g, or 10^{-16} to 10^{-11} solar masses). In this range, PBHs are too light to cause detectable microlensing of starlight, yet massive enough that they have not evaporated via Hawking radiation over the age of the universe. Within this narrow window, PBHs could still account for 100% of the universe's dark matter.

  Outside this window, PBHs can likely only make up a fraction (from a fraction of a percent to a few percent) of total dark matter.

Even if dark matter was densely packed during or shortly after the Big Bang, two fundamental physical mechanisms prevent it from collapsing into black, dark-matter, holes everywhere:

For matter to collapse into a black hole (a Kerr ring singularity if rotating), it cannot simply be densely packed; it must lose energy and angular momentum. Normal (Baryonic) Matter can form stars and black holes because it interacts via the electromagnetic force. Protons and electrons collide, emit radiation (photons), cool down, lose kinetic energy, and shed angular momentum, allowing them to fall inward under gravity.

Dark Matter is collisionless (or virtually collisionless). It only interacts via gravity. When dark matter particles fall toward a common center of gravity, they accelerate, overshoot the center, and orbit back. Because they cannot dissipate energy through electromagnetic radiation or collisions, they form extended, hot, diffuse clouds called dark matter halos rather than collapsing into black holes.

In the early universe, the high density of matter was balanced by the rapid expansion of spacetime. A primordial black hole could only form if a localized region had an unusually large density fluctuation (a peak in the primordial power spectrum) where gravity momentarily overcame the cosmological expansion. Without these specific overdensities, the expanding background kept the dark matter spread out.

If PBHs formed in the early universe, their theoretical mass depends directly on when they collapsed after the Big Bang, as a PBH's mass is roughly bounded by the mass enclosed within the cosmological horizon at the moment of creation:

- ***Ultra-small*** (10^{-5} g to 10^{15} g): Formed within the first 10^{-23} seconds. Anything lighter than 10^{15} g (about the mass of an asteroid) would have completely evaporated by today due to Hawking radiation.
- ***Sub-planetary / Asteroid-mass*** (10^{17} g to 10^{22} g): Formed around 10^{-18} to 10^{-14} seconds after the Big Bang. These are the size of an atom or proton physically, but possess the mass of a large asteroid or moon. This is the main surviving candidate range for 100% dark matter.
- ***Stellar-mass*** ($1 to 100\ \mathrm{M}_{\odot}$): Formed around 10^{-4} seconds after the Big Bang (during the QCD phase transition).
- ***Supermassive*** (10^2 to 10^5 $M_{\odot}$ or higher)**:** Formed around 1 second after the Big Bang.

What happens if dark matter is drawn into a black hole? A black hole is defined entirely by its spacetime geometry (described by mass, spin, and electric charge – the "no-hair theorem"). Once matter falls inside an event horizon, the universe retains no memory of whether the collapsing material was made of photons, baryonic matter, or dark matter. A black hole formed with mostly dark matter would be physically identical to a black hole formed from normal matter.

Could Primordial Black Holes act as the initial "seeds" for Supermassive Black Holes (SMBHs) at the centers of galaxies? Research suggests this is a real possibility:

1. ***The JWST High-Redshift Puzzle:*** Observations from the James Webb Space Telescope (JWST) have revealed fully formed supermassive black holes (billions of solar masses) early in the universe (redshifts $z > 7–10$). Standard accretion models of normal matter struggle to explain how stellar-mass black holes could grow that quickly.
2. ***PBHs as Direct Seeds:*** Massive primordial black holes formed in the early universe ($10^3–10^5\ M_{\odot}$) could serve as "seeds" at the centers of early dark matter halos. These seeds would then rapidly pull in surrounding gas and dark matter to build the supermassive black holes observed today.

In summary, while primordial black holes are unlikely to account for all dark matter across the entire mass spectrum, asteroid-mass PBHs remain a candidate. Moreover, primordial seeds created in the early universe may explain the rapid formation of supermassive black holes at galactic centers.

Dark Force

A "*dark force*" is a hypothetical force in the universe that scientists believe exists but cannot directly detect. It is thought to mediate interactions among dark matter particles, a mysterious substance that makes up most of the universe's mass and is invisible to ordinary light. Essentially, it's a proposed force that explains how dark matter might interact with itself, despite its unknown composition.

As with dark matter, we can't directly observe a "dark force" with current technology, but we infer its existence from gravitational effects in the universe. Some scientists theorize that it could account for the unusual distribution of dark matter around galaxies, a phenomenon that current models struggle to explain fully.

One proposal is a "dark force" of self-interacting dark matter or dark energy, the mysterious force driving the universe's accelerated expansion. Theories of self-interacting dark matter propose a new fundamental force that governs interactions among dark matter particles, potentially explaining discrepancies with the Standard Model of Cosmology. Since we already covered dark energy, we'll cover the dark matter idea.

Self-interacting dark matter is a hypothetical "dark force" that mediates interactions between dark matter particles, unlike the standard "cold dark matter" (CDM) theory, which assumes particles are non-interacting except through gravity. This theory helps explain two major puzzles: the unexpectedly high density of dark matter in the centers of some massive galaxies and the strong clustering of older, diffuse dwarf galaxies, which the Standard Model does not predict. If dark matter particles can interact via a "dark force," they could collide and transfer energy, altering the shapes and densities of dark matter halos around galaxies.

Summary

Chapter Seven addresses deep cosmological anomalies and alternative models designed to resolve issues left open by the Standard Model. It touches upon the "Hubble Tension" – the baffling statistical discrepancy between how fast the early universe expanded and how fast the local modern universe is expanding. It outlines alternative gravity models such as Modified Newtonian Dynamics (MOND), which seeks to explain galactic rotation without dark matter, and Double Special Relativity (DSR), which introduces the Planck length as a universal floor.

It features speculative physics like tachyons (hypothetical faster-than-light particles), details candidate particles for dark matter (WIMPs, axions, and primordial black holes. It explores the fascinating Strange Matter Hypothesis regarding ultra-dense "strangelets."

Chapter Eight: About Time

Ψ Tμν

"There was a young spaceman named Bright, whose speed was far faster than light. He departed one day, in a relative way, and returned on the previous night."

How We Got Here

For centuries, time was viewed as Isaac Newton did: a universal, absolute clock ticking away at the same rate for everyone, everywhere. But modern physics has shattered this clock. In General Relativity, time is not a separate background; it is woven into the very fabric of spacetime. It is flexible. We know that gravity and speed both stretch time – a phenomenon known as time dilation. A clock on a mountaintop ticks faster than one at sea level, and a clock on a speeding rocket slows down relative to those left on Earth.

It's difficult to get a good grasp of how time and space work under the rules of relativity. When you first hear the details, it is hardly intuitive, and it flies in the face of your day-to-day view of time – the ticking of clocks, days changing to nights, days into weeks, and so on. *Back to the Future, Doctor Who, The Time Machine*, and many other stories are fascinating for the possibilities they explore.

Light Cones

Let's start with reality about time using a famous illustration. A light cone is a geometric construct in four-dimensional spacetime that visualizes all possible paths that light can take through time. In a simplified two-dimensional representation, it looks like an hourglass. The top cone is the "future light cone," and the bottom cone is the "past light cone."

Every event – every point in spacetime – has its own unique light cone. The "future light cone" of an event is the set of all events that could be influenced by it. Because nothing can travel faster than light, everything we do can only affect the events within this future cone. The "past light cone" of an event consists of every prior event that could have influenced it. Our entire existence is shaped by the history that falls within our specific past cone.

Light cones were discovered as a direct consequence of Special Relativity. Einstein's theory showed that the speed of light is constant in a vacuum and is the universal speed limit. This fundamental constraint defines the boundaries of light cones. The concept was formalized and popularized by the mathematician Hermann Minkowski in 1907-1908. He recognized that special relativity required the unification of space and time into a single four-dimensional continuum, now known as Minkowski spacetime.

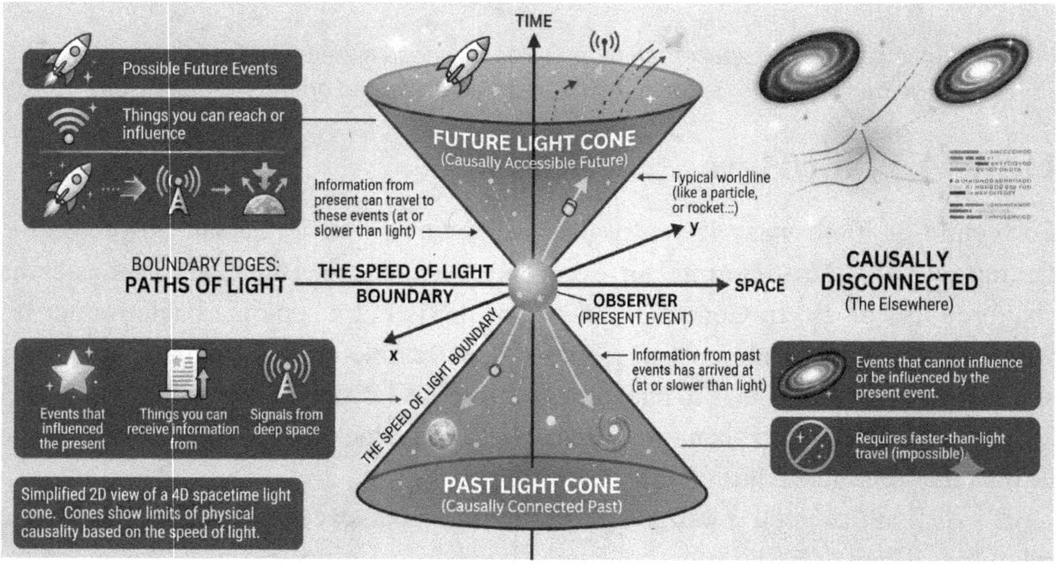

The light cone formalism became the perfect mathematical tool for understanding causality and information flow in this new 4D reality.

For an observer, light cones are the ultimate limits of perception and action. When we look at a distant star, we are not seeing it as it is now in our reference frame; we are seeing its light as it crosses our past light cone. The information that reaches us has taken time to travel, meaning we are essentially looking back in time to the event that originated that light.

Similarly, our actions can only affect events that lie within our future light cone. If we attempt to send a signal to a distant planet, it will follow a path that stays within this cone, and the soonest possible moment a recipient could respond is when our signal crosses their past light cone.

The defining principle of our universe is strict causality. Light cones are the physical manifestation of this rule, ensuring that a cause must always precede its effect. This linear, forward-moving causation is fundamental to the architecture of our cosmos.

In our universe, light cones maintain strict chronological protection. This means they ensure that information cannot travel from an event's future light cone back to its own past. Such a scenario, called a "closed timelike curve," would create logical contradictions, such as time-travel paradoxes in which an effect could alter its own cause. Light cones are the cosmic safeguards that prevent these causal loops, guaranteeing that the timeline of our universe remains rational, predictable, and fundamentally sequential.

I found some interesting convergence with some of the theories we've covered, so I want to cover them all here, rather than add to each of those theories (I won't cover Gödel's Universe here).

In physics, the light cone defines the path that light (or any massless particle) takes through spacetime. Mathematically, for two events separated by a spacetime interval ds^2, light rays follow paths where the interval is exactly zero:

$$ds^2 = -c^2dt^2 + dx^2 + dy^2 + dz^2 = 0$$

When Conformal Gravity Theory refers to the "symmetry of the light cone," it's describing the complete set of mathematical transformations that leave this zero-interval condition unchanged ($\mathrm{ds}^2 = 0$). In standard physics, spacetime operations are governed by:

- ***The Lorentz Group (6 generators):*** 3 spatial rotations and 3 velocity boosts.
- ***The Poincaré Group (10 generators):*** Adds 4 spacetime translations (moving forward/backward in time or space) to the 6 Lorentz transformations.

While the Poincaré group leaves all spacetime intervals invariant ($\mathrm{ds}^2 = \mathrm{constant}$), light rays only care about the condition $\mathrm{ds}^2 = 0$. Because 0 multiplied by any non-zero scaling factor is still 0, light rays permit extra transformations that standard massive physics does not:

If $ds^2 = 0 \implies \Omega^2(x) \cdot ds^2 = 0$

This introduces 5 additional symmetry operations, bringing the total from 10 to 15 generators:

- ***Dilatations (1 generator):*** Uniformly stretching or shrinking space and time (scale invariance).
- ***Special Conformal Transformations (4 generators):*** Non-linear transformations involving coordinate reciprocals that bend and scale grid lines locally while preserving angles.

Together, these 15 operations form the Conformal Group $\mathrm{SO}(4,2)$, which represents the full symmetry group under which the light cone remains invariant.

Since massless particles move along light cones ($\mathrm{ds}^2 = 0$), conformal symmetry is the natural, maximal symmetry of massless fields and light. Philip Mannheim argues that fundamental equations should possess full conformal symmetry (zero fundamental mass scales). Mass is then generated dynamically through the vacuum state rather than being hardcoded into the initial laws.

Because a 4-component Dirac spinor serves as the fundamental representation of this 15-generator conformal group, the theory naturally links left-handed and right-handed states. This requires the existence of right-handed neutrinos to complete the representation, resolving long-standing questions about parity and particle structures.

Twistors and Conformal Geometry

Roger Penrose's work centers on the idea that light rays–and the light cones they for m–are more fundamental than individual points in spacetime.

Standard relativity defines spacetime as a container of individual points (t, x, y, z). Penrose flipped this perspective in 1967 by inventing Twistor Space. In twistor theory, a single point in complex twistor space $\mathbb{CP}^3$ maps to an entire light ray in physical spacetime. A point in regular spacetime is merely an event where a family of light cones intersect.

In his Conformal Cyclic Cosmology (CCC), Penrose applies light-cone geometry to the origin and end of the universe. In the far future, when all matter has decayed or been swallowed by black holes, only massless photons remain. Because massless particles travel strictly along light cones ($ds^2 = 0$), they possess no internal clocks and cannot measure scale, distance, or time. By applying a conformal rescaling factor $\Omega(x)$, Penrose smooths out this infinitely expanded, massless future, turning it into the low-entropy Big Bang of a subsequent aeon.

CPT-Symmetric "Mirror" Universe

Neil Turok and Latham Boyle use light-cone geometry and conformal transformations to eliminate the Big Bang singularity without introducing an inflationary phase.

Standard cosmology encounters a mathematical breakdown at $t=0$, where density and curvature diverge to infinity. Boyle and Turok show that if you enforce CPT (Charge, Parity, and Time-reversal) symmetry across the origin, the Big Bang acts as a mirror boundary rather than a creation point. A mirror universe extends backward in time prior to $t=0$, where parity is inverted, charges are flipped, and anti-matter dominates.

Radiation dominates the early universe. Because relativistic radiation consists of effectively massless particles, its equations respect conformal symmetry. Rescaling the expanding scale factor $a(t)$ near $t=0$ reveals that the light cones pass smoothly through the Big Bang into the mirror universe without breaking.

All three approaches – Mannheim's conformal gravity, Penrose's twistor and CCC models, and Turok/Boyle's CPT mirror universe – share a foundational principle:

when mass is zero or negligible, physical scale disappears, and the conformal geometry of the light cone dictates the behavior of physics.

A Problem of Time

Our daily experience confirms that, for us, time flows strictly in one direction, moving from the past into the future. This directional flow is known as the arrow of time. However, our deepest mathematical equations describing the cosmos do not include an external ticking clock. These foundational equations, many of which we looked at earlier, are perfectly symmetrical; they work just as well whether time runs forward or backward. This means there is a disconnect between what we observe in daily life and what our theories propose about reality.

To understand why this contradiction exists, we must look at a cornerstone of quantum gravity: the Wheeler-DeWitt equation. Developed by physicists John Wheeler and Bryce DeWitt in the late 1960s, this equation attempts to merge quantum mechanics with general relativity.[74] When you combine these two frameworks to calculate the state of the entire cosmos, something odd happens. The time parameter, represented as t in physics, completely vanishes from the mathematics. The Wheeler-DeWitt equation describes the entire universe as a single, static, and frozen state.

This creates the "Problem of Time." If the universe encompasses everything that exists, there cannot be a clock sitting outside of it to mark its passing. If the ultimate equation of reality has no built-in clock, how do we experience a clear sequence of "before" and "after" every single day?

When physicists attempt to merge General Relativity and Quantum Mechanics into a single framework of quantum gravity, time mathematically disappears altogether. The conflict stems from how each pillar of modern physics treats time:

- ***In Quantum Mechanics, time is an external background stage.*** Time is an absolute, classical parameter (t) that ticks steadily outside the physical system. It drives change via the Schrödinger equation:

  $$\hat{H}\Psi = i\hbar\frac{\partial\Psi}{\partial t}$$

  Here, quantum events happen *in* time, but time itself is never subject to quantum uncertainty or dynamical change.
- ***In General Relativity, time is a dynamic actor.*** Time is woven directly into the 4D spacetime manifold. It has no universal "tick" – velocity and gravity warp the flow

of time depending on observer location. Spacetime is malleable and interacts dynamically with energy and matter.

When physicists apply standard quantum rules to General Relativity (canonical quantization), they combine the total energy operator (the Hamiltonian constraint, ($\hat{H}$) with the wave function of the entire universe (Ψ).

Because General Relativity is reparametrization-invariant – meaning physical reality does not depend on the arbitrary coordinate labels we assign to spacetime – the total Hamiltonian constraint must equal zero: $\Psi = 0$. Notice what is missing: there is no time derivative ($\partial / \partial t$) anywhere in the equation.

Known as the Wheeler–DeWitt equation, this foundational formula of quantum cosmology describes a completely static universe. The quantum state of the entire universe is frozen in place; it does not evolve or change. This creates a paradox: if the fundamental equation of quantum gravity contains no time variable, how do we experience an evolving universe with past, present, and future?

Physicists approach this paradox from several directions:

- ***Relational Time (Internal Clock):*** Time is not an absolute background, but a measure of change between correlations inside the universe. One subsystem (e.g., a clock or expanding space scale factor) serves as a local reference mark for another subsystem (e.g., a particle's position).
- ***Emergent Time:*** Time is an illusion or macro-level property, much like temperature. Atoms do not have a temperature; temperature emerges only when billions of atoms interact. Similarly, fundamental reality may be timeless, with time emerging only at coarse-grained classical scales.
- ***Fundamental Unification (String Theory / Loop Quantum Gravity):*** Alternative quantum gravity frameworks attempt to bypass the frozen-time state by reframing space and time as discrete quantum networks or holographic projections.

In 2026, a research team led by Professor Giovanni Bartini at the University of Birmingham constructed a laboratory experiment to test how time can emerge from a timeless system.[75] Instead of observing deep space with massive instruments like the James Webb Space Telescope, the team built a simplified, microscopic analog of the universe on a laboratory bench.

The team used a Bose-Einstein condensate (BEC) consisting of approximately 24,000 rubidium-87 atoms. A Bose-Einstein condensate represents a distinct state of matter that occurs when a gas of atoms is cooled to temperatures a fraction of a degree above absolute zero. At this extreme temperature, the individual atoms lose their separate identities. They overlap and merge into a single, macroscopic quantum wave – effectively acting as a single "super-atom".

This state allows researchers to observe quantum behaviors directly using standard laboratory equipment. The researchers used magnetic fields and lasers to trap this atomic cloud. They then projected a razor-thin barrier of light through the cloud, partitioning their mini-universe into two distinct zones:

- ***The Bright Sector:*** A region open to direct observation and measurement.
- ***The Dark Sector:*** A region that remained completely unobserved.

The rubidium atoms were free to tunnel back and forth across this light barrier. As the particles shifted between the bright and dark zones, the volume of the bright region expanded and contracted. This physical movement closely mirrored the cosmic expansion and contraction cycles associated with Big Bang and Big Crunch models.

The core objective of the experiment was to see if the internal events within the bright sector could be organized into a chronological sequence without looking at any laboratory wall clocks. To achieve this, the team relied on the concept of entropy. (See the *Background Materials* for more information on Entropy)

Entropy is a thermodynamic quantity that measures the specific spread of energy and disorder within an isolated system. According to the Second Law of Thermodynamics, the total entropy of an isolated system must increase over time. Because this regular accumulation of disorder is the only law in fundamental physics that possesses a strict direction, physicists use the growth of entropy to define the forward arrow of time.

By tracking the dispersion and exact exchange of atoms across the light barrier, the Birmingham team calculated an "entropic time" internal to the system. The results revealed several critical properties of relational time:

- ***Time is Relational and Internal:*** The internal calculation of entropy successfully tracked the progression of events without an external clock. This demonstrates that time can be defined solely by the interactions and energy exchanges taking place between different subparts of a system.
- ***The Arrow is Irreversible:*** Even when the atomic cloud physically contracted – mimicking a collapsing universe – the internal entropic time continued to move forward in a single, unwavering direction.
- ***Time Can Vary and Stop:*** The rate of entropic time fluctuated based on particle activity. It flowed faster during rapid atomic exchanges and stopped entirely when

the system reached a uniform state of balance known as heat death. Without internal change, time ceased to exist!

To verify the validity of this internal clock, the researchers rewrote the Schrödinger equation by replacing the traditional external time parameter with their newly derived internal entropic time. This modified quantum equation successfully predicted the physical behavior of the rubidium atoms. This structural consistency indicates that quantum mechanics remains perfectly functional when treated as an entirely internal, relational system.

Imagine a physical movie reel where the entire film – containing the beginning, the middle, and the end of the story – exists simultaneously as a static, unchanging object inside the metal canister. This represents the frozen universe described by the Wheeler-DeWitt equation.

However, when a beam of light passes through the individual frames of that reel, an observer inside the theater experiences a dynamic story that unfolds frame by frame. The flow of time is not an absolute property of the film canister itself; it is a relational property that emerges through the active projection and interaction of the internal components.

The Block Universe

The Block Universe theory, also known as Eternalism, is a cosmological framework derived from the literal interpretation of Albert Einstein's theory of relativity. In this model, you treat space and time as an integrated, unchanging four-dimensional continuum known as spacetime. Eternalism provides the philosophical backing for this view, asserting that the past, the present, and the future exist with equal reality.

There is no objective, universally moving moment that separates what has happened from what will happen. Instead, every event possesses a fixed, immutable coordinate within the four-dimensional structure. Your birth, your present reading of this sentence, and the final stages of the cosmos exist simultaneously within the geometry of the universe.

The physical justification for the block universe stems from the relativity of simultaneity in special relativity. When two observers move at different velocities relative to one another, they do not agree on which events occur at the same moment. An event that you perceive as occurring right now might belong to the past of an observer traveling at high speed, or it could sit in the future of an observer moving in the opposite direction. Because no physical experiment can establish one observer's frame of reference as the absolute truth, you must conclude that all moments are equally real. If the future of one observer already belongs to the past of another, then the future must exist unconditionally.

In this geometric view, the passage of time is a subjective experience rather than an intrinsic feature of physical reality. Physicists describe your life as a continuous path through the block, called a worldline. This worldline maps your entire trajectory through three spatial dimensions and one temporal dimension. Every single state of your body and every position of every atom you comprise remains eternally fixed at its specific coordinates. The perceived movement through sequential seconds arises from how consciousness processes sequential information along a worldline, not from any structural change within the universe itself.

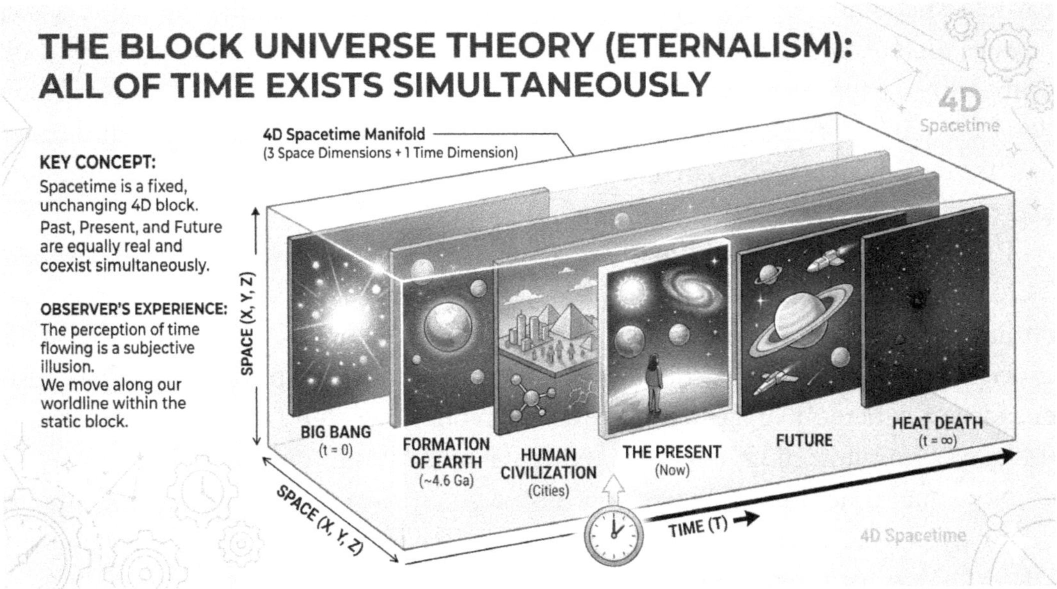

There is no special "now" moving through time; instead, every moment is equally real, much like every point on a map exists even if you aren't currently standing on it. Yet, even as relativity makes time flexible, the fundamental laws of physics – from Newton to Quantum Mechanics – are time-reversible. If you play a video of a subatomic interaction backward, it perfectly obeys the laws of physics. The only reason we see an "Arrow of Time" is Entropy. Like a shattered teacup that never reassembles, time appears to flow forward only because the universe is moving from a state of low-entropy order to high-entropy disorder.

Negative Time: Testing the Clock

We aren't just guessing about some of the strange properties of time; we have tested them. Atomic Clocks are so precise today that they can measure the difference in the flow of time between your head and your feet. These experiments confirm that time is a local experience, not a universal one.

More bizarrely, the theoretical and experimental exploration now extends into the concept of Negative Time. This is not a science-fiction, but a strange implication arising from quantum mechanics, specifically in the context of quantum tunneling. In quantum tunneling experiments, one of the most perplexing observations is the Hartman Effect – a phenomenon in which particles appear to exit a barrier before fully entering it.

In 1962, Physicist Thomas Hartman discovered this behavior through mathematical calculations using the Schrödinger equation.[76] When a quantum particle encounters an energy barrier higher than its own kinetic energy, classical physics dictates it must bounce back. In quantum mechanics, the particle has a finite probability of appearing on the opposite side. You might expect that a thicker barrier would require more time for the particle to cross. However, Hartman demonstrated that for sufficiently thick barriers, the transmission delay, or group delay, saturates and reaches a constant value.

Because the tunneling time T remains constant as the barrier thickness L increases, calculating the effective velocity using $v = L/T$ yields speeds that scale up indefinitely. For a thick enough barrier, the calculation results in a velocity that exceeds the speed of light (c). This apparent violation of special relativity caused decades of debate. It suggested that quantum tunneling could allow faster-than-light travel or communication, which would violate the laws of causality.

No actual time travel is occurring. Physicists resolved the paradox by analyzing how waves store energy. Tunneling is a wave phenomenon, meaning it applies to matter, electromagnetic, and sound waves. When a wave packet hits an opaque barrier, the incident and reflected waves interfere to create a non-propagating, evanescent (ephemeral) standing wave inside the barrier.

This standing wave oscillates in place and stores a finite amount of reactive energy. The group delay measured in laboratory experiments does not represent the transit time of a moving particle traversing the space. Instead, it measures the time required for the barrier to store this energy and release it on the transmission side. Because the maximum stored energy capacity saturates as the barrier thickens, the time delay saturates as well. The process does not involve actual information traveling faster than light, meaning it strictly obeys the constraints of relativity.

Negative time can refer to several things, depending on the context, ranging from a simple mathematical convenience to a mind-bending theoretical concept. The most straightforward answer is that negative time is simply the time *before* an arbitrarily chosen "zero" point. Think of a number line. The number zero isn't special; it's just a reference point. Negative numbers (-1,-2,-3) are just as valid and real as positive numbers (1, 2, 3).

Time plays a similar role in many physics and engineering problems. If we are launching a rocket from here on Oak Island at noon, we might set t=0 to noon. The countdown – "T-minus 10 seconds, 9, 8..." – is using negative time. In this context, t = -10 seconds is simply the point in time 10 seconds before launch. In this sense, negative time is completely real, and we use it all the time. It's just a label for a moment before a defined event.

This is where the question gets interesting. We experience time moving in only one direction – forward. We remember the past, but not the future. A shattered teacup never spontaneously reassembles itself and leaps back onto the table. Why? This forward direction is often called the "Arrow of Time," and the Second Law of Thermodynamics primarily governs it. This law states that the total entropy (a measure of disorder or randomness) of a closed system will always increase over time. The universe is moving from a state of low entropy (the highly ordered, hot, dense state right after the Big Bang) to a state of high entropy (a cold, disorganized, spread-out state in the future). A whole teacup is an ordered system. A shattered cup is a disordered system. The laws of probability make it vastly more likely for things to become more disordered than more ordered. From this perspective, negative time would mean running the universe's clock backward, a direction in which entropy decreases. We never see this happen on a large scale, which is why time has a built-in, one-way street sign.

At the most fundamental level, most laws of physics are time-symmetric. If you watch a video of a planet orbiting the sun, and then watch it in reverse, the reversed video still perfectly obeys the laws of gravity. You couldn't tell if it was being played forward or backward. The same is true for electromagnetism.

So, if the fundamental laws don't favor forward time, why do we? Besides entropy, there is one place where a subtle preference exists: the weak nuclear force (which governs radioactive decay) is not perfectly time-symmetric. But the overwhelming reason for the arrow of time is thermodynamics. This leads to two fascinating and more speculative ideas about negative time:

If we set the Big Bang as t = 0, then "negative time" would refer to a time *before* our universe began. The problem is that our current laws of physics, including general relativity, break down at the (assumed) Big Bang singularity.

Did time begin at t = 0? Some theories suggest that time itself was created in the Big Bang. Asking what happened "before" would be like asking what is north of the

North Pole – a meaningless question. Other speculative models, such as the "Big Bounce," propose that our universe originated from the collapse of a previous one. In this scenario, a universe existed at t < 0, and our concept of time would continue backward indefinitely.

The Feynman-Stückelberg interpretation explains antiparticles as particles moving backward in time, thereby resolving negative-energy solutions in the Dirac equation. In this view, a positron (antielectron) is simply an electron moving backward in time, but with positive charge, allowing for a consistent picture in which particle creation and annihilation are time-reversal events, making antiparticles equivalent to particles with opposite charge and momentum.

Instead of negative energy states, it's more convenient to picture antiparticles (like positrons) as their normal counterparts (electrons) moving backward along the time axis in a Feynman diagram. A particle with negative energy ($E < 0$) moving backward in time ($t \to -t$) has the same time dependence as a positive-energy particle ($E > 0$) moving forward in time, because $e^{-i(-E)(-t)} = e^{-iEt}$.

This interpretation makes Feynman diagrams work: antiparticles are shown using backward-pointing arrows, while particles are represented by forward-pointing arrows, thereby naturally accounting for charge and energy conservation in interactions. This is a powerful tool used by physicists to calculate particle interactions.

So, what is negative time? Conventionally, it's just any point in time before a chosen "zero," like in a countdown. Physically, it's a hypothetical direction in which entropy would decrease, a path the universe doesn't take on a large scale. In Quantum Physics, it's a powerful mathematical idea used to describe the behavior of antimatter.

Time Crystals

To understand a time crystal, it is helpful first to consider a standard crystal, such as salt or quartz. These materials exhibit a repeating three-dimensional structural pattern. A time crystal extends this concept into the fourth dimension by exhibiting a structure that repeats at regular intervals in time.

While a normal object in its lowest energy state remains static, the atoms in a time crystal continue to oscillate, flip, or rotate in a periodic cycle even when they possess no thermal energy. This behavior is unique because the system breaks what physicists call time-translation symmetry. Under normal circumstances, if you "kick" a system with a pulse of energy every second, you expect the system to respond every second. A time crystal, however, might only respond every two seconds, maintaining its own internal rhythm regardless of the timing of the external force.

Physicists view time crystals as a monumental discovery because they represent a new phase of matter that exists outside thermal equilibrium. For over a century, the study of matter was largely confined to systems that eventually settle down and stop moving as they reach a stable temperature. Time crystals prove that matter can be organized and "alive" with motion without ever heating up or losing energy to its surroundings. This discovery challenges the traditional understanding of the laws of thermodynamics and expands the map of possible states of matter. It demonstrates that many-body localization – a phenomenon in which particles become "stuck," preventing them from reaching equilibrium – can create stable, complex structures previously thought to be mathematically impossible.

The potential applications of time crystals focus primarily on the future of quantum technology. One of the greatest hurdles in quantum computing is decoherence, where the fragile quantum states of a computer are destroyed by environmental noise. Because time crystals are naturally stable and tend to "lock" into their rhythmic patterns, they could be used to create more stable quantum bits, or qubits, that are shielded from outside interference. Their inherent periodicity makes them ideal candidates for extremely precise measurements. They could be used to develop a new generation of atomic clocks or highly sensitive magnetic-field detectors. While we are still in the early stages of exploring this phase of matter, the ability to maintain perpetual, organized motion at the atomic scale provides a foundation for technologies that require perfect timing and extreme stability.

In 2021, a team of researchers using Google's Sycamore quantum processor achieved a major milestone by creating a stable time crystal within a quantum system.[77] This experiment was important because it moved the concept from a theoretical curiosity to a physical reality that could be observed and measured over hundreds of cycles. The researchers used a chain of 20 superconducting qubits – the quantum equivalent of computer bits – as the medium for the crystal.

To create the time crystal, the team first put the qubits into a random state. They then hit the system with periodic pulses of microwave radiation. In a normal system, this constant energy input would quickly turn the qubits into a chaotic, overheated mess, eventually destroying any organization. However, due to a phenomenon known as many-body localization, the qubits did not absorb heat. Instead, they became "locked" into a pattern in which they flipped their orientation, then flipped back exactly twice the microwave pulses' period.

The hallmark of this experiment was pattern stability. The qubits demonstrated "spatiotemporal order" that was remarkably resilient to errors. Even when the researchers slightly tweaked the timing or strength of the microwave pulses, the time crystal maintained its internal rhythm, refusing to sync with the external drive. This proved that the time crystal was a genuine phase of matter rather than just a fleeting simulation.

This experiment proved that quantum computers are not just tools for solving mathematical problems but also powerful laboratories for discovering new physics. By successfully shielding the system from the "noise" of the outside world, the Google team showed that we can engineer complex quantum states that remain stable over time. This breakthrough is a foundational step toward building quantum systems capable of performing long, complex calculations without degrading.

An assumption might be that a time crystal is a perpetual motion machine because both involve objects that move forever without an external power source. However, there is a fundamental distinction in physics that separates the two: the ability to perform work. A perpetual motion machine is a theoretical device that would violate the laws of thermodynamics by producing more energy than it consumes, or by perfectly converting heat into work without any loss. If such a machine existed, you could use it to turn a wheel or power a light bulb indefinitely.

A time crystal does not violate these laws because it is in its ground state, which is the lowest possible energy level a system can have. Because it is already at the "bottom" of its energy well, there is no energy to be extracted from its movement. To take energy out of the system, you would have to stop its motion, but that motion is the natural, lowest-energy configuration of the crystal. Trying to "plug" a time crystal into a battery to draw power would be like trying to extract energy from a rock sitting on the ground; it is already at rest in its own way.

The motion within a time crystal is more akin to "superconductivity" or "superfluidity." In a superconductor, an electric current can flow through a loop of wire forever without any resistance. As soon as you try to use that current to power an external device, you are drawing energy out and must eventually replace it. The time crystal's oscillation is a passive, inherent property of its existence. It moves because, in this strange phase of matter, remaining still would actually require more energy than staying in motion.

In short, a perpetual motion machine is a "free energy" scheme that physics forbids, whereas a time crystal is a "perpetual rhythm" that physics allows. The crystal is not creating energy out of nothing; it is simply refusing to stop moving because it has no way to lose the tiny amount of energy it already possesses.

In a typical physical system, order is usually the enemy of chaos; if you want something to stay organized, you expect it to be pure and uniform. However, for a time crystal, a specific type of internal "disorder" is actually the secret ingredient that prevents it from falling apart. Without this messiness, the time crystal would absorb energy from its surroundings, heat up, and eventually reach thermal equilibrium (through many-body localization). In the Google Sycamore experiment, the researchers did not make all the qubits identical. Instead, they introduced intentional "randomness" into the interactions between the qubits. This disorder acts as an internal friction or a series of barriers that prevent energy from flowing easily through the system. In a standard material, energy spreads out like a drop of ink in water until everything is uniform and warm. In a time crystal, the disorder "localizes" the energy, trapping it in place and preventing the particles from ever reaching a shared, static temperature.

Because energy cannot spread out and dissipate as heat, the system remains "cold" and stuck in its quantum state, even as external pulses hit it. This allows the time crystal to maintain its rhythmic flipping forever. The disorder effectively creates a permanent state of "stasis" for the energy while allowing the "motion" of the clock-like ticking to continue. It is a profound irony of physics: by making the material's internal structure chaotic and disordered, scientists can create a higher form of temporal order that is nearly impossible to break.

This reliance on disorder is what makes time crystals so robust. Because the pattern doesn't depend on perfect conditions, the crystal's "ticking" is shielded from minor fluctuations and external noise. If you slightly adjust the strength of the laser or the microwave pulse, the system's disorder acts as a buffer, ensuring the crystal maintains its own time regardless of the interference.

Does the “tick” provide the time crystal with the energy needed to move in the fourth dimension? In the original theoretical proposal for time crystals, the answer was intended to be yes. Nobel laureate Frank Wilczek first imagined a system that would spontaneously begin to oscillate on its own in its lowest energy state, without any external influence at all. This is called a continuous-time crystal. However, shortly after he proposed this, other physicists proved a "no-go theorem" which suggested that a truly static, isolated system in equilibrium cannot spontaneously start ticking. This led many to believe time crystals were a physical impossibility.

The breakthrough came when researchers shifted their focus to discrete-time crystals, which are the ones created in laboratories today. These do require a "kick," such as a periodic laser or microwave pulse. While this external drive provides the timing, it is not "fueling" the motion in the traditional sense. The time crystal does not absorb energy from the kick to keep moving; rather, the kick acts as a background field that allows the material to break symmetry. The system remains in a non-equilibrium state, ignoring the drive's specific energy and choosing its own motion frequency.

However, scientists have begun to observe versions that more closely resemble the original "no-kick" idea. In 2022, researchers at the University of Hamburg demonstrated a time crystal inside a quantum gas within an optical cavity.[78] In this setup, the "kick" wasn't a timed external pulse from a scientist, but rather a continuous beam of light. The system spontaneously organized itself into a repeating, rhythmic pulse even though the input was perfectly steady. While energy continues to enter the system, the crystal's "timing" is entirely self-generated.

The ultimate goal for many physicists remains the discovery of a passive time crystal. This material would sit in a vacuum, perfectly isolated from any light or energy, and still exhibit a rhythmic pulse forever. While we have not yet found a way to create this without violating the laws of equilibrium, the progress from "timed kicks" to "steady light" suggests that we are moving closer to understanding how matter can create its own internal clock entirely from its dynamics.

Time at the Microscopic Scale

A scientific study, published in *Nature Physics* in 2024 by researchers at the Technical University of Darmstadt and Roskilde University, provides a fascinating look at the "arrow of time" within complex materials.[79]

Unlike a crystal, which has a perfectly ordered atomic structure, glass is essentially a "frozen liquid." Its molecules are disordered, but they are packed so tightly that they can barely move. Even though glass appears solid, it is not in equilibrium. Over decades or centuries, the molecules in glass very slowly "settle" into a more stable arrangement. This process is called molecular aging. Because the glass is constantly moving toward a more stable state, this process traditionally has a clear "arrow of time" – the material is always getting "older" and never "younger."

The researchers used ultra-sensitive video cameras and laser light scattering to observe the molecular movements in glass and plastic. By analyzing these tiny fluctuations, they sought to determine whether the movements were time-reversible. The study showed that at the molecular level, these fluctuations are statistically time-reversible. If you were to watch a video of a single molecule's internal jitters, you would have no way of knowing if the video was playing forward or in reverse.

This creates a strange physical paradox. The material as a whole has a clear arrow of time (it is aging), but the individual molecular fluctuations driving that aging do not.

The Macroscopic View (Irreversible): The glass is settling and moving toward a state of higher entropy. You can tell the difference between "before" and "after."

The Microscopic View (Reversible): The specific molecular dance appears identical in both directions.

The researchers described this by saying that time does not "run" in the same way for these molecules as it does for the object they compose. The "arrow of time" is effectively absent from the molecular jitters themselves. This discovery challenges our fundamental understanding of how entropy and the arrow of time emerge in complex materials. Traditionally, we assume that the "arrow of time" is baked into every level of a system that is moving toward entropy. This study suggests that the "arrow" might be an emergent property that only appears when you look at the system as a whole, rather than something present in every individual interaction.

In summary, the study doesn't suggest that you can turn "old" glass back into "new" glass. Instead, it reveals that the microscopic "ticks" of the clock for a glass molecule are indifferent to whether the clock is moving forward or backward, even while the glass itself continues to grow old.

Simultaneity and Synchronicity

Simultaneity forms a crucial puzzle in modern physics. Einstein's work in Special Relativity demonstrated conclusively that the perception of two events occurring at the same time is not universal; what appears simultaneous to one observer can, and often must, appear to happen at different times for another observer in a different frame of reference. This fundamental realization, often termed the "death of Now," shows that there is no single, absolute "Observable Time" or universal timeline that everyone across the cosmos can agree on. Time is not an independent background but is inextricably linked to the motion and position of the observer.

This scientific fact occasionally finds a curious, speculative parallel in the philosophical concept of Synchronicity. Synchronicity, a term popularized by psychologist Carl Jung, describes the occurrence of two or more events that appear meaningfully related but lack a causal physical connection. Instead, the events seem linked by an underlying, non-physical principle or a shared meaning. While modern physics firmly grounds its framework in causality, the idea of acausal, meaningful connections–a concept that intellectually fascinated brilliant minds, including the

Nobel laureate physicist Wolfgang Pauli, who collaborated extensively with Jung–suggests a deeper layer to reality.

The contrast between the scientifically proven relativity of simultaneity and the speculative nature of synchronicity highlights the boundary between established physics and philosophical exploration.

The New Frontiers: Speculative Theories

Is Time an Illusion?

Some theories suggest that time is not a fundamental, linear backdrop but rather an emergent, multidimensional, or even illusory phenomenon. Key ideas include that time has three dimensions, that time arises from quantum entanglement, or that time results from the irreversible imprinting of information. These approaches aim to merge quantum mechanics and general relativity.

Research conducted at the University of Surrey in the UK is fundamentally challenging one of the most ingrained concepts in physics: the unidirectional, universal flow of time. Specifically, their investigations into quantum systems have yielded compelling evidence that these systems can exhibit not one but *two* distinct arrows of time.[80]

This finding introduces complexity to our understanding of temporality, moving beyond the traditional thermodynamic arrow of time – the irreversible progression from order to disorder (an increase in entropy). The existence of a second, potentially opposing, temporal direction within the quantum realm opens up new theoretical avenues for reconciling quantum mechanics with general relativity, a long-standing challenge in modern physics, by positing that the arrow of time is not a fixed cosmological constant, but rather a variable property that emerges from the underlying quantum dynamics of a system.

Another compelling idea is the Thermal Time Hypothesis (TTH), championed by Carlo Rovelli and Alain Connes. It proposes that time is not a fundamental part of the universe's "hardware" at all. Instead, it is a statistical illusion–an emergent "blur" that appears only because we are unable to see the full, infinite complexity of the quantum world.

TTH is linked to information theory, which Claude Shannon founded in 1948. Its core idea is that information is a physical, measurable quantity. The basic unit of information is the "bit." A bit represents the resolution of uncertainty between two equal possibilities (like a coin flip, heads or tails). In information theory, (Shannon) entropy is a measure of uncertainty or "surprise." A message with low entropy is predictable (e.g., "AAAAAA"). It contains little new information. A message with high entropy is random and unpredictable (e.g., "Zg8(k!"). Every character is a surprise and carries a lot of information. The main point is that information resolves uncertainty.

The Thermal Time Hypothesis proposes that time is not a fundamental part of reality, but instead *emerges* from a more basic concept: information. It suggests that our experience of "the flow of time" is just a manifestation of the flow of information and the laws of thermodynamics.

In Einstein's General Relativity, time is relative and part of a flexible four-dimensional "spacetime." In quantum mechanics, time is a rigid, absolute, universal "metronome" that everything ticks along to. When you try to combine the two theories (to create a theory of quantum gravity), the equations "freeze" and the variable for time (t) vanishes. This suggests that at the most fundamental level, time may not exist.

The TTH proposes that time is not fundamental. Instead, it is an emergent property that "flows" from the universe's thermal state. It states that the "flow of time" *is* the same as the flow of thermal equilibrium. Just as heat naturally flows from a hot object to a cold one, the universe's statistical state naturally evolves. This statistical, thermodynamic evolution is what we perceive and measure as the passage of time.

This is where it all comes together, as Information Theory and the Thermal Time Hypothesis share the concept of entropy. In physics, "entropy" is a measure of disorder (a key part of thermodynamics). In the 1950s and 1960s, physicists proved that Shannon's "information entropy" and the "thermodynamic entropy" of physics are essentially the same thing.

1. A high-entropy system requires more information (more bits) to describe.
2. A low-entropy (ordered) system requires less information to describe.

This was demonstrated physically by Landauer's principle, which states that erasing one bit of information in a computer *must* release a minimum, fixed amount of heat (entropy) into its environment. Information is physical. Under the Thermal Time Hypothesis, time ceases to be a fundamental background and instead becomes a byproduct of thermodynamic flux. Because thermal entropy is fundamentally a tally of 'missing' information regarding a system's microstates, the arrow of time is revealed as the macroscopic unfolding of information itself.

In this view, the universe is a giant information-processing system. The fundamental "stuff" of reality isn't matter or energy, but information (bits). The "Big

Bang" was a state of incredibly low entropy – a highly ordered state that could be described with very little information. The "Arrow of Time" (the reason time flows forward and not backward) is simply the universe evolving from this simple, low-entropy state to a more complex, high-entropy (disordered) state, as required by the second law of thermodynamics. Our perception of "now" is just a snapshot of this unfolding process. "Time" is the label we give to the changing state of the universe's information.

Gunther Kletetschka, a physicist at the University of Alaska Fairbanks, proposes a theoretical framework featuring three dimensions of time alongside three spatial dimensions.[81] Published in 2025, this model suggests that 3D time is the fundamental "canvas" of reality, with space a byproduct. It aims to unify quantum mechanics and general relativity, explaining particle masses and allowing for alternative, non-linear timelines.

Instead of one linear time, this model proposes three separate, orthogonal axes of time. Kletetschka suggests that time is the primary fabric of the universe, and space is a byproduct, reversing the traditional "spacetime" view. The theory purports to explain the specific masses of fundamental particles, such as electrons and muons, by treating them as eigenmodes of the temporal metric. It aims to connect quantum mechanics and general relativity by providing a six-dimensional (6D) framework (3D space + 3D time). The model predicts particle masses with percent-level accuracy and suggests potential deviations in the speed of gravitational waves, providing testable, falsifiable claims. While interesting, the theory remains speculative and has not yet achieved widespread acceptance in the broader scientific community.

The Page-Wootters (PaW) mechanism proposes that time is not fundamental but emerges from quantum entanglement within a timeless, stationary universe by treating an internal system and an "internal clock" as an entangled pair, showing how apparent evolution arises as a conditional equation of motion.

This idea, developed by Don Page and William Wootters in 1983, suggests that observers within the universe experience time as correlations between system states and clock states, thereby resolving issues in quantum gravity in which time often disappears. Experiments have supported the idea that time is an emergent phenomenon for internal observers, consistent with the Wheeler-DeWitt equation. That is to say, the entire universe, including all matter and clocks, is described by a single, unchanging quantum wave function, as per the Wheeler-DeWitt equation in quantum gravity. The universe is conceptually divided into the "system of interest" (everything we observe) and an "internal clock" (such as a particle or field). These two parts become entangled, linking the system's state to specific clock states. An observer measuring the clock's state sees the system's state change, creating the appearance of time evolution (e.g., t = 1, t = 2, …) from a timeless whole. What we perceive as flowing time is a manifestation of correlations within the quantum state, not a fundamental flow. An external observer of the whole system sees no time, while an internal observer (using the clock) experiences it. The mechanism helps tackle the "problem of time" in quantum gravity and has inspired quantum algorithms for parallel-in-time simulations.

Other theories suggest that time emerges from the irreversible imprinting of information, with thermodynamic laws preventing its erasure. This implies that time is a consequence of change rather than a container for it, or that the flow of entropy drives that time and measures change in open systems.

The Wolfram Physics Project models time computationally, with the state of the universe computed from the previous state according to specific rules. Stephen Wolfram launched the project in 2020. It attempts to find a fundamental theory of physics by modeling the universe through computer programs rather than traditional mathematical equations. Traditional physics relies on continuous space and differential equations. This project proposes that the universe is fundamentally discrete, digital, and computational.

At the lowest level, the project models space as a collection of abstract points connected by relations. This structure forms a hypergraph, a network in which connections can bind any number of points together, rather than just pairs. The model does not place these points in a pre-existing space. Instead, the connections in hypergraph define space itself.

To simulate the evolution of the universe, you repeatedly apply simple, abstract rewriting rules to the hypergraph. For example, a rule might take any two connected nodes and replace them with a specific arrangement of new nodes and connections. When you run these computer experiments millions of times, complex structures emerge from these simple rules. The project attempts to derive the major pillars of modern physics from these computational rules.

Space and Gravity: As the hypergraph grows, the dense collection of points begins to resemble continuous, multi-dimensional space. By tracking how information spreads through the network, researchers construct a causal graph. The project demonstrates that for certain classes of rules, the large-scale behavior of this causal graph obeys Einstein's equations of General Relativity.

Quantum Mechanics: When a rule can be applied to a hypergraph in multiple ways, the system splits into a network of histories called a multiway graph. This branching and merging of paths represents quantum superposition. The project models quantum entanglement as paths that share a common ancestral state in this multiway space.

These theories seek to resolve the "problem of time": general relativity treats time as flexible, while quantum mechanics treats it as a rigid, external parameter. We are at a strange crossroads. We have a "Cosmic Realm" where time is a smooth, warped manifold, and a "Quantum Realm" where it might be pixelated, emergent, or even non-existent.

The future of time research lies in reconciling these views. Projects like LIGO are looking for ripples in time (gravitational waves) that might carry the "fingerprints" of the Big Bang. As our computers become "quantum" and our clocks become more precise, we may finally reach the point where we can "see" the grains of time at the Planck scale. We trade the infinite precision of the quantum soup for the ability to have a history, a narrative, and a life. We are the mechanism by which a timeless universe experiences the wonder of a "now".

Summary

Chapter Eight looked at the nature of time across multiple views. It contrasts the absolute universal clock of Isaac Newton with Einstein's flexible, dilating time coordinate. It considered the "Block Universe" theory (Eternalism), which posits that the past, present, and future exist simultaneously on a static 4D map. It reviews how the one-way "Arrow of Time" is fundamentally dictated by entropy and thermodynamic irreversibility, but that the "Quantum Arrow of Time" may allow for two arrows of time.

Finally, it introduces concepts like Planck time limits, time crystals (phases of matter that structurally oscillate without absorbing or losing energy), Wolfram Physics, and the Thermal Time Hypothesis (TTH) – which proposes that time isn't fundamental at all, but rather a macro-statistical illusion emergent from a timeless quantum soup.

Chapter Nine: The Amazing Electron

"I hate being around electrons. They're always so negative!"

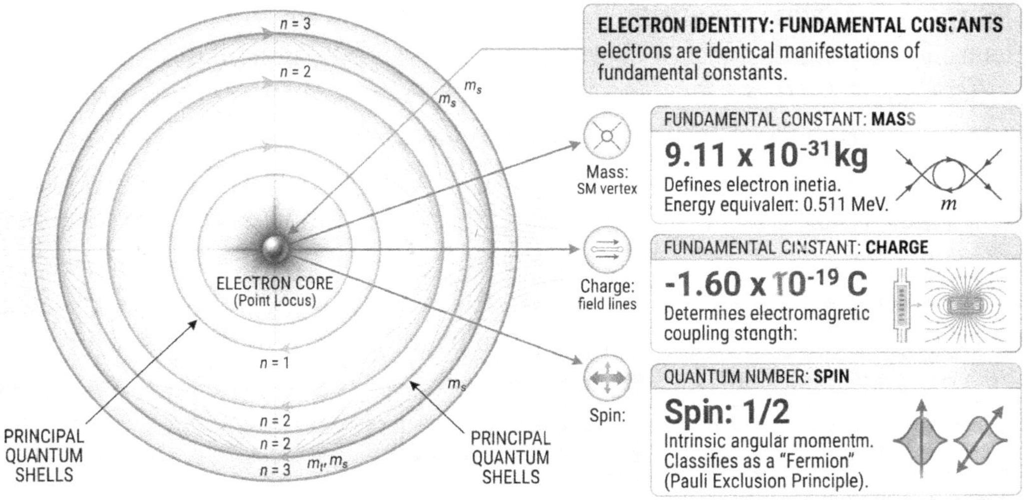

Electrons are considered "special" because they are the primary drivers of almost every physical and chemical interaction in our daily lives. While protons and neutrons provide the mass and "identity" of an atom, electrons provide its behavior. Electrons are the "glue" that holds atoms together.

Whether through sharing (covalent bonds) or transfer (ionic bonds), all chemical reactions and the formation of molecules such as water and DNA depend entirely on electron movement. Only the outermost electrons (valence electrons) participate in reactions. The number of these electrons determines whether an element is highly reactive, like fluorine, or completely inert, like neon.

Electricity is essentially the flow of electrons through a conductor. Without this flow, we would have no power for lights, computers, or any electronic device. Moving electrons generate magnetic fields. Every permanent magnet or electromagnet relies on the spin and alignment of electrons to function. Devices such as transistors and microchips work by precisely controlling the flow of electrons through semiconductors.

Electrons do not exist as simple "dots" orbiting a nucleus. Their quantum behavior makes them appear as both particles and waves, existing in "probability clouds" (orbitals) where they are effectively "smeared out" around the atom. Electrons can "quantum tunnel", passing through physical barriers that should be impossible to cross–a phenomenon used in modern flash memory and electron microscopes. No two electrons in an atom can occupy the same quantum state. This Pauli exclusion principle prevents atoms from collapsing and gives matter its volume and "solidity". Electrons can pass "through physical barriers that should be impossible to cross?"

Electrons are incredibly small, with a mass roughly 1/1,836 that of a proton. Despite having almost no mass, they occupy nearly all the volume of an atom. The electron is the lightest stable charged particle. Because there is no lighter charged particle for it to decay into, it is believed to be essentially permanent. The most precise experimental estimate puts the minimum longevity of an electron at 6.6×10^{28} years (or 66,000 yottayears – 66 billion trillion trillion years) at a 90% confidence level. This duration is roughly 5 quintillion times longer than the universe's age.

Wow, electrons are immortal! When we "touch" something, our electrons push against the electrons in that "something", and we "feel" it – is that just wild or what? We could swap out all the electrons in our solar system for the ones in Alpha Centauri, and there would be no difference – all electrons are identical (so maybe there's only one–and, no, that's not as unbelievable as it might sound).

History of the Electron

The history of the electron is foundational to modern physics, tracing a path from the earliest understanding of electricity to the development of quantum mechanics and the latest technological applications, such as spintronics.

In the 1830s, Michael Faraday, through experiments with electrolysis, established the laws governing how electric current affects chemical solutions, suggesting that discrete units of electrical charge were involved.

In 1874, George Johnstone Stoney first proposed the existence of a fundamental unit of electrical charge and, in 1891, named this theoretical unit the "electron." Scientists such as William Crookes and Jean Perrin studied cathode ray tubes (evacuated glass tubes). They observed mysterious rays traveling from the cathode (negative electrode) to the anode (positive electrode), but the nature of these rays was debated: were they waves or particles?

In 1897, J.J. Thomson, working at the Cavendish Laboratory, performed a series of definitive experiments using cathode ray tubes. He demonstrated that both electric and magnetic fields could deflect cathode rays. This proved they were composed of charged particles. By measuring the amount of deflection, he was able to calculate the ratio of the particle's charge (e) to its mass (m). The value he found was over a thousand times smaller than that of the hydrogen ion (the lightest known atom). Thomson concluded that these particles, which he called "corpuscles", were fundamental components of the atom, establishing that the atom was not indivisible, as previously thought.

Following the discovery of the nucleus, scientists struggled to develop a stable atomic model that accounted for electron orbits. The following illustrates the various models proposed, with Louis de Broglie making the leap that enabled the first complete formulation of quantum theory.

In 1904, J.J. Thomson proposed the Plum Pudding Model, which held that electrons were embedded in a uniformly positive sphere, but it failed to explain why positively charged alpha particles were scattered at large angles by the gold foil. In 1909, Robert Millikan performed his famous Oil Drop Experiment, which measured the fundamental charge of the electron (e) with unprecedented accuracy. Combining this with Thomson's e/m ratio, he calculated the electron's mass, confirming its status as an incredibly light particle.

In 1911, Ernest Rutherford's Planetary Model proposed a small, dense, positively charged nucleus with electrons orbiting it like planets. It failed because, according to classical electromagnetism, an orbiting electron must constantly radiate energy and spiral into the nucleus almost instantly (classical catastrophe). Then, in 1913, Niels Bohr postulated that electrons could only exist in specific, discrete, stable orbits (stationary states) without radiating energy. Electrons emit or absorb energy only when jumping between these orbits. This explained the stability of the atom and successfully predicted the spectral lines of Hydrogen, introducing the concept of quantized energy levels. To add to Bohr's ideas, Louis de Broglie, in 1924, proposed that electrons, being particles, also possess wave-like properties, with an associated wavelength $(\lambda = h/p)$. This provided the physical justification for Bohr's quantization: only orbits whose circumference was an integer multiple of the electron wavelength were stable.

In 1925, Werner Heisenberg developed Matrix Mechanics, the first complete formulation of quantum theory. This formalism naturally led to the Uncertainty Principle (1927), which states that the position and momentum of an electron cannot both be known simultaneously. In 1926, Erwin Schrödinger developed Wave Mechanics. The solution (ψ) describes the electron's wave function. The square of the wave function $(|\psi|^2)$ defines the orbital – the probability cloud where the electron is most likely to be found.

In 1928, Paul Dirac formulated the Dirac Equation, which combined special relativity with quantum mechanics. This equation not only correctly described the electron's behavior but also automatically predicted the electron's intrinsic property of spin and the existence of its antiparticle, the positron (antimatter). The years immediately following the Dirac equation were defined by filling in the blanks of the atomic nucleus and confirming Dirac's "spooky" predictions.

Current research focuses on exploiting the electron's quantum properties, particularly its spin, to develop new technologies. Spintronics uses the electron's spin state (↑↓) to store and process data, in addition to its charge. This has led to the commercialization of MRAM (Magnetoresistive Random-Access Memory), a non-volatile (retaining data when power is off), faster, and more energy-efficient form of memory.

For Quantum Computing, the electron's two spin states naturally form a qubit (quantum bit). Current research involves trapping and manipulating single electron spins in silicon quantum dots to build highly stable quantum processors.

Everyday Object Spin

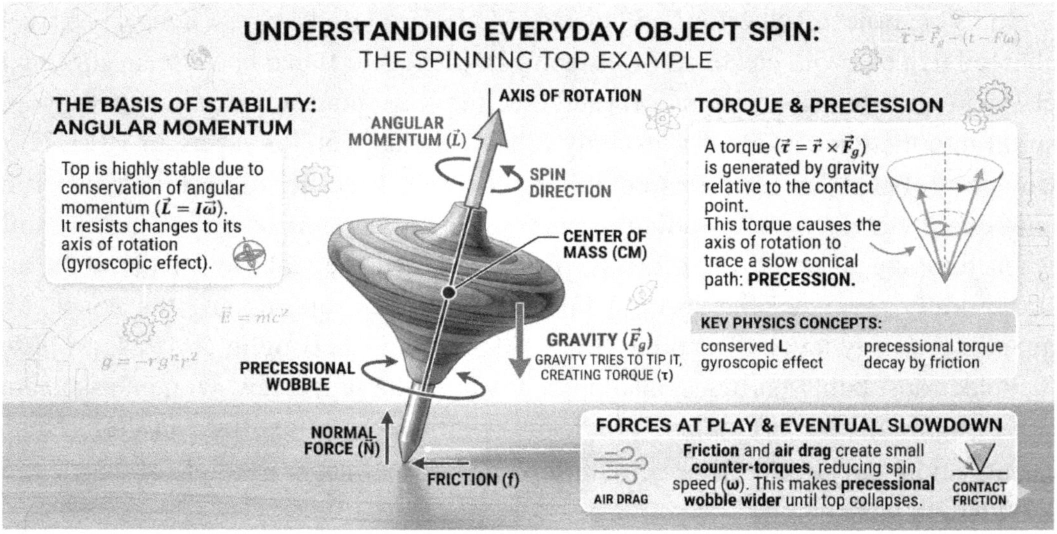

To begin, let's look at what spin is in a regular object (like a child's top, the Earth, or a gyroscope). When we understand regular spin and angular momentum, we can then apply this knowledge to the electron. Here are my observations from various sources. I avoided extensive editing to share my thought processes.

What does" spin" mean for an everyday object, say, a child's top? What does it mean when we say it is spinning? After some research, my explanation is that the top's mass is rotating around a vertical axis (ignoring angular momentum, gravity, friction, and precession for the moment). You impart kinetic energy to the top, and because the mass is constrained around its axis, the energy is converted into rotational energy.

So, it's the "mass" of the top that is spinning, meaning what? The top is made of "stuff," perhaps plastic. Plastics are composed of molecules, which are made up of atoms. We know atoms are not indivisible, but they are good enough for this explanation.

Here's the first important point: the congregated atoms themselves are not spinning; they are moving in a circle around an axis – they are turning, not spinning. It's the aggregate of all the atoms that are said to be spinning, that is, all of the atoms that comprise the top. In other words, for a thing to be able to spin, that thing must have constituent parts that rotate around the axis together.

"My 'spinning' means I am rapidly rotating around my pointed tip, creating a force called angular momentum that keeps me upright despite gravity pulling me down; as I spin, my axis of rotation remains stable due to a phenomenon called gyroscopic precession, allowing me to seemingly defy gravity and stay balanced on my point until friction eventually slows me down and I topple over."

I admit, I still have misgivings about the nomenclature; maybe spin should be renamed "*quantum gyration*" and "*wobbly angular stuff*".

Everyday Object Angular Momentum

Let's look at a regular "object" in our classical world. The most straightforward definition of angular momentum is *the measure of an object's tendency to continue rotating.* It is the rotational analog of linear momentum (mass in motion). Just as linear momentum depends on mass and velocity, angular momentum depends on mass, velocity, and how that mass is distributed relative to the center of rotation (the object's shape or moment of inertia). In any closed system (with no external torque), angular momentum is always conserved; it cannot be created or destroyed.

A good example is when an ice skater pulls their arms and legs inward; they decrease their moment of inertia (how mass is distributed). Since angular momentum is conserved, rotational speed (angular velocity) increases dramatically.

Broadly speaking, a normal physical object (e.g., the Earth) can have two types of angular momentum. The first type is due to the object's center of mass rotation about some fixed external point (e.g., the Sun) – generally known as *orbital angular momentum*. The second type is due to the object's internal motion – known as *spin angular momentum* (for a rigid object, the internal motion consists of spinning about an axis passing through the center of mass).

When a spinning top is set in motion, its "angular momentum" measures its rotational inertia, essentially how much it wants to maintain its spin in the same direction. It appears to defy gravity and stay upright as long as it's spinning fast enough. The amount of angular momentum depends on inertia, a measure of how its mass is distributed around the axis of rotation, and on angular velocity, which determines how fast it is spinning. This is because angular momentum acts as a vector pointing along the axis of the spinning top, resisting any change in its rotation unless an external force (such as friction) acts on it (a vector has both magnitude and direction). Angular momentum is conserved. When no external torque (i.e., a twisting force) is acting on the top, its angular momentum remains constant.

When the top starts to slow down, it leans because gravity exerts a torque that slowly precesses its rotational axis about the vertical axis. You can visualize this precession by imagining a spinning top on a flat surface. As long as it spins quickly, it maintains high angular momentum and stays upright. When the top slows down, its angular momentum decreases, and it begins to wobble, eventually falling over due to gravity. A spinning top demonstrates angular momentum by maintaining its spin due to its inherent rotational inertia, resisting changes in its rotational direction unless an external torque acts on it, and exhibiting precession when gravity creates a torque, causing its axis to rotate slowly about a vertical axis.

So, momentum is a physical quantity that an object might possess by virtue of its motion. Angular momentum occurs when the motion is rotational. Angular momentum is "the *quantity of rotation of a body, which is the product of its moment of inertia and its angular velocity*."

Electron Quantum Spin

Going back to our explanation of the spinning top, you may still think it's a strange description, but if we reduce the top to a single atom, can it spin? What is its axis of rotation? Now we look further at the electron, which is considered fundamental – it is not made of smaller components, such as atoms. Physicists often refer to an electron as a point particle because its diameter is too small to measure – about 10,000 times smaller than a hydrogen atom.

As we said earlier, to spin, an object has to be a composite so that its parts can rotate around its axis, and an electron has no "parts"! Knowing this, why do quantum physicists say an electron spins? Unfortunately, they used that label because it relates to other properties the electron does have, such as angular momentum and precession, and it seemed appropriate given the Stern-Gerlach experiment. So, an electron does not spin in the conventional sense, so what is it doing? This is where intrinsic angular momentum comes in.

An electron spin refers to a *form* of angular momentum of electrons, so don't assume that the regular angular momentum is an exact description of an electron. Physicists say the spin quantum number provides information about an electron's unique quantum state. This makes it easier to compute the properties of elementary particles, such as electrons. An electron's angular momentum is a quantum property whose magnitude is permanent.

In the 1920s, Otto Stern and Walther Gerlach of the University of Hamburg in Germany, knowing that all moving charges produce magnetic fields, proposed measuring the magnetic fields generated by electrons orbiting atomic nuclei. To everyone's surprise, they found that electrons themselves appear to spin very rapidly, producing tiny magnetic fields independent of their orbital motion. The terminology 'spin' was used to describe this apparent rotation of subatomic particles – a name we are stuck with.

And that explains why we struggle with the concept of non-spinning spin – it's a unique physical quantity in quantum particles. It is analogous to a planet's spin, giving a particle angular momentum and a tiny magnetic field, known as a magnetic moment. However, based on the known sizes of subatomic particles, the "surfaces" of charged particles would have to move faster than the speed of light to produce the measured magnetic moments. Spin is quantized, meaning only discrete spin values are allowed. This situation creates complications that make spin one of the more challenging aspects of quantum mechanics.

In a sense, spin is an essential property that influences the ordering of electrons and nuclei in atoms and molecules, lending it great physical significance in chemistry and solid-state physics. Spin is likewise a vital consideration for all interactions among subatomic particles, whether in high-energy particle beams, low-temperature fluids, or the tenuous flow of particles from the sun known as the solar wind. Indeed, many, if not most, physical processes, ranging from the smallest nuclear scales to the largest astrophysical distances, depend on the interactions of subatomic particles and their spins.

Quantum Spin and Intrinsic Angular Momentum

All elementary particles exhibit spin, a basic quantum property akin to intrinsic angular momentum, and are categorized as fermions (half-integer spin: electrons, quarks, neutrinos, protons, neutrons) or bosons (integer spin: photons, gluons, Higgs). Spin is misleading because it isn't literally spinning but a quantum state with discrete values (e.g., $\pm\frac{1}{2}$, ±1) that determines particle behavior, governing matter structure or force mediation.

Fermions (Half-Integer Spin, e.g., $\frac{1}{2}$, $\frac{3}{2}$)

- ***Electrons, Muons:*** Key components of atoms, responsible for chemistry.
- ***Quarks:*** The building blocks of protons and neutrons (protons/neutrons also have half-integer spin).
- ***Neutrinos:*** Elusive particles involved in the weak nuclear force.

Bosons (Integer Spin, e.g., 0, 1, 2)

- ***Photons:*** Carry the electromagnetic force (spin 1).
- ***Gluons:*** Carry the strong nuclear force (spin 1).
- ***W and Z Bosons:*** Mediate the weak nuclear force (spin 1).
- ***Higgs Boson:*** The only known elementary particle with spin 0 (no intrinsic spin).

Spin is an intrinsic Property and fundamental to a particle's identity, not just movement. Only specific (quantized) values (multiples of $\frac{1}{2}$) are allowed. Half-integer spin fermions obey the Pauli Exclusion Principle. The table below categorizes particles by spin and role in the universe.

Spin Number	Particle Type	Examples	Meaning / Property
0	Scalar Boson	Higgs Boson	Rotationally invariant; looks the same from every direction.
$\frac{1}{2}$	Fermions	Quarks, Leptons (Electrons, Neutrinos), Protons, Neutrons	Matter constituents; must be rotated twice to return to the same state.
1	Vector Bosons	Photons, Gluons, W and Z Bosons	Force carriers; behave like vectors under rotation.
$\frac{3}{2}$	Composite Fermions	Delta Baryons (excited states)	Highly symmetrical matter states; rare in stable matter.
2	Tensor Boson	Graviton (theoretical)	Proposed mediator of gravity; requires rotation for symmetry.

***Fermions vs. Bosons**:* Particles with half-integer spins ($\frac{1}{2}$, $\frac{3}{2}$) are Fermions, which obey the Pauli Exclusion Principle–no two can occupy the same quantum state. Particles with integer spins (0, 1, 2) are Bosons, which can "pile up" in the same state (like photons in a laser).

***Rotation and Symmetry**:* The spin number relates to how much you must rotate a particle to make it look the same again. For example, a spin 1 particle looks the same after a full turn, while a spin 2 particle looks the same after only

Intrinsic Property: Unlike the spin of a basketball, particle spin is an inherent property that cannot be sped up or slowed down.

The simplest definition of spin in an electron is that it is an intrinsic form of angular momentum, much like a tiny top spinning on its axis. However, there are two crucial differences:

1. It is not a literal rotation: The electron is considered a point-like particle with no internal structure, so it can't physically be "spinning." Spin is a fundamental, fixed property – like its mass or charge – that is a purely quantum-mechanical phenomenon.
2. It is quantized: For an electron, spin can only point in one of two opposite directions when measured along any axis (usually called "up" or "down"). It cannot point sideways or at any angle in between.

Think of it as the particle's internal magnetic dipole moment. This property dictates how electrons interact with magnetic fields. An internal magnetic dipole moment (often called an "intrinsic" magnetic moment) is a basic quantum property that describes the magnetic moment of subatomic particles, such as electrons and protons. It acts like a tiny bar magnet inside an atom, dictating how particles align and interact in magnetic fields.

Spin Angular Momentum

An object's internal motion is known as *Spin angular momentum* because, for a rigid object, the internal motion consists of spinning about an axis passing through the center of mass. By analogy, quantum particles can possess both orbital angular momentum from their motion through space and spin angular momentum from their internal motion.

While electrons have a magnetic field similar to a bar magnet, the fundamental difference is that electrons possess intrinsic angular momentum, so the analogy with classical extended objects is not entirely accurate (because electrons are structureless point particles). I still find it better to think of an electron as having no constituent parts (like quarks in a proton), so it has nothing to rotate around an axis. Thus, in quantum mechanics, it is best to think of spin angular momentum as a kind of intrinsic angular momentum possessed by particles. Each type of elementary particle has a characteristic spin angular momentum, just as each type has a characteristic charge and mass. Again, they are not physically spinning, but we know they have an analogue of angular momentum because the Stern-Gerlach apparatus causes them to precess. Bar magnets would pass through the magnetic field, producing a spatial distribution at the detector. Electrons, on the other hand, cannot rotate and create the same distribution because they precess.

I started to feel as though I *migh*t understand quantum spin, but it needed time to sink in. The fundamental step in understanding spin is to understand how torque causes a particle with angular momentum to precess. Without intrinsic angular momentum, electrons passing through the Stern-Gerlach setup would produce a distribution of positions at the detector, rather than discrete accumulation points.

Quantum spin is an intrinsic angular momentum. It is a property that particles possess because they are particles – I don't get it either. This angular momentum would be produced if a particle were to spin physically, but the measured values are too high. The necessary rotation speed would exceed the speed of light. This is why spin is an intrinsic angular momentum. The particles are not spinning, but they possess angular momentum as though they were.

We know that intrinsic angular momentum exists because applying a torque to a particle with spin causes it to precess rather than rotate. The brilliance of the Stern-Gerlach experiment lay in its ability to induce this effect.

Interim Summary

Spin angular momentum is an inherent characteristic of the particle, much like its mass or electric charge, and does not depend on the particle's motion through space. The amount of intrinsic angular momentum is a constant for a given particle type. Despite the name, it is not caused by the particle physically spinning. The electron is considered point-like and has no structure to rotate. When measured along any axis, the spin's orientation can only take one of a few discrete (quantized) values (e.g., "spin up" or "spin down" for an electron).

Spin Up, Spin Down, and Quantized Spin Numbers

The spin of an electron is a *label* because it's really intrinsic angular momentum. Bohr's atomic model states that this angular momentum is quantized, meaning it can take only specific, fixed values. Bohr also specified that the angular momentum of an electron must be a multiple of $\frac{h}{\pi}$, (hbar - $\hbar$), where h is Planck's constant. This limitation confines electrons to specific stable paths around the nucleus. Electrons are either oriented with angular momentum pointing up or down; hence, "spin up" and "spin down". The angular momentum vector cannot change orientation, i.e., it cannot point up or down. The torque causes the angular momentum vector to trace out a circle while pointing either up or down.

Building on Bohr's research, Louis de Broglie later proposed that electrons exhibit both particle and wave characteristics in motion. This duality was crucial to understanding how electrons behave on the quantum scale. De Broglie said that if electrons behave like waves, then for them to exist in a stable orbit around the nucleus, the circumference of their orbit must be an integer multiple of the wavelength.

Electrons in free space can carry quantized orbital angular momentum projected along the direction of propagation. This orbital angular momentum corresponds to helical wavefronts, or, equivalently, a phase proportional to the azimuthal angle. Electron beams with quantized orbital angular momentum are also called electron vortex beams.

The Stern-Gerlach Experiment

The Stern-Gerlach experiment is famous as it demonstrated the quantization of an electron's spin, meaning it can only be oriented in specific directions (like "up" or "down"), by showing that a beam of silver atoms when passed through a non-uniform magnetic field, splits into two distinct beams, indicating that their magnetic moments (related to their spin) are quantized and not randomly oriented as classical physics would predict; essentially, it proved that quantum mechanics is at play at the atomic level.

The Stern-Gerlach experiment is one of the most important and foundational experiments in physics. It elegantly and unexpectedly revealed a bizarre property of the quantum world: ***spin***.

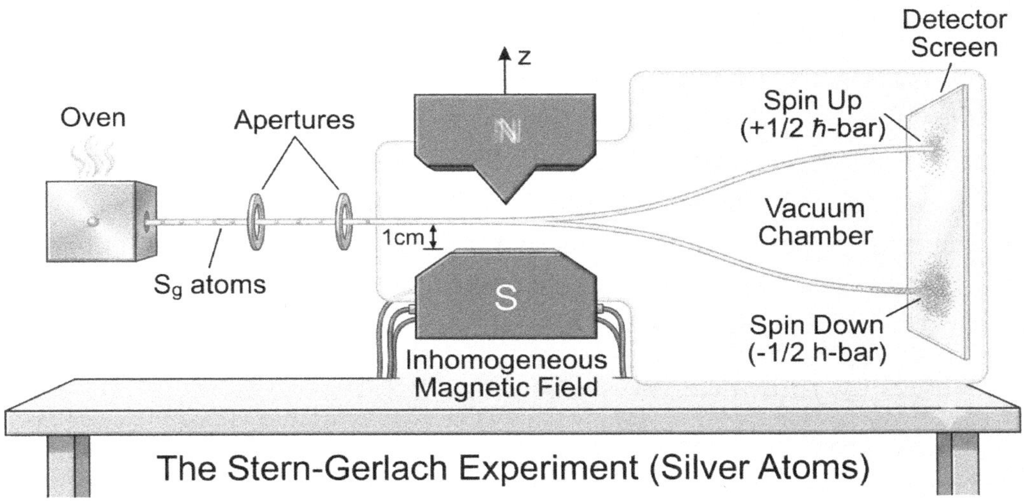

The Stern-Gerlach Experiment (Silver Atoms)

The Goal: What Were They Trying to Prove?

In the early 1920s, physicists were grappling with new quantum ideas. One idea was "space quantization," which suggested that an atom's angular momentum could only point in specific, discrete directions relative to a magnetic field. This was a radical departure from classical physics, which said it should be able to point in *any* direction.

Otto Stern and Walther Gerlach designed an experiment to test the idea. They fired a beam of atoms through a specially designed magnetic field to see what happens to them. They heated silver in an oven until it vaporized. This created a gas of individual silver atoms. This gas was allowed to escape through a small hole and then passed through a series of slits. This process, called collimation, produced a straight, thin, well-defined beam of silver atoms traveling in the same direction.

This is the most crucial part. The beam was sent through a powerful, *non-uniform* magnetic field. One pole of the magnet was knife-edged, making the magnetic field much stronger near it than near the other, U-shaped pole. This gradient is key – it means the magnet doesn't just twist the atoms; it actively pushes on them. A simple glass plate was placed at the end of the apparatus. As silver atoms from the beam struck the plate, they would build up a visible deposit, showing where they landed.

This is where the experiment becomes legendary. The classical prediction and the actual quantum result were utterly different.

The Classical Prediction

A silver atom has a single electron in its outermost shell, which gives the atom a magnetic moment, essentially making it a tiny bar magnet. Classically, these tiny atomic magnets should be randomly oriented in space – some pointing up, some down, some sideways, and every possible angle in between.

When this random collection of magnets passes through the magnetic field gradient, each one should be deflected by a different amount depending on its orientation. The atoms pointing mostly "up" would be pushed up the most. The ones pointing mostly "down" would be pushed down the most. The ones on their sides wouldn't be pushed at all. The classical prediction was that the beam would spread out into a single, continuous, smeared vertical line on the detector plate.

The Actual Quantum Result

When Stern and Gerlach ran the experiment and developed the detector plate, they did not see a continuous line. Instead, they saw two distinct, separate spots. The beam had been split cleanly in two. The result was so startling that it took time to fully understand its implications. The two spots provided the first direct, stunning proof that angular momentum is quantized. The atoms' magnetic moments could not point in any direction they wanted. They were forced to choose one of two possible discrete orientations relative to the magnetic field: "up" or "down," and nothing in between.

Initially, this quantization was attributed to the electron's orbital angular momentum. However, calculations showed that this was incorrect. It was later understood that this property was the electron's intrinsic angular momentum, something it just *has*, like mass or charge. This new quantum property was named Spin. The electron is a $spin\frac{1}{2}$ particle, meaning its spin can only be measured as "spin-up" $(+\frac{1}{2})$ or "spin-down" $(-\frac{1}{2})$.

In summary, the Stern-Gerlach experiment, designed to test a known quantum hypothesis, accidentally discovered a fundamental, previously unknown property of the universe – spin. In doing so, they changed the course of physics forever.

While the *magnitude* of spin angular momentum is fixed, its orientation is quantized when you measure it along a particular axis. If you try to measure the spin of an electron along, say, the z-axis, you will only ever find it to be in one of two possible states:

Spin Up (or $spin + \frac{1}{2}$) Often denoted as "↑" or $|\uparrow\rangle$. The spin angular momentum component along the chosen axis is measured to be $+\frac{\hbar}{2}$.

Spin Down (or $spin - \frac{1}{2}$): Often denoted as "↓" or $|\downarrow\rangle$. The spin angular momentum component along the chosen axis is measured to be $-\frac{\hbar}{2}$.

These are often called spin states. It's crucial to understand that "spin up" and "spin down" are *relative to the axis of measurement*. If you change the axis you're measuring along, the "up" and "down" directions change accordingly. Imagine a coin. It can land "heads" or "tails." Electron spin is somewhat similar – when you measure it along an axis, you get one of two "outcomes" ("spin up" or "spin down").

One of the consequences of electron spin is that it confers an intrinsic magnetic moment on electrons. Imagine a tiny bar magnet associated with each electron. A magnetic moment is a measure of how strongly an object interacts with a magnetic field. Classical electric currents create magnetic moments. For electrons, even though they aren't literally "spinning charges" in the classical sense, their spin *behaves as if* it were a tiny circulating current, creating a magnetic dipole moment. Electron spin is the fundamental origin of magnetism in materials. The magnetic properties of magnets, magnetic materials, and even atoms and molecules are determined mainly by the spins of their electrons.

Before you measure the spin of an electron, it's not necessarily in a definite "spin up" or "spin down" state. Instead, it is said to be in a quantum superposition of both states. This means it's in a combination of "spin up" and "spin down" simultaneously until a measurement is made. The act of measurement forces the electron to "choose" one of the definite spin states (either "spin up" or "spin down") along the axis of measurement. That is, the wave function collapses.

Spinors

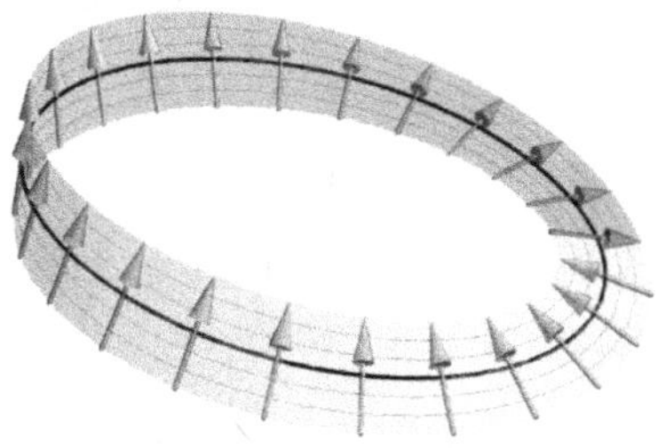

A spinor is visualized as a vector pointing along the Möbius band, exhibiting a sign inversion when the circle is continuously rotated through a complete turn of 360°. In physics and mathematics, spinors describe the intrinsic angular momentum (spin) of subatomic particles such as electrons.

While they might seem similar to vectors at first glance, they have a unique geometric property: they require 720° rotation (two full turns) to return to their original state. To understand a spinor, it helps to compare it to a standard Euclidean vector.

If you rotate a vector (like an arrow) by 360°, it points exactly where it started, but if you rotate a spinor by 360°, it actually flips its sign (it becomes negative). You must rotate it another 360° (for a total of 720°) to return it to its starting configuration.

The Plate Trick illustrates this property. If you hold a plate in your hand and rotate it 360 degrees by twisting your arm, you end up with your arm in a tangled, uncomfortable position. However, continue rotating it in the same direction for another 360° (a total of 720°), and your arm will untangle and return to its original state.

Mathematically, spinors are elements of a representation space of a Clifford Algebra. In quantum mechanics, we most commonly encounter them as Pauli Spinors (for non-relativistic physics) or Dirac Spinors (for relativistic physics). They are defined as elements of the representation space of Clifford algebras, which are crucial for understanding Lorentz transformations and the behavior of particles in quantum mechanics and relativity. For example, spinors describe electrons, which have $spin - \frac{1}{2}$ and the Dirac equation, which governs their behavior, is inherently a spinor equation.

Spinors are elements of a vector space that form a Clifford algebra. These algebras naturally incorporate the concepts of spin and rotational symmetry, making them a fundamental tool in advanced physics. They are essential for describing fundamental fermions, such as electrons, which have a spin $\frac{1}{2}$. They are also integral to relativistic quantum mechanics, enabling the development of equations such as the Dirac equation to accurately describe these particles.

Note: In 1878, William Kingdon Clifford combined Sir William Rowan Hamilton's work on quaternions with Hermann Grassmann's theory of linear extension. This led to an original system of geometric algebra that describes rotations, reflections, and scaling in multidimensional spaces – Clifford Algebras.

Spinors are foundational to the Dirac equation, which describes the motion of relativistic $spin\frac{1}{2}$ particles, such as electrons. In quantum field theory, spinors represent fundamental matter fields and are crucial for understanding particle interactions and properties. Spinors are the most fundamental objects that transform under the Lorentz group, the group of transformations in special relativity. The Pauli Spinor is a non-relativistic electron described by a two-component complex column vector:

$$\chi = \binom{\alpha}{\beta}$$

Where $|\alpha|^2$ is the probability of the electron being "spin up," and $|\beta|^2$ is the probability of it being "spin down." In relativistic quantum mechanics, the Dirac equation requires four components to account for both spin (up/down) and the existence of antimatter (particle/antiparticle). The Dirac spinor ψ satisfies:

$$(iy^{\mu}\delta_{\mu})\psi = 0$$

As promised, here's a version of the above that should be easier to understand:

How Physicists Use Spinors: Physicists use spinors to describe the behavior of matter. Every elementary matter particle has an intrinsic property called spin. Spin does not mean the particle physically rotates like a toy top. Instead, it represents a permanent, built-in type of angular momentum. Physicists categorize spinors based on how fast a particle moves:

Pauli Spinors for Slow Particles: When electrons move at standard, non-relativistic speeds, physicists use Pauli spinors. This tool uses a two-number column. One number tracks whether the electron's spin points up, and the second tracks whether it points down.

Dirac Spinors for Fast Particles: When electrons move near the speed of light, Einstein's theory of relativity applies. Fast particles follow the Dirac equation, which relies on Dirac spinors. This version requires a four-number column. Two numbers track the electron's spin-up and spin-down states. The remaining two numbers track the spin states of its antimatter counterpart, the positron.

The Mathematical Origin: In 1878, mathematician William Kingdon Clifford combined existing theories about 3D rotations and multi-dimensional spaces. His work created Clifford Algebras. This mathematical system provides the precise framework for describing how particles change, rotate, and interact in quantum mechanics and relativity.

Technical Details for Advanced Reference: Formally, spinors are elements of a representation space of a Clifford algebra. They serve as the most fundamental objects that transform under the Lorentz group in special relativity.

- ***Non-relativistic framework:*** The Pauli spinor uses a two-component complex column vector to describe a spin-1/2 particle.
- ***Relativistic framework:*** The Dirac equation demands a four-component spinor. This structure handles the two degrees of freedom for spin alongside the two degrees of freedom required for particle-antiparticle states.

The spinor nature of fermions enables the Pauli Exclusion Principle to operate, preventing matter from collapsing and explaining the structure of the periodic table. In three-dimensional space, vectors transform under the group SO(3). Spinors, however, transform under SU(2), which is the "double cover" of SO(3). A "double cover" describes a 3D physical rotations using two distinct mathematical "spins" (or states) for each rotation. This means a particle must rotate 720° (two full circles) to return to its original quantum state.

Electron spin is mathematically described using spin angular momentum operators (S), usually represented by Pauli matrices (σ) acting on a two-component spinor. It is defined by the spin quantum number $s = \frac{1}{2}$, resulting in a quantized magnitude of $\sqrt{s(s+1)}\hbar = \frac{\sqrt{3}}{2}\hbar$, and a magnetic quantum number $m_s = \pm\frac{1}{2}$ for spin-up/down states.

Spinor/Wave Function: Represented $|\psi\rangle = \begin{pmatrix} c_1 \\ c_2 \end{pmatrix}$ as a column vector (spinor) $F = m\sqrt{\frac{Gma_0}{r^2}}$, where $|c_1|^2$ and $|c_2|^2$ represent the probabilities of being in the spin-up or spin-down state.
Operators: The spin operator is $S = \frac{\hbar}{2}\sigma$, where σ are the Pauli matrices.
Eigenvalues: The z-component S_z is $m_s\hbar = \pm\frac{1}{2}\hbar$.
Framework: Arises naturally from the Dirac equation in relativistic quantum mechanics. SO(3) is the Special Orthogonal Group in three dimensions, representing all possible rotations about the origin in 3D Euclidean space.

The Pauli Exclusion Principle, based on electron spin and statistics, is fundamental to the structure of atoms and the periodic table. It dictates how electrons fill atomic orbitals. Covalent bonds are formed by pairing electrons with opposite spins. Ferromagnetism, paramagnetism, and diamagnetism all arise from the collective behavior of electron spins in materials. Spin-orbit coupling (interaction between spin and orbital angular momentum) affects atomic spectra, leading to fine structure in spectral lines.

In summary, electron spin is an intrinsic quantum-mechanical form of angular momentum. It's not literally spinning. Spin magnitude is fixed at spin-$\frac{1}{2}$, and spin orientation along a measured axis is also quantized (spin up or spin down), giving electrons a magnetic moment – the basis of magnetism. Quantum numbers describe spin states and can be in superposition. Electron spin is fundamental to atomic structure, chemical bonding, magnetism, and many modern technologies.

Electron Orbitals

Electron orbitals are central to the quantum mechanical model of atoms. An electron orbital is a 3-dimensional region around an atom's nucleus where an electron is most likely to be found. Governed by quantum mechanics, orbitals describe wave-like behaviors rather than fixed, planetary orbits. Each distinct orbital can hold a maximum of two electrons.

A set of three quantum numbers defines the specific characteristics of an orbital:

Quantum Number	Symbol	Name	Description
1.	n	Principal	Defines the electron's energy level and overall size. n can be 1, 2, 3, etc. Higher n means higher energy and larger orbital size.
2.	l	Angular Momentum	Defines the shape of the orbital. l can range from 0 to n-1. These values correspond to the orbital types: l = 0 is s, l = 1 is p, l = 2 is d, and l = 3 is f.
3.	ml	Magnetic	Defines the orbital's orientation in space. ml can range from -l to +1. If l=1 (a p orbital), ml can be -1, 0, 1, meaning there are 3 distinct p orbitals oriented along the x, y and z axes.

There are four primary orbital shapes, and each is determined by the angular momentum quantum number (l).

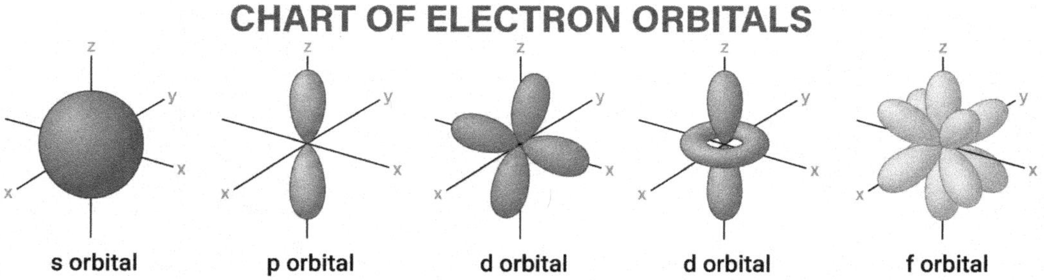

- **s** Orbitals $(l = 0)$. ***Spherical***, and the probability of finding the electron is highest at the nucleus and decreases symmetrically as the distance from the nucleus increases. Only one s orbital exists for any energy level $(ml = 0)$.
- **p** Orbitals $(l = 1)$. ***Dumbbell-shaped*** (two lobes on opposite sides of the nucleus). There are three distinct p orbitals for each energy level $(n \geq 2)$, oriented along the three spatial axes: $p + x, p_y$ and p_y.
- **d** Orbitals $(l = 2)$. Primarily ***cloverleaf-shaped*** (four lobes) or a dumbbell with a donut around the middle. There are five distinct d orbitals for each energy level $(n \geq 3)$. These are important for transition metals.
- **f** Orbitals $(l = 3)$. ***Very complex***, usually consisting of six to eight lobes. There are seven distinct f orbitals for each energy level $(n \geq 4)$. These are important for lanthanides and actinides.

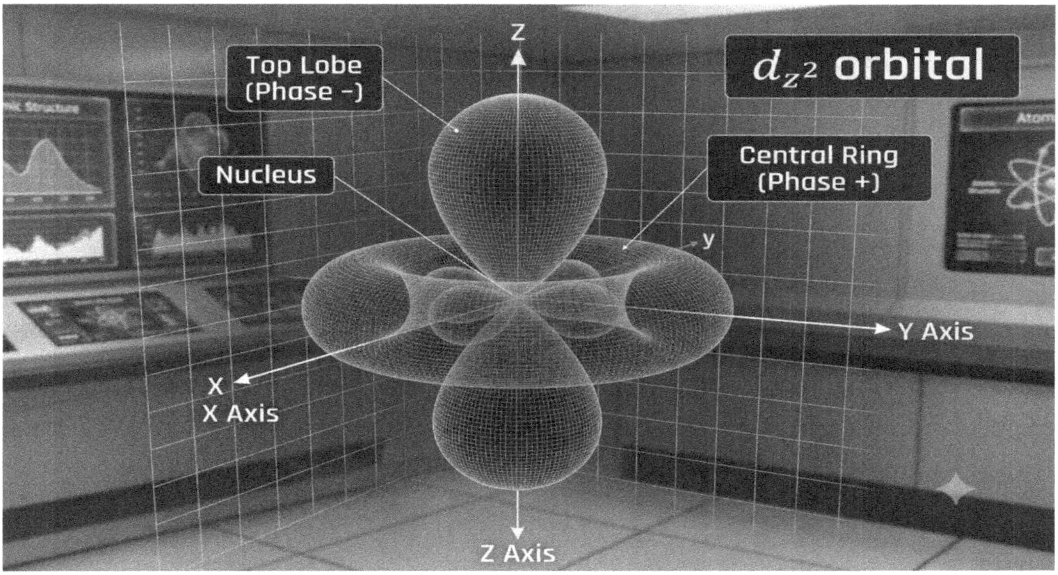

The Pauli Exclusion Principle states that no two electrons in an atom can have the same set of four quantum numbers (n, l, m). The fourth number, m_s (the spin quantum number) can only be $\frac{1}{2}$ (spin-up) or $\frac{1}{2}$ (spin-down). Each orbital (n, l, m) can hold one electron with spin up and one with spin down.

***l* Value**	**Orbital Type**	**Number of Orientations (*ml*) values**	**Total Electron Capacity (2*ml*)**
0	s	1	2
1	p	3	6
2	d	5	10
3	f	7	14

Cooper Pairs

Before we get into Cooper Pairs, let's take a quick look at superconductivity. No doubt you have heard that it usually only shows up in some special supercooled materials – with scientists striving to find "room temperature" superconductors. Even under these conditions, there are practical uses – such as quantum computing (to reduce decoherence). I didn't realize that there were so many other applications:

Magnetic Confinement Fusion: Energy companies build compact fusion reactors, known as Tokamaks, using powerful superconducting magnets. The HH70 Tokamak in China uses advanced superconducting components to confine plasma heated to millions of degrees.

Particle Accelerators: Large scientific facilities use superconducting magnets to steer and focus particle beams. The Large Hadron Collider (LHC) at CERN uses thousands of superconducting magnets chilled to near absolute zero to accelerate subatomic particles to nearly the speed of light before smashing them together.

Data Centers and Power Grids: Data facilities require immense amounts of electricity, and standard copper cables lose a lot of energy as heat. Replacing copper with superconducting cables in data centers cuts power transmission losses by up to 90%.

Medical Diagnostics: Hospitals rely on magnetic resonance imaging (MRI) scanners daily. This application represents the largest commercial use of superconductivity. These machines use superconducting coils to generate strong, perfectly stable magnetic fields that align hydrogen atoms in your body, enabling radiologists to capture detailed images of soft tissue.

Materials and Operating Temperatures: Superconducting materials require cooling to a specific threshold, known as the critical temperature, before their electrical resistance drops to zero. Engineers divide these materials into low-temperature and high-temperature categories based on the coolant required to keep them operational.

Traditional Low-Temperature Superconductors (LTS) require complex cooling systems filled with liquid helium. Despite the name, High-Temperature Superconductors (HTS) are still extremely cold. However, they operate at temperatures warm enough to allow the use of liquid nitrogen or mechanical cryocoolers, which dramatically lower operating costs.

And now back to Cooper Pairs. A Cooper pair consists of two electrons bound together at low temperatures. Physicist Leon Cooper first described this phenomenon in 1956, establishing the foundation for the Bardeen-Cooper-Schrieffer (BCS) theory of superconductivity.[82] This pairing allows electrical current to flow through certain materials with zero resistance.

In a vacuum or a standard conductor, two electrons repel each other because they carry identical negative electric charges. Inside a superconducting crystal lattice, however, an attractive force overcomes this Coulomb repulsion.

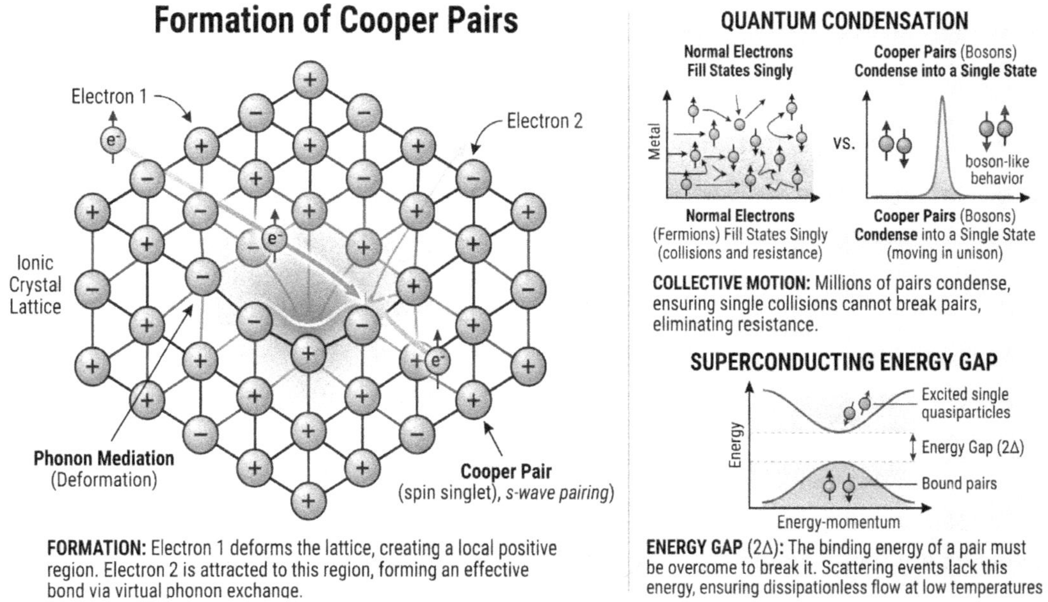

When an electron travels through the lattice, its negative charge attracts the surrounding positive metal ions. This attraction deforms the lattice, creating a localized region of higher positive charge density. Before the ions bounce back to their original positions, a second electron is drawn toward this temporary positive concentration. The quantized lattice vibration, known as a phonon, acts as the mediator that binds the two electrons together.

Since individual spin $\frac{1}{2}$ electrons obey the Pauli exclusion principle, this forces electrons in a normal metal to fill up energy levels sequentially, leading to collisions and electrical resistance. When two electrons form a Cooper pair, their spins most often align in opposite directions, creating a singlet state with a total spin of 0. This integer spin alters their basic quantum statistics, transforming the pair into a boson. Unlike fermions, unlimited bosons can occupy the same quantum state. When the temperature drops below a specific threshold, thousands of Cooper pairs condense into a single, coherent quantum ground state.

Because the condensed Cooper pairs share a single quantum wave function, they move through the crystal lattice in unison. A defect or thermal vibration that would normally scatter a single electron cannot readily disrupt the collective motion of the entire condensate. To cause resistance, an obstacle must deliver enough energy to break a Cooper pair apart. The energy required to break a pair and create two independent quasi-particles is known as the superconducting energy gap, denoted as Δ. This binding energy is remarkably weak, measuring around a few millielectronvolts (10^{-3} eV). Because thermal agitation easily disrupts this bond, conventional superconductivity occurs only below a strict critical temperature (T_c). For instance, aluminum must be cooled below 1.2 K to form Cooper pairs, while niobium requires a temperature below 9.25 K.

In March 2025, research from a group at Caltech identified a new quantum state in thin flakes of an iron-based superconductor ($FeTe_{0.55}Se_{0.45}$).[83] Using scanning tunneling microscopy on a clean, contaminant-free surface, the physicists mapped the material's superconducting energy gap at the atomic scale. They discovered that the energy gap strength modulates periodically, matching the exact spacing of the atoms in the crystal lattice. The team named this the Cooper-pair density modulation state. The variation in gap measurements reached up to 40%, indicating that Cooper-pair concentrations fluctuate at a sub-nanometer scale.

In September 2025, an experiment focused on quantum entanglement achieved the first real-time observation of Cooper pairs splitting into individual electrons.[84] The researchers connected a superconducting aluminum electrode to two isolated copper islands. When a Cooper pair split, one spin-entangled electron tunneled into the left island while the second electron tunneled into the right island. By utilizing single-electron transistors as real-time charge detectors, the team recorded these double-tunneling events on a microsecond timescale. This setup isolates the split pairs from background electron noise with 99% fidelity, offering a practical method for generating localized electron entanglement.

The Wheeler-Feynman "One Electron" Theory

The "One-Electron Universe" is a thought experiment that emerged from a 1940 conversation between physicists John Wheeler and Richard Feynman. It proposes a radical solution to one of the most fundamental mysteries in physics: why every electron in the universe has the same mass, charge, and spin.

The theory began with a phone call from John Wheeler to his graduate student, Richard Feynman. Wheeler suggested that the reason all electrons are identical is that they are all, in fact, the same particle. In standard physics, we view the universe as being populated by trillions of individual electrons. Wheeler's insight was to apply the geometry of spacetime to this problem. He proposed that a single electron traces a massive, complex path (a "worldline") that zigzags through time, spanning the entire history of the universe.

The key to making this work mathematically is interpreting antimatter. According to the Dirac equation, a positron (the antimatter counterpart of an electron) can be mathematically modeled as an electron moving backward in time. Wheeler's theory suggests that when an electron reaches a certain point in the future, it "turns around" and travels toward the past. To an observer moving forward in time, this backward-traveling electron appears as a positron. When it eventually reaches the past, it turns around and moves forward again, reappearing as an electron. The single electron's worldline crosses our "now" millions of trillions of times. Each intersection point is what we perceive as an individual electron or positron.

The primary beauty of the One-Electron Universe is that it perfectly explains the principle of indistinguishability. In quantum mechanics, electrons are not just "similar"; they are identical. You cannot label them or tell them apart. Wheeler's theory provides a physical reason for this: you cannot tell them apart because there is nothing to distinguish–they are the same object encountered at different stages of its existence.

While mathematically elegant, the theory faces an empirical hurdle. If every electron moving forward is matched by the same electron moving backward as a positron, the universe should contain an equal number of electrons and positrons. However, the universe we observe exhibits a massive "baryon asymmetry." There are vastly more electrons than positrons. When Feynman pointed this out to Wheeler, Wheeler jokingly suggested that perhaps the missing positrons were "hidden inside protons." While the theory of a single electron is not accepted as a literal description of reality today, it laid the groundwork for Feynman's development of Feynman Diagrams, where antiparticles are still shown as particles moving backward in time.

Spintronics

Spintronics exploits the intrinsic angular momentum (spin) of electrons, in addition to their fundamental electric charge, to create new forms of electronic devices. While conventional electronics use the electron's charge to store and process information (representing a bit as a flow of current or a voltage level), spintronics uses the electron's spin state – spin-up (↑) or spin-down (↓) - to represent the 1s and 0s of binary data.

It relies on three fundamental processes: The first is to create an electron current in which the spins are polarized (i.e., most electrons have their spins aligned in the same direction (↑ or ↓). This can be achieved using ferromagnetic materials. Once the spin is polarized, the spin state must be controlled. This can be achieved using electric or magnetic fields, or by using specific quantum materials that exhibit strong spin-orbit coupling (the interaction between an electron's spin and its motion). The device must be able to read spin information, using a magnetoresistive device. This means that its electrical resistance changes with the orientation of the electron's spin and the material's internal magnetization.

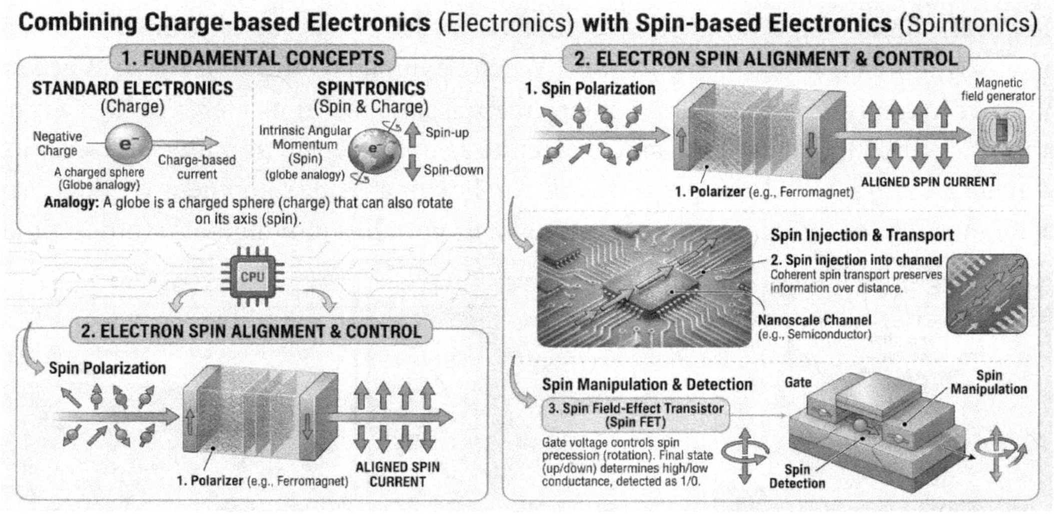

Spintronics offers advantages that could dramatically change computing – not least by reducing power consumption for server farms needed for the expansion of Artificial Intelligence. Information stored in spin orientation remains stable even when power is turned off (like a magnetic hard drive), eliminating the need for constant power to refresh memory (like DRAM). Moving and manipulating spin requires less energy than moving charge, leading to faster switching speeds and much lower power consumption. Spintronic devices can store a lot more data in a smaller area than current charge-based memory cells.

The most successful commercial application is Magnetoresistive Random-Access Memory (MRAM), which uses the change in resistance observed in a magnetic tunnel junction. Companies are now mass-producing STT-MRAM (Spin-Transfer Torque MRAM), which uses a spin-polarized current to switch the magnetic orientation, making it extremely fast, energy-efficient, and non-volatile.

Spintronics is a critical component in the race to build practical quantum computers. The two spin states (↑ or ↓) naturally serve as a qubit. Scientists are working on silicon quantum dots and nitrogen-vacancy (NV) centers in diamond to trap and manipulate single electron spins. These systems are highly coherent, making them promising candidates for quantum hardware.

The first likely consumer devices to adopt spintronics will be those that require high-speed, low-power, non-volatile memory and advanced processing capabilities – which means integrating MRAM and spintronic-enabled logic chips.

As technology matures, spintronics will move into logic and sensor applications, enabling new device capabilities. Current processors separate memory and logic, which creates a performance bottleneck known as the Von Neumann bottleneck. Spintronic logic aims to integrate these functions into next-generation computing and specialized AI accelerators. This will enable computation to occur directly where the data is stored (in-memory computing or spin-based logic gates), resulting in much faster processing and a substantial reduction in energy loss due to data traffic across the chip.

Spintronic principles are already improving magnetic sensors used in industrial settings, and these improvements will filter down to consumer technology. Spintronic sensors are exceptionally sensitive to magnetic fields, enabling more precise, smaller, and lower-power magnetometers for applications like advanced navigation without a clear satellite signal.

In the long term, using electron spin as a qubit in quantum computing should lead to commercial quantum machines, such as specialized quantum processors offered as cloud services or built into research-grade computers.

Summary

Chapter Nine attempted to explain how amazing the electron really is. It explained how fundamental it is to our world of matter and how it interacts. It explained angular momentum, spin, and other characteristics of the electron. We saw that you need to "rotate" an electron 720° a – two full turns – to return it to its starting position.

We discovered that electrons are not plums in an atomic pudding, nor do they orbit the nucleus like planets around a star. Instead, they exist in orbitals – a 3-dimensional region around an atom's nucleus where an electron is most likely to be found.

We learned how Cooper pairs allow electrical current to flow through certain materials with zero resistance. We saw the emerging field of Spintronics exploits electron spin to create new forms of electronic devices. And finally, we read about the strange and somewhat amusing "one-electron theory".

Chapter Ten: The Great Divide

A Tale of Two Universes

At the most immediate level, our experience consists of tangible objects – tables, stars, animals, and familiar everyday artifacts – distributed across what appears to be empty space and unfolding through an apparently uniform, linear progression of time. Yet, when we look closely using the frameworks of both general relativity and quantum mechanics, nature reveals itself: our intuitive conceptions of matter, space, and time are not elementary features of reality, but rather emergent estimates that break down at the fundamental scales. The great divide between the flawless shape of spacetime described by relativity and the probabilistic, discontinuous behaviors of quantum particles exemplifies what many see as a continuing gap within modern physics.

This divide not only confronts the unity of our scientific worldview, but also compels us to recognize the limitations of common sense as an interpretive tool, revealing the necessity of finding a more integrated conceptual framework that reconciles the large-scale and microscopic orders of nature.

The first lesson on our journey proved that "empty space" is not empty. Quantum Field Theory showed us that the vacuum is a quantum froth of virtual particles constantly being created and annihilated. What we perceive as solid matter – the electrons in your hand, the photons hitting your eye – are temporary excitations in those underlying fields. Like waves in the sea, particles emerge, interact, and disappear back into the sea. We are not solid creatures moving through a static universe; we are ripples moving through a world that is fundamentally interconnected and full of activity.

The second lesson was that "certainty" is an illusion. Heisenberg's Uncertainty Principle and the double-slit experiment showed us that the universe refuses to be pinned down. It exists in a state of superposition, a blur of possibilities that only becomes a definite reality when it is compelled to do so. We are not neutral observers watching the universe from outside. Asking nature a question – measuring a position or checking a spin – forces nature to provide a specific answer, collapsing the wave of probability into the hard fact of existence.

The third lesson is that time is not absolute. As I leave you behind on my trip to Alpha Centauri at near light-speed, our clocks will not agree. For me, only a few years have passed, but on Earth, decades have passed, and you are no longer young. We cannot even agree on when events occur. I may see two simultaneous lightning bolts flash from the sky, whereas you, in a different reference frame (the moving train), see one, followed by the next. Although we disagree, we are both correct. And then we find that gravity, the weakest of the four fundamental forces, can become so strong that nothing, not even light, can escape its grip – we are witnessing black holes. Time and space (spacetime) warp and twist in ways we can't imagine, and gravity and quantum physics need to come together to fix the Great Divide.

The quantum and relativistic nature of time, as best we can understand them today, is not a clear, straightforward "whole". Luckily, we experience time from one moment to the next; anything else would (for us) be extremely strange and confusing. Our world-view may prevent us from ever grasping the true nature of time. Is it smooth and continuous, or discrete and quantized? Imagine looking at a glass of water. It appears to be a smooth, clear liquid. But if you could see every individual molecule, you wouldn't see smoothness but a frenzied blur of bouncing particles. We see the "smooth" water because we aren't capable of seeing the details at the quantum level.

Because we cannot know the exact state of every field in the universe, we perceive the "blur" of their tiny interactions as entropy (a simplistic view, but one we can work with). We experience this statistical blurring as the forward advance of time. In this view, the past is simply the low-entropy state we can remember, and the future is the high-entropy fog we are moving toward.

So what is the cosmic reason, and what are the rules? The rules, as we saw, are the symmetries of the fields – the gauge invariances that tell the ocean waves how to move. But the reality we live in – the world of cause and effect, of "before" and "after", of solid ground and shining stars – is a perspective. It is the specific way our consciousness interacts with the complexity of the quantum soup.

The quantum and relativistic universe shatter our "common sense" assumptions, forcing us to rethink the relationship between the observer and the observed entity. Before these theories, the prevailing philosophy held that the universe was a Newtonian machine in which every cause had a predictable effect. Relativity challenged this by removing the "objective" center of the universe. There isn't a "God's eye view" of time or space; every perspective is equally valid, meaning "now" is a local concept rather than a universal truth.

Quantum Mechanics introduced indeterminacy – that nature is fundamentally probabilistic, and the future is not "written" in the present. This has reopened age-old debates about free will versus a predetermined reality. Or, as we saw, perhaps we live in a fully deterministic universe where free will is an illusion. Perhaps quantum probability is resolved by the hidden variables so beloved by the physicist Albert Einstein.

What is Reality?

Perhaps the deepest philosophical shift is the realization that we cannot describe the world without including ourselves in the description. In the quantum realm, the properties of a particle (like its position) may not even exist until a measurement occurs. This challenges Realism – the belief that an objective world exists independently of our perception.

Classic Western philosophy is built on Atomism: the idea that you can understand a system by breaking it into its smallest parts, whereas entanglement suggests a Holistic universe. If two particles on opposite sides of the galaxy are linked (entangled) instantaneously, the universe is "non-local." However, Conformal Gravity theories and the Pilot-Wave theory may be correct in assuming that Quantum Mechanics is incomplete, or we are interpreting it incorrectly.

We evolved to understand the world at the "middle scale" – the world of hunting, gathering, and solid objects. Our language (subject-verb-object) and our logic (A cannot be B) struggle to describe a reality in which a thing can be both a wave and a particle, in which space and time are one and the same, and in which the vacuum, "nothingness," is actually a boiling sea of energy.

Philosophically, we have come to understand that human intuition is a local tool, not a universal one. We are like ants trying to understand the internet; we can measure its effects, but the "true" nature of the system might be beyond the limits of our minds to comprehend.

Is There A *Great Divide*?

Is there truly a Great Divide, or are we just hoping that gravity needs to be quantized? I'm open to either possibility, depending on what the best theories and evidence show. Maybe we don't need gravitons at all, and gravity and quantum mechanics could fit together in a single unified theory. There are good arguments on both sides, so let's look at some ideas from people who think gravity doesn't have to be quantum.

The idea that gravity is fundamentally classical, sometimes called the "non-quantum gravity" or "postquantum" hypothesis, is becoming a more popular alternative to theories like string theory or loop quantum gravity. Supporters of this view say that space and time should stay smooth and classical, while matter stays quantum.

Several theoretical physicists support this idea, including Jonathan Oppenheim from University College London. He is one of the main creators of the modern "postquantum theory of classical gravity." He believes gravity is not quantized, but instead has its own randomness. This randomness lets gravity interact with quantum matter without leading to mathematical problems.

Lajos Diósi, a theoretical physicist at the Wigner Research Center for Physics in Hungary, has supported the idea of classical, random (stochastic) gravity for many years.

Freeman Dyson, the well-known theoretical physicist who passed away recently, argued that gravity and quantum mechanics might actually describe two separate parts of physics. He often questioned whether we really need to unify them and doubted that anyone would ever detect a single graviton. I'm a fan of Dyson, so I take his opinions seriously.

This idea goes back to the famous physicist Niels Bohr. He believed that measuring devices and the background of space and time are always classical, so he argued against the need to quantize gravity. Like with Einstein, you should be cautious if you want to disagree with Niels Bohr – or if he would disagree with you.

Philip Mannheim's Conformal Gravity also suggests that both theories can fit together in a single, unified framework. It might even show that our ideas about non-locality are not quite right.

The physics community is still divided. Many researchers are searching for a quantum theory of gravity, but supporters of classical gravity have suggested real, small-scale experiments to test if gravity shows quantum effects like entanglement.

Personally, I hope there is still much more to discover, and that we will eventually bridge the gap between quantum theory and general relativity. I also hope you weren't expecting a clear answer or disappointed that we didn't find one. I hope you enjoyed learning about the discoveries, experiments, and the ongoing search to close this gap.

The Future

We have come a long way in the last two hundred years, but I don't have a crystal ball for what's to come. Scientists aren't great at predicting the future. Here are two great examples of not appreciating human capabilities:

a) As we saw, Einstein predicted the existence of gravitational waves in 1916 as part of his general theory of relativity. He later doubted they could ever be detected, believing they were too weak to measure. In 2015, LIGO successfully detected gravitational waves from a black hole merger, proving Einstein's original 1916 theory correct and his later doubts wrong.

b) The neutrino was proposed by Wolfgang Pauli in 1930. Because the particle had no electric charge and virtually no mass, Pauli believed it would interact too weakly with matter ever to be seen. He famously lamented: "*I have done a terrible thing, I have postulated a particle that cannot be detected.*" It was finally discovered in 1956 by American physicists Clyde Cowan and Frederick Reines.

Two geniuses for sure, but they couldn't predict the incredible advances in science, technology, and our thirst for answers. Many of us will have seen bold predictions about household robots, bases on the Moon and Mars by the 1970s or 1980s. After fifty years of expectations, robots are starting to fulfill the promise, and fusion energy might actually be only twenty years away. So I will avoid using my crystal ball too much when we look at the future of quantum theory and gravity.

The Challenges Ahead

1. Quantum Mechanics & Quantum Gravity
2. Black Holes
3. Dark Matter and Dark Energy
4. Hubble Tension
5. String Theory & Supersymmetry
6. The Restaurant at the End of the Universe

Quantum Mechanics & Quantum Gravity

Which interpretation is correct, and will we ever find definitive proof? This is a tough one because most of the theories have something going for them, but finding the experiment or the evidence to say "this is the correct one" is frustratingly difficult.

While the Many-Worlds Interpretation has become very popular, I am not a fan. It seems very unlikely that it can ever be proven, and it's overkill to solve the problem because we don't understand the nature of wave-function collapse. Having said that, all kudos to Hugh Everett III, who proposed it in 1957 and called it the "relative state formulation". If you recall, the MWI eliminates the need for a "collapse", treating the entire universe as a single, deterministic, quantum-mechanical system governed solely by the Schrödinger equation. As much as I still don't like it, I can't argue that it was a very novel concept and deserves respect.

I admit I am a fan of the less accepted Pilot-Wave Interpretation, but that's based more on my view of how clever and elegant it seems, and my desire to see non-locality explained in different ways.

The only other interpretation I am going to mention is the Copenhagen Interpretation, because it was first, is still taught in schools, and many physicists find it easier to go with it than to lose sleep over what it all really means.

Developed primarily by Niels Bohr and Werner Heisenberg in the 1920s, it posits that a quantum particle doesn't have a definite, fixed state (like position or momentum) until it is directly measured or observed. Instead, it exists simultaneously in a "cloud" of multiple probabilities, known as a superposition. However, it relies on the somewhat ambiguous nature of what constitutes an "observation." It suggests that a fundamental, physical boundary exists between a probabilistic quantum system and the measuring device (which is classical and produces definitive results).

Neither Albert Einstein nor Erwin Schrödinger were big fans. Einstein famously stated in a letter that he refused to believe "God plays dice with the universe." Schrödinger's (who created the wave equation) proposed his famous cat thought experiment to highlight the absurdity of the Copenhagen interpretation. There are still those that believe quantum mechanics is incomplete, and that it can be combined with general relativity without requiring gravity to be quantum.

So what about quantum gravity, the graviton, those little strings, or loops? I would be thrilled to see solutions that demonstrate that General Relativity is indeed quantum in nature. The problem, as I see it, is that we are now dealing with the Planck scale and the impossibility of observing what is happening at those extreme scales.

What gives me some optimism is that, while detecting gravitons seems out of our reach, as was assumed to be the case with gravitational wave detection, ideas are emerging that could provide evidence of their existence. Before that eventuality, reading books on Loop Quantum Gravity, Gravity from Qubits, and other theories remains fascinating.

Ψ

I think there is plenty of scope for new theories, new experiments, and opportunities to prove (or disprove) interpretations. Does it matter? Why yes, it does. Who knows what the discovery might reveal about the world we live in and how it may affect the human race?

Whether gravity is quantum or not, the Great Divide will remain until we can pin down which one it is, why, and how it operates in our universe. It's all very exciting, in my humble opinion, and I hope we haven't hit too many dead ends (but I like philosophy too!).

Black Holes

I am optimistic about future findings because black hole astrophysics has moved from a theoretical science to an observational era powered by next-generation global networks and advanced computing. Just as Einstein doubted we could ever detect gravitational waves, technology has proven such previous skepticism wrong.

The Event Horizon Telescope (EHT) no longer just takes static photos. Multi-year imaging campaigns have revealed chaotic, shifting magnetic field flips around the supermassive black hole *M87**. The EHT data connects the glowing ring of a black hole directly to the base of its powerful plasma jets, providing a deep testing ground for physics. The upcoming Africa Millimeter Telescope (AMT) in Namibia will add critical baselines to Earth's network, enabling the creation of real-time, high-definition black hole movies. LIGO-India officially broke ground, adding a crucial fifth detector to the planet's gravitational-wave array to drastically improve sky localization and pinpoint exactly where black holes collide.

Black holes sit at the intersection where the General Theory of Relativity and quantum mechanics clash. The influx of data on Hawking radiation and event horizons brings physicists closer than ever to a unified "Theory of Everything".

New studies suggest supermassive black holes naturally mimic the high energies of particle colliders. Scientists can use them as natural laboratories to potentially hunt for elusive dark matter particles. Researchers are optimizing tools to locate ancient, "primordial" black holes born seconds after the Big Bang, which might entirely account for the universe's missing dark matter. If Dark Matter exists as a unique "particle", I think it's likely that it will somehow be linked with black holes. If it interacted with primordial black holes (PBHs), it might even tell us something new about why we see such large black holes before the onset of galaxies?

In the next edition, I suspect there will be a new chapter on how dark matter and normal matter black holes come together under the influence of intense gravitational forces such that the other fundamental forces can't influence the outcome of them forming extremely powerful black holes. There will be many questions, I'm sure. I was tempted to spend more time on this, but I'd like to see more results from the physics community.

Dark Matter and Dark Energy

As astronomical data and particle theory improve, I feel optimistic that we will solve these puzzling problems. I know they've been around for a long time, with no resolution, but sometimes all it takes is a new idea or observation.

My optimism is helped by the fact that we are approaching the problems from multiple angles. A discovery in one area that narrows down the possibilities and is likely to lead to better, more focused theories in those other avenues. It seems possible that dark matter could be detected experimentally, and that would be real science – observations and theories validated by experiment. I am hoping that, somehow, the elusive right-handed neutrinos will play a role in the puzzle of dark matter and the matter/antimatter asymmetry after the Big Bang.

Dark Energy is a little more intangible, but it seems likely that we will continue to hone observations and theories until we have at least a plausible addition to the Standard Model of Cosmology. How Dark Matter and Dark Energy intersect with Quantum Mechanics and General Relativity (and I assume they must) will likely be another incredible step forward in our understanding of how our universe operates.

The Hubble Tension

It has been less than thirty years since the discovery that the expansion of the universe is accelerating. That has created the Hubble Tension puzzle, but I foresee scientists finding a way to reconcile the disparate observations and calculations, bringing them into agreement. Once that part of the puzzle is solved, they can focus on what dark energy is and why it exists as it does.

Quantum Field Theory suggests that the vacuum is teeming with virtual particles and zero-point energy. If you add up all this energy, the density is gargantuan. Based on the observed expansion of the universe (governed by Dark Energy), the actual energy density of the vacuum is incredibly small. The prediction and the observation differ by approximately 120 orders of magnitude.

It's even possible that we may find new physics and that Einstein's General Relativity equations may require some modifications (though I am dubious).

Neutrinos

The world of neutrino physics is still busy with theories and experiments to improve our knowledge of these "ghost particles". While we have moved past the initial discovery of neutrino oscillation, we are now grappling with fundamental questions about the particle's role in the origin of matter and its strange relationship to the laws of spacetime.

We know from oscillation experiments that there are three neutrinos, each with a different mass. However, we don't know the ordering of these masses. Resolving this is crucial because it influences our understanding of how galaxies formed and how much "weight" neutrinos contribute to the universe's total mass. Every other fermion (like the electron or quark) is a Dirac particle, meaning it has a distinct antimatter counterpart. However, because neutrinos are electrically neutral, they could be Majorana particles – meaning the neutrino and the antineutrino are actually the same thing.

Physicists are looking for Neutrinoless Double Beta Decay, and, if observed, it would prove that neutrinos are their own antiparticles and that Lepton numbers are not conserved. If neutrinos are Majorana particles, this provides a natural explanation for their tiny masses via the See Mechanism (heavy right-handed neutrinos "balancing" the light left-handed ones).

In the Standard Model, we only ever observe "left-handed" neutrinos. If neutrinos are Majorana particles, "right-handed" neutrinos must exist, but they would be incredibly heavy – perhaps near the GUT scale (10^{16} GeV). These particles are "unknown" because they are far too heavy to be produced in any collider. We can only "see" them indirectly through the way they influence the masses of the light neutrinos we can measure.

The Cosmic Neutrino Background is the "holy grail" of observational cosmology. While the CMB gives us a snapshot of the universe at 380,000 years old, the CvB was formed just one second after the Big Bang. Even though it’s an incredibly difficult task, I have some hope that we will be able to detect it and transform cosmology into a precision science for the first second of existence. Even if we cannot directly observe “right-handed” neutrinos, new data may provide indirect evidence of their existence.

Supersymmetry (SUSY)

While it remains one of the most elegant solutions to the Standard Model's shortcomings, the fact that the Large Hadron Collider has not yet detected a single "sparticle" has forced a massive re-evaluation. The primary reason physicists expected to find supersymmetry at relatively low energies, but did not, is the Hierarchy Problem.

In the Standard Model, the Higgs boson is "unnatural." Because it interacts with every particle that has mass, quantum corrections from those particles should mathematically drive the Higgs mass up to 10^{19} GeV. For the Higgs to remain at its observed 125 GeV, the universe's constants would have to be "fine-tuned" to a staggering degree – like balancing a needle on its tip for 13.8 billion years.

Supersymmetry solves this by introducing "superpartners" (such as photinos, gluinos, and squarks) that provide equal and opposite quantum corrections, naturally canceling the runaway mass and keeping the Higgs light. If SUSY were "natural," these partners should have appeared at roughly the same mass scale as the Higgs and the top quark.

So why do I remain optimistic about supersymmetry and the chances of detecting supersymmetric particles? Maybe the discovery will not be made, at least not in my lifetime. Still I think something will show up, either in a new theory or an experiment, and provide a way forward – perhaps helping string theory move forward once more. It's an elegant theory, but nature is not obliged to be "pretty".

SUSY is a mathematical framework, not a single model. Verifying SUSY might require us to move beyond the TeV scale into the "High-Energy Frontier" of exploration. I assume, optimistically, that such powerful colliders would make many discoveries, advancing our overall knowledge.

The End of the Universe?

The Big Freeze: Currently favored by most cosmologists, this scenario occurs if the universe continues to expand indefinitely at its current accelerated rate. But that may change if we do confirm the expansion rate is slowing.

The Big Rip: If the repulsive force of dark energy grows stronger, it could trigger a violent conclusion. The accelerating cosmic expansion becomes so aggressive that it overpowers all other forces of nature. Dark energy would pull apart clusters of galaxies, dismantle galaxies, and tear solar systems away from their host stars. In the final moments, the outward tearing force would rip planets apart, break molecular bonds, and completely shred individual atoms and spacetime itself.

The Big Crunch: If dark energy fluctuates or fades over time, gravity could once again become the dominant force in the universe. If the outward expansion slows down, halts, and reverses, the universe will contract. Galaxies will collide and pack closely together, causing cosmic radiation to heat the universe to catastrophic temperatures. In this reverse of the Big Bang, all matter, stars, and planets will be compressed into a single, infinitely dense singularity. One possibility is a Big Bounce, which could trigger a brand-new Big Bang.

Vacuum Decay: Unlike the other endings, an unpredictable quantum event could happen at any moment. The Higgs field might not be in its lowest possible energy state, meaning our universe rests in a temporary, "metastable" vacuum. If a random quantum tunneling event causes a tiny patch of space to drop into a true, lower energy state, it would create a bubble of "true vacuum", instantly expanding outward in all directions at the speed of light, and obliterate everything. Ouch!

Conformal Cyclic Cosmology: Sir Roger Penrose posits that the universe has neither a beginning nor an end. Instead, it undergoes an infinite succession of cosmic cycles called aeons, where the far-future "heat death" of one universe seamlessly becomes the "Big Bang" of the next. Unlike traditional cyclical models that require a physical contraction like the Big Crunch, Penrose's theory relies entirely on a clever geometric transition based on the physics of scale.

Summary

I remain optimistic. The world's space programs are more vibrant. We have the Hubble and James Webb space telescopes showing us the wonders of the universe, and we can expect even more from the Vera Rubin Observatory, the Giant Magellan Telescope, the ELT, and the Argus Array. We can look forward to the Nancy Grace Roman space telescope, the PLATO and SPHEREx projects, and the Compton Spectrometer and Imager. And Artemis will take us back to the Moon.

In theoretical physics, we can expect to see advances in Black Hole theories, quantum gravity, graviton detection, new ways to revisit string theory, and complex Dark Matter/Dark Energy models.

We should see new gravitational wave data from LIGO undergoing analysis, and tech companies shifting focus from simply having many qubits to demonstrating "logical qubits" - multiple physical qubits working together to correct their own errors, which could be critical for future quantum computers. There will certainly be significant technological advances, including spintronics and photonics.

I don't see any reason to be disappointed with the progress in cosmology or with the idea that we're nearing the end of new theoretical physics. We have no idea when (or where) the next Da Vinci, Newton, Einstein, or Hawking might show up, or what paradigm shift they may bring to the table.

Ψ

The future looks pretty exciting from where I stand.

Final Thoughts

In this investigation I have aimed to give a thorough account of quantum theory and general relativity, explaining both what we currently understand and the areas that are still unresolved in overcoming The Great Divide. Our effort to gain knowledge in Quantum and the Cosmic Realms is in the end about discovering patterns and coherence in what at first appears to be disorder. It is now evident that even beneath the surface of quantum uncertainty there are deep structural regularities which govern the behavior of the universe. Although the desire to unify physics usually focuses on the quantization of gravity, real advances might require entirely new points of view or conceptual frameworks that have not yet been imagined. In the end, facing the unknown forces us not only to improve our theories, but also to consider our own intellectual limitations – and maybe to stay open to the possibility that the answers we are looking for could challenge the very basis of our understanding of reality.

We are just like Valentine Michael Smith – people who are strangers in a strange land. Gravity gives us with the entire universe so that we can live in it and explore it. We act as a mechanism through which the universe can observe itself. The macroscopic world comes about as a result of blurring the quantum details. Instead of having infinite precision we have the possibility of possessing a history, a narrative, and a life. The mystery of the quantum world is not a hindrance to understanding; it is in fact the basis of our existence. We are the ripples that have learned to look at the ocean and wonder where the waves originate.

If you have enjoyed this book, you should definitely go on to read some of the other excellent works mentioned in the bibliography.

Background Materials

Abelian

An "Abelian" (named after Niels Abel) group is a group in abstract algebra in which the operation is commutative, meaning the order of elements does not affect the result. This contrasts with a non-abelian group, such as the group of 2×2 matrices under multiplication, where order matters. Examples of Abelian groups include the set of integers under addition or the set of non-zero rational numbers under multiplication.

The crucial property is that the group operation is commutative $(a \times b = b \times a)$ for all elements a and b in the group. All subgroups of an abelian group are normal, and every element forms its own conjugacy (a fundamental equivalence relation) class. Examples include:

- ***Additive Integers*** $(\mathbb{Z})$***:*** The set of all integers $\cdots, -2, -1, 0, 1, 2, \cdots$ under addition. Since $x + y = y + x$ it is the most foundational example of a countably infinite cyclic group.
- ***Multiplicative Rationals*** $(\mathbb{Q}^x)$***:*** The set of non-zero rational numbers under multiplication. Because the product of two fractions is the same regardless of order, it satisfies the commutative property.
- ***Integers Modulo*** n $(\mathbb{Z}_n)$***:*** A finite group consisting of the remainders when dividing by n (e.g., hours on a clock for $n = 12$). Addition performed modulo n is always commutative.

A non-example is the general linear group $GL_n(\mathbb{R})$ of $n \times n$ invertible matrices with real entries for $n \geq 2$, which is not abelian because matrix multiplication is not generally commutative. While most commonly associated with groups, the term "abelian" also appears in other mathematical contexts. For instance, an Abelian category is a category where morphisms can be added, and where kernels and cokernels exhibit specific, desirable structural properties; the category of abelian groups $(\mathbf{Ab})$ is the prototypical example. Similarly, an Abelian variety is a projective algebraic variety that also functions as an algebraic group, standing as a fundamental object of study in algebraic geometry. Finally, an elementary abelian group is a finite abelian group where every non-identity element has the same prime order p.

A classic non-example is the group of $n \times n$ invertible matrices with real numbers, written as $GL_n(\mathbb{R})$ for sizes of 2×2 or larger. This group is not abelian because changing the order of matrix multiplication usually changes the final answer. While the word "abelian" is most famous for describing commutative groups, it is also used in other areas of advanced math:

- ***Abelian Category:*** A mathematical framework where maps can be added together, and where pieces can be broken down cleanly using generalized versions of "kernels" (finding what maps to zero).
- ***Abelian Variety:*** A smooth, geometric shape defined by polynomial equations that also works like an abelian group. You can add points on the shape together, and it is a key concept in geometry.
- ***Elementary Abelian Group:*** A finite abelian group where every element (except the identity/zero) has the same prime-number order. This means if you repeat the group's operation a specific prime number of times on any element, you always land back at the starting point.

An elementary abelian group is a finite group where every non-identity element has a prime order$p$. Mathematically, it is always constructed by taking the direct product of a finite number of cyclic groups of order p. We write this structure as:

$$(\mathbb{Z}_p)^k \quad \text{or} \quad \prod_{i=1}^{k} \mathbb{Z}_p$$

- p is a prime number (like (2, 3, 5)).
- $\mathbb{Z}_p$ is the cyclic group of integers modulo (p).
- k is a positive integer representing the number of copies being multiplied together.
- *Total Size*: The number of elements (cardinality) in the group is exactly p^k.

Example: If $p = 2$ and $k = 2$, you get $(\mathbb{Z}_2)^2$, which has $2^2 = 4$ elements. This is famously known as the Klein Four-Group.

In abstract algebra and geometry, "Abelian" structures are defined by their predictability, balance, and order. Whether looking at the multiplicative rationals or abelian categories, the ability to swap the order of operations without changing the outcome provides a highly stable mathematical environment. Conversely, non-abelian structures like the matrix group $GL_n(\mathbb{R})$ introduce complexity because order matters. Understanding the constraints of these groups forms the backbone of modern algebra and geometric analysis.

Alpha Particle

An alpha particle (symbol α) is a positively charged particle made of two protons and two neutrons, identical to the nucleus of a helium-4 atom. Alpha particles have a net spin of zero. Their kinetic energy varies, with higher energy for larger nuclei (~7 MeV) and lower energy for smaller nuclei (~3 MeV). The average kinetic energy is around 5 MeV, and the velocity is approximately 4% the speed of light. Because it lacks electrons, an alpha particle has a net electrical charge of 2+. However, the particles readily accept electrons from other matter and quickly become regular electrically neutral helium atoms ($^4_2$He).

Alpha particles produced by radioactive decay have a lower penetration depth than beta or gamma radiation. All it takes is a few centimeters of air, a sheet of paper, or a layer of skin to stop an alpha particle. Alpha particles that form from ternary fission are much more energetic and penetrate about three times further. Alpha particles, which are cosmic rays, are so energetic that they can pass through the human body or several meters of radiation shielding. Similarly, alpha particles produced in particle accelerators are highly energetic and highly penetrating.

An alpha particle is a helium nucleus $\frac{4}{2}He^{2+}$ with a mass number of 4 and an electrical charge of +2. They are relatively heavy and slow compared to other forms of radiation, like beta and gamma rays. Alpha particles are highly ionizing, meaning they can easily knock electrons out of atoms they interact with. The most common source of alpha particles is the radioactive decay of heavy, unstable atomic nuclei, such as uranium, radium, and plutonium. This process transforms the original nucleus into a new element with an atomic number two less than the original. Alpha particles are also produced in nuclear fusion reactions in stars and are found in cosmic rays.

As with other particles, there is an antimatter version of an alpha particle, which is the anti-alpha. An International team of researchers detected the anti-alpha using the STAR detector at Brookhaven National Laboratory in 2011.[85] The experiment involved colliding gold ions moving nearly at the speed of light. At the time, the anti-alpha or antimatter helium-4 nucleus was the heaviest antimatter particle.

Alpha particles are used in several applications. Many household smoke detectors use a small amount of an alpha-emitting isotope (Americium-241) to ionize the air in a chamber. Smoke disrupts this ionization, triggering an alarm. Alpha-emitting radioactive materials, such as Plutonium-238, are used as power sources in radioisotope thermoelectric generators for some spacecraft. Some static eliminators use alpha emitters to ionize air and reduce static electricity.

Beta Decay

Beta decay is a fundamental type of radioactive decay in which an atomic nucleus spontaneously transforms, changing its composition and emitting a beta particle (either an electron or a positron) and a neutrino or antineutrino. It's one of the primary ways unstable isotopes achieve a more stable configuration.

The weak nuclear force governs the process of radioactive decay. There are two main types of beta decay: Beta-minus (β−) decay and Beta-plus (β+) decay.

Beta-Minus (β−) Decay is the most common form of beta decay. It occurs in nuclei that have an excess of neutrons relative to protons, making them unstable. To achieve stability, one of these excess neutrons transforms into a proton. A neutron (n0) within the nucleus effectively "decays" into a proton (p+). To conserve electric charge (a neutral neutron becomes a positive proton), a negatively charged electron (e -) is created and immediately expelled from the nucleus. This electron is the beta-minus particle. An electron antineutrino (v¯e) is also emitted. This particle has no electric charge and very little (or zero) mass, and it carries away energy and momentum, thereby conserving them.

The fundamental reaction is: n0 ⟶ p++e−+v¯e. The Atomic Number (Z) increases by 1 (because there's one more proton). The Mass Number (A) remains unchanged (because a proton replaces a neutron, so the total number of nucleons stays the same). Since the atomic number changes, the atom transforms into an atom of the next element up in the periodic table.

Example: Carbon-14 Decay

Carbon-14 (^{14}C), with 6 protons and 8 neutrons, is an unstable isotope used in carbon dating. It undergoes beta-minus decay to become Nitrogen-14 (^{14}N).

$$\frac{14}{6}C \rightarrow \frac{14}{7}N + e^{-} + v + 0.1565MeV$$

Here, a neutron in C-14 decayed into a proton, turning it into N-14, and an electron (beta particle) and an antineutrino were emitted.

Beta-Plus (β+) Decay (Positron Emission) occurs in nuclei with an excess of protons over neutrons, making them unstable. To achieve stability, one of these excess protons transforms into a neutron. A proton (p+) within the nucleus transforms into a neutron (n0). To conserve electric charge (a positive proton becomes a neutral neutron), a positively charged positron $(e+)$is created and immediately expelled from the nucleus. This positron is the beta-plus particle. A positron is the antimatter counterpart of an electron. An electron neutrino (*ve*) is also emitted, carrying away energy and momentum. The fundamental reaction is: p+⟶n0+e++ve. The Atomic Number (Z) decreases by 1 (because there's one less proton). The Mass Number (A) remains unchanged. The atom transforms into an atom of the previous element down in the periodic table.

Example: Carbon-11 Decay

Carbon-11 ($^{6}{}^{11}C$), often used in medical imaging, undergoes beta-plus decay to become Boron-11 ($^{5}{}^{11}B$). $^{6}11C \longrightarrow 115B+e++ve$ Here, a proton in C-11 became a neutron, turning it into B-11, and a positron (beta particle) and neutrino were emitted.

In Beta Decay, the Weak Nuclear Force is the only fundamental force capable of changing the "flavor" of quarks (the constituent particles of protons and neutrons). In beta decay, a down quark transforms into an up quark (for β− decay) or an up quark transforms into a down quark (for β+ decay). Despite the dramatic transformations, several fundamental quantities are always conserved: The total electric charge before and after the decay is the same. The total mass number of protons and neutrons (nucleons) remains the same. The Lepton Number, a specific quantum number related to electrons, neutrinos, and their antiparticles, is conserved. Energy and momentum are also conserved, which is why the neutrino/antineutrino is necessary; without it, the emitted electron/positron would have a fixed energy, but experiments show a continuous energy spectrum.

Beta decay is a crucial process in astrophysics (e.g., the life cycles of stars) and in practical applications such as nuclear medicine (PET scans, radiotherapy) and radioactive dating.

Bose-Einstein Condensate

Sometimes called the "fifth state of matter", a Bose-Einstein Condensate (BEC) only exists in extremely cold environments. In our everyday world, atoms in a gas behave like tiny, distinct billiard balls, moving around and bouncing off one another. But strip away their thermal energy – cooling them to within a few billionths of a degree above absolute zero – and their behavior shifts dramatically. Their individual "wave-packets" begin to stretch and expand. Eventually, these waves become so large that they overlap and merge, causing thousands, or even millions, of individual atoms to lose their separate identities and settle into the same quantum state. They cease to be a collection of particles and instead become a single, unified "super-atom."

The Indian physicist Satyendra Nath Bose and Albert Einstein predicted the possibility of this state in the mid-1920s. They realized that the subatomic world is divided into two classes of particles: Fermions and Bosons. Fermions are governed by the Pauli Exclusion Principle, which prevents any two of them from occupying the same state. Given enough cold and enough quiet, bosons like helium-4 atoms or photons will happily pile on top of one another. When this happens, the microscopic quantum world, which is usually hidden from our eyes, suddenly expands to a macroscopic scale. We can actually see the collective wave behavior of millions of atoms acting in perfect unison.

This behavior leads to strange properties, such as Superfluidity, in which the condensate can flow without any friction. If you were to place a Bose-Einstein Condensate in a cup and start it spinning, it would theoretically never stop. It could even "crawl" up the sides of the container and over the rim to escape, driven by its own internal quantum pressure. In a Bose-Einstein Condensate, the distinction between "this atom" and "that atom" disappears. There is only the field, behaving as a single, harmonious entity.

Cherenkov Radiation

Cherenkov radiation is the phenomenon that occurs when a charged particle travels through a transparent medium faster than the speed of light. While nothing can travel faster than the speed of light in a vacuum, light slows down when it passes through substances like water, glass, or plastic. This creates a "speed limit" within the material that high-energy particles can sometimes exceed.

When a charged particle, such as an electron, moves through an insulating material (a dielectric), it polarizes the molecules along its path. As the particle passes, the molecules quickly return to their original state, emitting a pulse of electromagnetic radiation. Under normal circumstances, these pulses interfere destructively, canceling each other out. However, if the particle is moving faster than the speed at which light propagates in that medium, the electromagnetic waves "pile up" and interfere constructively. This creates a coherent shock wave of light, much like an aircraft creates a sonic boom when it exceeds the speed of sound.

The most famous visual manifestation of this effect is the blue glow seen in the cooling pools of nuclear reactors. This happens because high-energy electrons (beta particles) are emitted during nuclear fission and travel through water at speeds exceeding the speed of light in water (roughly 75% of c). The blue color results from the radiation's frequency. The intensity of Cherenkov radiation is proportional to the frequency of the light. More energy is emitted at shorter wavelengths (blue and ultraviolet) than at longer wavelengths (red), making the light appear distinctly blue to the human eye.

The condition for Cherenkov radiation is defined by the particle's velocity (v) and the refractive index (n) of the medium:

$$v > \frac{c}{n}$$

The angle at which the light is emitted (θ) is also related to the velocity of the particle, forming a cone of light behind the particle:

$$cos(\theta) = \frac{c}{nv}$$

Cherenkov radiation is more than just a visual curiosity; it is a critical tool in experimental physics:

- ***Nuclear Safety:*** The intensity of the blue glow in a reactor pool is directly related to radioactivity, allowing operators to visually confirm whether a reactor is active or whether spent fuel rods remain highly radioactive.
- ***Particle Detectors:*** Physicists use "Cherenkov counters" to identify the speed and charge of subatomic particles. By measuring the angle of the produced light cone, they can calculate exactly how fast a particle is moving.
- ***Astrophysics:*** Large-scale experiments, such as the IceCube Neutrino Observatory in Antarctica, use ice to detect neutrinos. When a neutrino interacts with the ice, it can sometimes produce a high-speed muon which generates Cherenkov radiation, that allows sensors to track the neutrino's origin.

Compton Effect / Wavelength

The Compton Effect is a phenomenon that provided early evidence for the particle nature of light (photons) and confirmed that energy and momentum are conserved during particle-photon interactions. The effect occurs when a high-energy photon collides with a charged particle, usually a stationary electron.

An incoming photon strikes the electron, and the photon transfers some of its energy and momentum to the electron, causing the electron to recoil. The original photon is scattered at a new angle (θ) with reduced energy. Since a photon's energy (E) is related to its wavelength (λ) by $E\frac{hc}{\lambda}$, a loss of energy means the scattered photon must have a longer wavelength (λ') than the incident photon (λ). This is the Compton Shift.

Arthur Compton's 1923 experiment successfully measured this shift, which could not be explained by classical wave theory (which predicted the electron would absorb the wave energy and re-radiate light of the *same* wavelength). Compton's result provided definitive proof that light behaves as discrete particles (photons) with measurable momentum $(p = \frac{h}{\lambda})$ cementing the concept of wave-particle duality as central to quantum mechanics.

The Compton Wavelength is a physical constant that characterizes the shift in a particle's observable wavelength during Compton. The change in wavelength depends only on the scattering angle (θ) and is defined by the following formula:

$$\Delta\lambda = \lambda' - \lambda = \frac{h}{me^{c}}(1 - cos\theta)$$

The term $\frac{h}{m_e c}$ is the Compton Wavelength (λ_c) of the electron. The Compton Wavelength (λ_c) is the characteristic length scale for the quantum mechanical scattering of a photon off a particle of mass m. For an electron (m_e), the value is:

$$\lambda_c = \frac{h}{m_e c} \approx 2.43 \times 10^{-12} \text{ meters}$$

The Compton Wavelength represents the wavelength of a photon whose energy ($E = \frac{hc}{\lambda_c}$) is precisely equal to the rest-mass energy ($E = mc^2$) of the electron.

More broadly, it defines the scale at which a particle's mass becomes detectable in light scattering. If a photon has a wavelength significantly larger than λ_c, it behaves primarily as a wave (classical interaction); if its wavelength is similar to or smaller than λ_c, it behaves mainly as a particle (quantum interaction).

Degenerate Matter

While bosons are social particles that like sharing the same space, fermions – such as the electrons, protons, and neutrons that make up the bulk of our visible world – are governed by the Pauli Exclusion Principle. This principle states that no two fermions can occupy the same quantum state simultaneously – it forces atoms to maintain a certain distance and volume.

However, inside the dying remnants of stars, where gravity becomes a crushing force, this principle is pushed to its breaking point. It creates a state of matter so dense that our classical understanding of "solids" fails to describe it. When a star like our Sun exhausts its nuclear fuel, it can no longer generate the outward pressure necessary to hold back the pull of gravity. As the star collapses, its atoms are crushed until the electrons are stripped away and packed together as tightly as the laws of physics allow.

In this state, matter is packed so densely that a sphere the size of a city can contain the mass of two Suns. The entire star behaves like a single, giant atomic nucleus. If the star is massive enough, even the electron degeneracy pressure will eventually fail, leading to an even more extreme transition. Gravity overcomes the electrons and forces them to merge with protons, turning the entire core into a massive ball of neutrons – a Neutron Star, where Neutron Degeneracy Pressure will now support the core.

At this point, the star becomes a White Dwarf, and its stability is maintained by Electron Degeneracy Pressure. Because electrons cannot occupy the same space, they begin to move at incredible speeds to find "empty" quantum states to inhabit. This high-speed movement creates a "pressure" that halts the star's collapse. A teaspoon of white dwarf material would weigh as much as a mountain on Earth, yet it remains held apart because the electrons must obey the Pauli Exclusion Principle. If the mass of the star increases any further, even the neutrons cannot hold their ground, and the star will collapse into a black hole. The laws of degeneracy fail to support the core.

De Sitter Space

A "de Sitter space" is a theoretical model of spacetime with a positive curvature. It describes a universe that is rapidly expanding, driven by a "cosmological constant" that acts like a repulsive force, pushing everything apart. It can be described as resembling the surface of a balloon, where parallel lines eventually converge. It's relevant to understanding cosmic inflation and the universe's current accelerating expansion driven by dark energy.

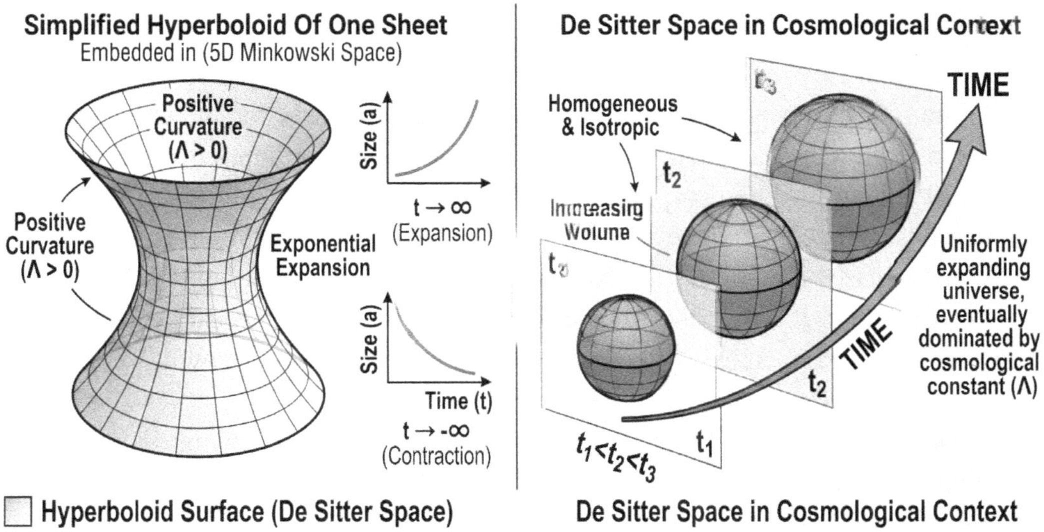

I think an explanation of flat spacetime might be helpful. When physicists say spacetime is flat, they don't mean the universe resembles a flat sheet of paper. They mean that the geometry of the universe follows standard, predictable rules because there is no gravity warping it.

Here's an analogy that might help: Think about a flat wooden tabletop. If you draw two parallel lines across the table, those lines will stay exactly the same distance apart forever. If you roll two marbles along those lines, they will travel straight ahead and never collide. This is Euclidean geometry – the standard math I was taught at school in England.

Contrast this with a curved surface – a globe, or a balloon. If you and a friend stand at the equator a few miles apart and both walk due north – perfectly parallel to each other – you will eventually bump into each other at the North Pole. The curvature of the Earth forces your paths to bend together, even though neither of you turned. Mass and energy bend the fabric of the universe. When a massive object like the Sun sits in space, it creates a dip in this fabric, much like a heavy bowling ball resting on a trampoline – we looked at this earlier. This warping is what we perceive as gravity. When spacetime is flat, there are no massive objects nearby to cause these dips. The cosmic grid remains perfectly straight and uniform.

In a perfectly flat region of spacetime, the laws of physics operate without the complications of gravity. Light travels in absolute straight lines – a laser beam fired across the vacuum will never bend or deflect. Parallel paths remain parallel – two probes launched side-by-side into the void will maintain their exact spacing indefinitely. We cannot, of course, get away without time entering the picture: Time moves at a uniform rate. A clock sitting on a spaceship ticks at the same speed as an identical clock floating miles away.

We cannot find perfectly flat spacetime on Earth because the planet's mass warps the space around us, pulling objects toward the ground. To experience flat spacetime, you must travel into the deep void between galaxies. Imagine an astronaut floating in the empty space between the Milky Way and the Andromeda galaxy. If the astronaut releases a hammer, it will float completely still right next to them. If they push the hammer, it will travel in a perfectly straight line at a constant speed forever. Because there are no stars or planets nearby to warp the environment, the astronaut is experiencing flat spacetime.

Physicists use a specific mathematical framework called Minkowski space to describe this unwarped universe. It serves as the baseline for Albert Einstein's theory of special relativity, which calculates how speed and time interact when gravity is taken out of the equation.

A de Sitter universe is characterized by a positive cosmological constant and a positive energy density, whereas if both are negative, it's called an anti-de Sitter universe. The latter has been found useful in a variant of string theory, known as AdS/CFT (Anti-de Sitter/Conformal Field Theory), which features the Holographic Principle as its primary characteristic.

Eigenvalues & Eigenvectors (Eigenstates & Eigenfunctions)

Eigenvalues and eigenvectors are fundamental concepts in linear algebra that can seem abstract at first, but they have a very intuitive and powerful meaning. Imagine having a machine that can stretch, shrink, and rotate objects. In mathematics, this "machine" is a matrix, and the "things" it acts on are vectors (which you can think of as arrows with a direction and length).

When you put a vector into the machine (i.e., multiply it by a matrix), it usually comes out pointing in a new direction and having a new length. However, for any given transformation, there are almost always a few "special" vectors. When you put these special vectors into the machine, something amazing happens: they come out pointing in the same direction as they went in. They might get stretched, shrunk, or even flipped, but their direction along their original line is unchanged.

An eigenvector is one of these "special" vectors whose direction is not changed by the transformation. The corresponding eigenvalue is the simple number that tells you how much the eigenvector was stretched, shrunk, or flipped.

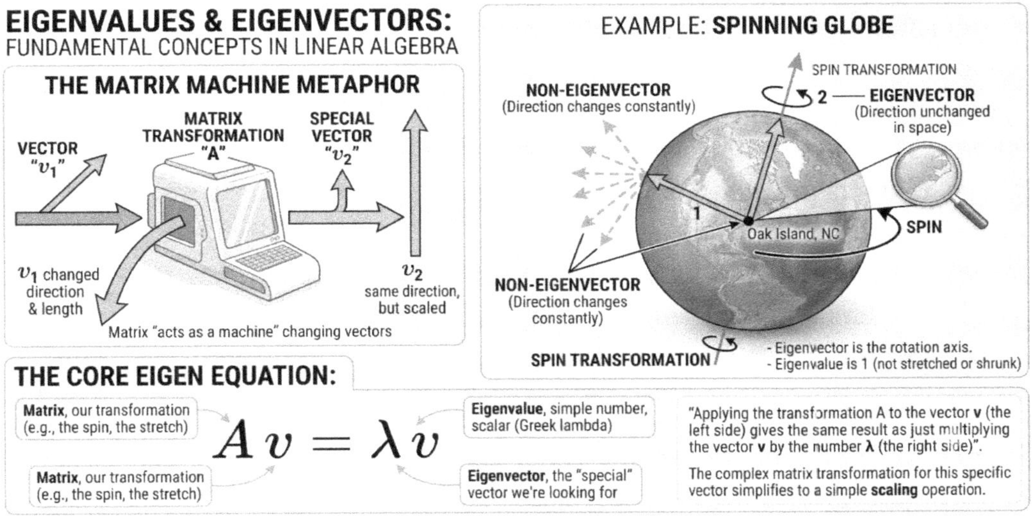

Imagine a globe of the Earth. Now, spin it around its axis (the line from the North Pole to the South Pole). This spin is your transformation. Let's look at a few vectors originating from the center of the globe: A vector to Oak Island, NC: As the globe spins, this vector is constantly changing direction. It points towards the wall, then the window, then the door, and so on. This is not an eigenvector. A vector pointing straight up to the North Pole: As the globe spins, this vector continues to point straight up. Its direction in space *remains unchanged*. It just spins on itself. This is an eigenvector. The eigenvalue is the stretching factor. Since the vector's length didn't change, the stretching factor is 1. So, for a spinning globe:

- An Eigenvector is the rotation axis.
- The Eigenvalue is 1 (because it's not stretched or shrunk).

This core idea is captured in a single, elegant equation: $A_v = \lambda_v$.

- A is the matrix, our transformation (e.g., the spin, the stretch).
- v is the eigenvector, the "special" vector we're looking for.
- λ (the Greek letter lambda) is the eigenvalue, just a simple number (a scalar).

This equation reads: *"Applying the transformation A to the vector v (the left side) gives the same result as just multiplying the vector v by the number λ (the right side)."* The complex matrix transformation for this specific vector simplifies to a simple scaling operation.

Eigenvalues and eigenvectors are crucial in almost every field of engineering and science. In Quantum Mechanics, the Schrödinger equation is an eigenvalue equation. The eigenvalues correspond to the allowed, quantized energy levels of an atom or molecule (like the energy levels of an electron in an atom). The eigenvectors (called eigenfunctions here) describe the probability of finding the electron in a particular location.

For those who like the gory details: In quantum mechanics, eigenstates and eigenfunctions form the mathematical framework connecting physical observables (like position, momentum, energy, and spin) to real-world measurements.

An eigenstate is a state vector ($|\psi\rangle$) in Hilbert space that remains unchanged in direction when acted upon by a physical operator ($\hat{A}$). An eigenfunction ($\psi(x)$) is the spatial wave function representation of that eigenstate. The eigenvalue (λ) produced by the operation represents the precise scalar value measured in a laboratory experiment: $\hat{A}\psi(x) = \lambda\psi(x)$.

Quantization of Measurements

- Every physical quantity corresponds to a Hermitian operator. (A Hermitian operator is a special type of linear operator that equals its own conjugate transpose: $\hat{A}^\dagger = \hat{A}$. In matrix notation, taking the conjugate transpose († or "dagger") means swapping rows with columns (transposition) and taking the complex conjugate of every entry (i $\rightarrow -i$). For a matrix A, this is written as A$^\dagger = (A^*)^T = A$.)

  When measuring an observable, the system collapses into one of that operator's eigenstates, and the measurement yields the corresponding eigenvalue.

Superposition and Expansion

Arbitrary quantum states $|\Psi\rangle$ are generally not pure eigenstates of a given operator. Instead, they exist as superpositions (linear combinations) of basis eigenstates:

$|\Psi\rangle = \sum_n c_n|\psi_n\rangle$. The squared modulus of the expansion coefficient ($|c_n|^2$) gives the exact probability of measuring eigenvalue a_n.

Stationary States (Energy Eigenstates)

Eigenfunctions of the Hamiltonian operator ($\psi = E\psi$) represent stationary energy states. Their spatial probability densities ($|\psi(x)|^2$) remain time-invariant, governing atomic stability.

Where They Are Used in Practice

- ***Atomic and Molecular Physics (Energy Levels):*** Solving the time-independent Schrödinger equation for hydrogen yields energy eigenfunctions labeled by quantum numbers (n, l, m). These dictate atomic orbital shapes (s, p, d, f) and electron shell energies. Transitions between energy eigenvalues produce discrete spectral lines.
- ***Quantum Computing and Information (Qubits):*** Qubit states are expressed as superpositions of the computational basis eigenstates $|0\rangle$ and $|1\rangle$, which are eigenvectors of the Pauli-Z operator ($\hat{\sigma}_z$) :

  $$\hat{\sigma}_z|0\rangle = +1|0\rangle, \quad \hat{\sigma}_z|1\rangle = -1|1\rangle$$
- ***Quantum Optics and Laser Physics (Coherent States):*** Laser light is described using coherent states, which are eigenstates of the photon annihilation operator ($\hat{a}$).
- ***Condensed Matter and Solid State Physics (Bloch Waves):*** Electrons moving through a periodic crystal lattice occupy Bloch states – eigenfunctions of the spatial translation operator. These determine band structures, distinguishing conductors, semiconductors, and insulators.
- ***Nuclear Magnetic Resonance (NMR) and MRI (Spin States):*** Nuclear spin states align as eigenstates of applied magnetic field operators. Radiofrequency pulses drive transitions between spin energy eigenvalues, forming the foundation of medical MRI scans.

Energy

Potential Energy: Stored energy and the energy of position.

Chemical energy is energy stored in the bonds of atoms and molecules. Batteries, biomass, petroleum, natural gas, and coal are examples of chemical energy.

Mechanical energy is energy stored in objects by tension. Compressed springs and stretched rubber bands are examples of stored mechanical energy.

Nuclear energy is energy stored in the nucleus of an atom–the energy that holds the nucleus together. Large amounts of energy can be released when the nuclei are combined or split apart.

Gravitational energy is energy stored in an object's height. The higher and heavier the object, the more gravitational energy is stored. When a person rides a bicycle down a steep hill and picks up speed, the gravitational energy is converted to motion energy.

Kinetic Energy: Motion of waves, electrons, atoms, molecules, substances, and objects.

Radiant energy is electromagnetic energy that travels in transverse waves. Radiant energy includes visible light, X-rays, gamma rays, and radio waves. Light is one type of radiant energy. Sunshine is radiant energy, which provides the fuel and warmth that make life on Earth possible.

Thermal energy, or heat, is the energy that comes from atoms and molecules moving in a substance. Heat increases when these particles move faster. Geothermal energy is the heat energy in the earth.

Motion energy is energy stored in moving objects. The faster an object moves, the more energy is stored. It takes energy to get an object moving, and energy is released when an object slows down. Wind is an example of motion energy. A dramatic example of motion energy is a car crash–a car comes to a total stop and releases all of its motion energy at once in an uncontrolled instant.

Sound is energy moving through substances in longitudinal (compression or rarefaction) waves. Sound is produced when a force causes an object or substance to vibrate. The energy is transferred through the substance in a wave. Usually, the energy in sound is lower than in other forms of energy.

Electrical energy is delivered by tiny, charged particles, called *electrons*, that move through a wire. Lightning is an example of electrical energy in nature.

Entropy

"Ah, entropy. It's just not what it used to be."

The Second Law of Thermodynamics states that systems naturally evolve toward states of maximum multiplicity (Ω), which is why disorder (high entropy) increases over time. The most current and widely accepted explanation of entropy comes from statistical mechanics and is fundamentally based on probability and information theory. Ludwig Boltzmann first formalized this explanation. The modern definition of entropy is not simply "disorder" but rather a measure of the number of possible microstates (specific arrangements of atoms and molecules) that correspond to a given macrostate (observable properties such as temperature and pressure).

The famous Boltzmann formula quantifies this: $S = kln\Omega$
Where:

- S is the entropy.

- k is the Boltzmann constant (the link between the microscopic and macroscopic worlds).
- ln is the natural logarithm.
- Ω (Omega) is the multiplicity or the number of micro-states corresponding to the observed macro-state.

The most abstract and current explanation connects entropy to information, as formalized by Shannon and applied to physics by Landauer. Entropy is a measure of the amount of information required to describe a system's micro-state perfectly. A high-entropy system (like a hot gas) requires more information (more bits) to specify the exact location and momentum of every single atom. The universe's evolution toward higher entropy is its evolution toward a state in which less information is observable or can be extracted.

UNDERSTANDING ENTROPY: THE INEVITABLE INCREASE IN DISORDER

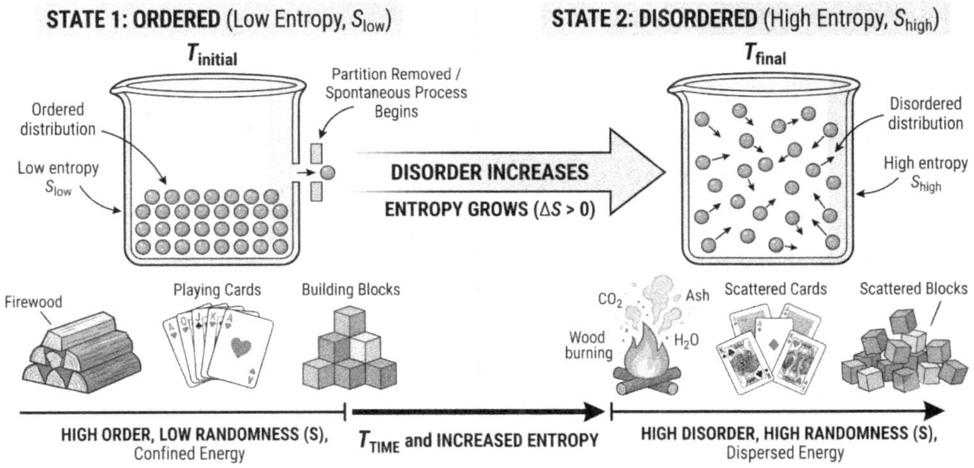

While the statistical-mechanical view is the consensus, other theoretical and conceptual perspectives complement it or seek to extend the understanding of entropy. One perspective argues that the universe's true entropy is dominated by gravity rather than by ordinary matter and radiation. Jacob Bekenstein and Stephen Hawking showed that the maximum entropy a region of space can have is proportional not to its volume but to the surface area of its event horizon. The early universe was incredibly smooth and low in entropy. The increase in entropy since the Big Bang is largely due to the formation of gravitationally bound structures, particularly black holes. A single black hole contains far more entropy than all the stars that formed it. In this view, the cosmological arrow of time is determined by the universe's increasing ability to form and grow black holes.

The Thermal Time Hypothesis is a speculative idea proposed by physicists such as Carlo Rovelli. It proposes that time is not fundamental but an emergent property that arises when we view the universe statistically (thermodynamically). The "flow" of time is identical to the flow toward thermal equilibrium, namely the increase in entropy. There is no fundamental "t" (time); there is only the statistical evolution of the system's micro-states. This relates the Second Law of Thermodynamics directly to the fundamental nature of time itself.

Although, in its strict form, it has been largely disproven by the Pusey–Barrett–Rudolph theorem, some perspectives argue that entropy has a subjective element. The Observer Dependence view suggests that the entropy we measure depends on the specific variables we choose to track. If an observer had perfect knowledge (infinite information) about every micro-state, they might conclude the universe's entropy never fundamentally changes. This view is generally considered incomplete because fundamental thermodynamics shows that the change in entropy is an objective physical reality, regardless of an observer's knowledge.

The idea that entropy provides the arrow of time – the fundamental reason why time flows forward and not backward–is one of the most profound concepts in physics, linking thermodynamics, cosmology, and philosophy. This concept evolved through several key historical stages:

The journey began with the establishment of the Second Law of Thermodynamics, based purely on macroscopic observations of heat engines and natural processes. Sadi Carnot (1824) laid the groundwork by studying the efficiency of heat engines. His work, focused on the idea that heat flows spontaneously from hot to cold, suggested a natural, irreversible direction for thermal processes. Rudolf Clausius (1850s) formalized the concept of entropy (ΔS) as a quantifiable property. He said that the total entropy of an isolated system can only increase over time $(\Delta S \geq 0)$. This provided the first formal, irreversible constraint on natural processes, making entropy a directional concept. Clausius later famously summarized: "The energy of the universe is constant. The entropy of the universe tends to a maximum." This established entropy as a measure of the disorder or unavailability of energy to do work.

The connection between entropy and the direction of time was solidified when Ludwig Boltzmann linked entropy to the probability of microscopic states. Boltzmann argued that the increase in entropy was not a strict, deterministic law, but a statistical inevitability. He defined entropy based on the number of possible microscopic arrangements (Ω micro-states) that correspond to a macroscopic state (macro-state) using his formula $S = kln\Omega$. Boltzmann explained that systems naturally evolve from low-probability (ordered) states to high-probability (disordered) states simply because there are vastly more ways for a system to be disordered than ordered. This provided the first physical mechanism for the arrow of time: time flows in the direction of increasing probability.

Physicist Ernst Zermelo raised a powerful objection, noting that since classical mechanics is time-reversible, any closed system must eventually return to its initial state (Poincaré Recurrence Theorem). If the system returns to its original (low-entropy) state, how can entropy always increase? Boltzmann countered that while recurrence is theoretically possible, the time required for a large system (like a box of gas) to return to its low-entropy state spontaneously is astronomically vast–far longer than the age of the universe. Thus, the Second Law holds for all practical purposes.

Lemaître and Hubble's 1920s discovery that the universe is expanding implied a definite starting point: the Big Bang. This raised the question: Why did the universe start in an extremely low-entropy state? If the universe were in thermal equilibrium (maximum entropy) at the start, there would have been no processes, no structures, and no arrow of time.

Roger Penrose (1970s–1980s) highlighted that the universe's initial low-entropy state is the most extreme fine-tuning problem in cosmology. He argued that the initial state was characterized by incredibly low gravitational entropy (high spatial order). The increase of gravitational entropy (the clumping of matter to form stars, galaxies, and eventually, black holes) is the dominant process driving the cosmic arrow of time.

The consensus today is that the Thermodynamic Arrow of Time is understood to be a consequence of the Cosmological Arrow of Time: Time flows forward because the universe started in an extremely low-entropy, highly ordered state (the Big Bang) and is still moving toward the state of maximum entropy (thermal equilibrium, or "heat death"). The irreversible increase in entropy provides the directionality that gives all our experiences – memory, aging, growth, and cause and effect–their unidirectional nature.

Gamma Ray (Radiation)

Gamma radiation (symbol γ) is a form of electromagnetic radiation, just like visible light, radio waves, microwaves, infrared, ultraviolet, and X-rays. What distinguishes it is its extremely high frequency, which yields very high energy and a very short wavelength. Unlike alpha and beta particles, which really are particles (helium nuclei and electrons/positrons, respectively), gamma-rays are pure energy (photons). They are produced during nuclear processes.

The two main scenarios for gamma ray emission are:

1. Nuclear De-excitation: After an atomic nucleus undergoes radioactive decay or is formed in a nuclear reaction (such as fusion or fission), it is often left in an "excited state" with excess energy. Just like an electron in an excited orbital will drop to a lower energy level and emit a photon of light, an excited nucleus will drop to a more stable, lower energy state by emitting a gamma ray photon.

Imagine a ball bouncing down a set of stairs. When it's at the top, it has potential energy. It drops to a lower step, releasing some of that energy as sound or heat. An excited nucleus is like the ball on a higher step. It emits a gamma ray to drop to a lower, more stable step.

The gamma ray emission doesn't change the number of protons or neutrons in the nucleus – the atom's atomic number and mass number remain the same, meaning the element does not change. It becomes a more stable version of itself.

2. Particle Annihilation: When a particle meets its corresponding antiparticle (e.g., an electron meets a positron), they annihilate each other, converting their entire mass into pure energy in the form of high-energy gamma ray photons.

Since they are photons, they have no mass and no electric charge. This is a crucial difference from alpha and beta particles. They have the highest frequencies and shortest wavelengths in the electromagnetic spectrum. Because they are chargeless and massless, gamma rays interact very weakly with matter. They can easily pass through human tissue and even thick layers of concrete or lead. Despite being uncharged, gamma rays are a form of ionizing radiation. When a high-energy gamma photon interacts with an atom, it can knock electrons out of their orbits, creating ions. This ionization can damage biological cells and DNA, leading to radiation sickness, cancer, or genetic mutations.

Applications of Gamma Radiation

- ***Medical Treatment (Radiotherapy):*** Precisely targeted gamma rays (e.g., from Cobalt-60 sources or linear accelerators) are used to destroy cancerous tumors.
- ***Medical Imaging (PET Scans):*** The annihilation of positrons produces gamma rays, which can be detected to create images of metabolic activity in the body.
- ***Astronomy:*** Gamma-ray telescopes detect gamma rays from cosmic events such as supernovae, black holes, and neutron-star collisions, providing insights into the universe's most energetic phenomena.

In summary, gamma radiation is pure, high-energy electromagnetic radiation emitted by nuclear processes. Its lack of mass and charge gives it immense penetrating power, making it both a dangerous form of radiation and an incredibly useful tool in medicine, industry, and scientific research.

Gauge Theory

To understand gauge theory, we need to step back and review how it became an essential foundation for physicists and mathematicians today. Unfortunately, I had to mix and match explanations to give the reader a clear understanding so some information is repeated.

Group Theory and Gauge Theory as the Language of Symmetry

In the modern physicist's view of the cosmos, symmetry is far more than an aesthetic quality; it is the structural scaffolding from which all physical laws are derived. In theoretical physics, forces are not arbitrarily "added" to a universe that was previously static. Instead, they are revealed as logical necessities – requirements imposed by the universe's insistence on internal consistency across space and time.

Group theory provides a rigorous mathematical framework for describing these symmetries, defining the rules for transforming a system while preserving its essential nature.

A mathematical group is defined as a set of elements G and a binary operation (often called the "circle operator" or group law). To form a logically rigorous group, the structure must satisfy four non-negotiable axioms:

1. ***Closure:*** Operating on any two elements in the group must produce another element within that same group.
2. ***Associativity:*** The order in which we group operations does not change the outcome: $(a \cdot b) \cdot c = a \cdot (b \cdot c)$.
3. ***Identity Element:*** There must exist a unique element e that leaves others unchanged $(a \cdot e = a)$. A classic counterexample is the set of positive real numbers under addition $(R+,+)$; because 0 is not included, the identity is missing, and the structure fails to be a group.
4. ***Inverse Element:*** Every element a must have a unique inverse a^{-1} that will return the identity. For example, the set of integers under multiplication fails here; while 1 is the identity, the inverse of 2 is 1/2, which is not an integer.

We further distinguish between Abelian groups, in which the order of operations does not matter $(a \cdot b = b \cdot a)$, and non-Abelian groups, in which it does. Electromagnetism, governed by the $U(1)$ group, is Abelian – a simple phase shift is commutative. However, the nuclear forces are governed by $SU(2)$ and $SU(3)$ non-Abelian groups, where the "knobs" we turn are matrices. Because matrix multiplication is generally non-commutative $(AB \neq BA)$, the sequence of transformations fundamentally alters physical interactions, leading to the complex, "sticky" behavior of the weak and strong forces.

Matter-Antimatter Symmetry

The fundamental difference between an electron and its antimatter counterpart, the positron, is the direction of complex phase evolution over time. While the electron's phase rotates clockwise in the complex plane, the positron's phase rotates counterclockwise. This reversal in the "direction of the clock" implies that matter and antimatter must possess equal and opposite charges. It is a huge "So What?" because charge is not an arbitrary label but a consequence of how a particle's internal phase vibrates over time.

Physics distinguishes between Global Symmetry (redefining the phase the same way everywhere) and Local Symmetry (changing the phase "here" but not "there"). While global symmetry is a simple relabeling, local symmetry is a violent act. It introduces gradients – spatial and temporal changes – that disrupt the field's energy and momentum.

The Lagrangian Breakdown

The Lagrangian (L) is the "seed" of physical law. When we apply a local $U(1)$ phase transformation to the Dirac Lagrangian, we create a mathematical crisis. The four gradients of the phase field $(\partial_\mu \theta)$ create a "mess" of extra energy and momentum terms that the original equation cannot balance. The symmetry is broken, and the theory appears to fall apart. To restore order, we must introduce the Photon field (A_μ). This field acts as the "picker-upper" of the mathematical mess. We redefine our derivative as a Covariant Derivative D_μ, defined as $D_\mu = \partial + i\frac{q}{\hbar c}A_\mu$. This new operator ensures that any "mess" created by the change in phase $(\partial_m u\theta)$ is exactly canceled out by a corresponding change in the gauge field A_μ. Thus, the electromagnetic force is not an "expansion pack" for the universe; it is a logical necessity for maintaining local phase symmetry.

Tell Me Again, What *Exactly* is Gauge Theory?

In physics, a gauge theory is a type of field theory in which the Lagrangian, and hence the system's dynamics, remain invariant under local transformations caused by certain smooth families of operations (Lie group). Formally, the Lagrangian is under these transformations. Lie groups are the mathematical framework used in quantum mechanics to describe continuous symmetries and their corresponding conservation laws. Their infinitesimal generators form Lie algebras, which correspond directly to the physical observables (like momentum and angular momentum) that dictate a system's quantum dynamics.

In quantum field theory, gauge theory describes fundamental forces and elementary particles by incorporating the concept of local gauge symmetry. This means the laws of physics remain unchanged even when a local transformation is applied to the fields describing the particles. The theory introduces gauge fields that mediate these forces and is a cornerstone of The Standard Model. The core idea is that the physical laws are invariant under transformations that can vary from point to point in spacetime. Imagine having different units for measuring temperature at different locations; gauge theory ensures that the underlying physical reality remains consistent despite these local changes in "gauge". To maintain this local symmetry, gauge fields are introduced. These fields are associated with the forces and mediate interactions between particles.

Gauge theories describe the fundamental forces of nature: Electromagnetism is described by Quantum Electrodynamics, where photons are the gauge bosons. The Weak Nuclear Force is responsible for certain types of radioactive decay. The Strong Nuclear Force binds quarks together to form protons and neutrons, mediated by gluons. A fundamental symmetry in the laws of physics allows it to be applied differently at every point in space and time. To keep the laws consistent under these local changes, new fields (gauge fields) were introduced. These gauge fields, and the particles they carry (gauge bosons), are then identified with the fundamental forces and the particles that mediate them.

The Global vs. Local Distinction

Imagine everyone on Earth agrees to turn their clocks forward by one hour at the exact same time. The "gap" between everyone stays the same, so nothing really changes. This is global symmetry. Now imagine every person on Earth can decide to change their clock whenever they want, by any amount. Suddenly, the world is in chaos. You can't coordinate a meeting because everyone's "gauge" is different (local symmetry). To restore order in this local chaos, you would need a "Correction Service" – a signal (the gauge field) that tells everyone how to translate their time to everyone else's. In physics, the "Correction Service" is a fundamental force like electromagnetism.

While "temperature" is a useful analogy, gauge theory in the Standard Model deals specifically with the phase of a particle's wave function. In Quantum Electrodynamics, an electron is a wave. You can shift the "phase" of that wave (where it starts its peak and trough) without changing the probability of where the electron is. If you shift the phase differently at every point in space (a local transformation), the math of the electron's motion breaks. To fix the math and keep the "physics" the same, you *must* introduce the photon field. The photon is the mathematical "glue" that allows electrons to have different local phases while still behaving consistently.

It is helpful to explain *why* we have these transformations in the first place. Gauge theory suggests that our mathematical descriptions of the universe contain redundancy. Think of it like a map. You can describe a location using latitude/longitude or a different coordinate system. The coordinates (the gauge) aren't "real," but the physical location is. Gauge theory is the tool that allows us to change "coordinates" at every single point in space without changing the physical reality. It ensures that the "fluff" in our math doesn't interfere with the "truth" of the physics.

Here's another example: If you have $100, it doesn't matter if it's in a single bill or twenty $5 bills – the "value" (the physical state) is the same, but the "gauge" (the denomination) has changed. A gauge theory ensures that the "value" of the universe stays the same even when we change how we "count" it at different points in space.

Special Unitary Groups

The notation $\mathrm{SU}(\mathrm{n})$ refers to the Special Unitary group of degree n. You define these groups using $\mathrm{n} \times n$ complex matrices that preserve the inner product of vectors. These matrices must satisfy two specific mathematical conditions. First, the matrix must be unitary. This means the conjugate transpose of the matrix multiplied by the matrix itself results in the identity matrix. The matrix must be special. This condition requires the determinant of the matrix to be exactly 1.

You determine the number of independent generators for these groups by counting the remaining degrees of freedom after you apply the constraints. A general $\mathrm{n} \times n$ complex matrix begins with n^2 complex entries, which provide $2\mathrm{n}^2$ real parameters. The unitary condition $\mathrm{U}^{\dagger} U = I$ creates n^2 independent real equations. When you subtract these, you are left with n^2 parameters. The requirement that the determinant equals 1 removes one more degree of freedom because it fixes the phase of the matrix. This subtraction results in the final formula for the number of generators: $\mathrm{n}^2 - 1$.

For SU(2), you use 2×2 matrices. When you apply the formula $2^2 - 1$, you get 3. These three generators correspond to the three gauge bosons that mediate the weak interaction. In your study of quantum mechanics, you often represent these generators using the Pauli matrices. For SU(3), you use 3×3 matrices. The calculation $3^2 - 1$ gives you 8. These eight generators represent the eight gluons that carry color charge in Quantum Chromodynamics.

The specific count of 8 gluons is a direct result of the SU(3) symmetry. If the symmetry were U(3) instead of SU(3), the determinant would not be restricted to 1. You would then have 3^2 or 9 generators. That ninth gluon would be color neutral and would interact with all quarks equally, which would create a second type of long range force similar to electromagnetism. Experimental evidence does not show such a force. This confirms that the strong force follows SU(3) symmetry and uses exactly 8 gluons. Gauge symmetry is the *reason* for conserved quantities. $U(1)$ gauge invariance is the fundamental reason for the conservation of electric charge. If the universe didn't have this specific "knob" (symmetry), electricity as we know it couldn't exist as a stable system.

Once more, let's categorize a Lie Group and add a simple breakdown of the "Symmetry Languages":

- **U(1)** : The simplest group, responsible for electromagnetism. It involves a single "phase" change. The force carrier is the photon.
- **SU(2)** : A more complex group for the Weak Force, involving three gauge bosons (W^+, W^- and Z).
- **SU(3)** : The most complex group for the Strong Force, involving 8 "colors" of gluons

In essence, gauge theory tells us that forces are nature's way of maintaining consistency. If the universe wants the freedom to let things vary from point to point (locality), it pays for that freedom by creating force-carrying particles (bosons) to act as the translators between those points.

In modern physics, $SU(4)$ doesn't represent a standalone fundamental force . Instead, it is a mathematical framework used in several "Beyond the Standard Model" theories to describe hypothetical unified or "dark" forces. An $SU(4)$ gauge theory is a Yang-Mills theory, which involves 15 gauge bosons. Unlike the Standard Model, which uses $SU(3) \times SU(2) \times U(1), SU(4)$ is a primary candidate for "Grand Unified Theories" (GUTs) that seek to unify the different fundamental forces into a single framework.

Hamiltonian

A Hamiltonian is a mathematical function or operator that represents the total energy of a system and is used in a variety of fields, including quantum mechanics, control theory, and classical mechanics. It is a fundamental concept for describing a system's behavior over time. In classical mechanics, it's a function of position and momentum, while in quantum mechanics, it's an operator.

In quantum mechanics, a Hamiltonian is an operator, represented by the symbol $\hat{H}$, that corresponds to the total energy of a quantum system. It's the sum of the kinetic energy operator (T) and the potential energy operator (V), $\hat{H} = T + V$. The Hamiltonian is used to solve the Schrödinger equation, which describes how a quantum system evolves in time. The Hamiltonian's "eigenvalues" determine the "energy spectrum" or set of possible energy values for the system.

An example of a Hamiltonian operator in quantum mechanics is that of a single particle with mass m moving in one dimension under a potential $V(x)$. The Hamiltonian represents the total energy (kinetic + potential) and is expressed as:

$$\hat{H} = \hat{T} + \hat{V} = -\frac{\hbar^2}{2m}\frac{d^2}{dx^2} + V(x)$$

- ***Kinetic Energy Operator*** $(\hat{T})$: $-c\hbar^2 2mcd^2dx^2$ (where $\hbar$) is the reduced Planck's constant).
- ***Potential Energy Operator*** $(\hat{V})$: $V(x)$, representing the environment, such as a harmonic oscillator $(V(x) = \frac{1}{2}m\omega^2x^2)$.
- ***Total Energy***: The Hamiltonian allows us to calculate the energy eigenvalues and time evolution using the Schrödinger equation.

In classical mechanics, a Hamiltonian is a function that represents the total energy of a system, expressed in terms of its generalized positions and momenta. It provides an alternative formulation of classical mechanics, known as Hamiltonian mechanics, developed by Sir William Rowan Hamilton. This formalism is closely related to geometry and serves as a bridge to quantum mechanics.

A non-abelian (e.g., $(a \times b)$ is not always the same as $(b \times a)$) group with normal subgroups (i.e., invariant under conjugation). The quaternion group (Q_8) is the smallest and best-known example of a Hamiltonian group. The quaternion group is important in physics and computer graphics for representing orientations and rotations in three-dimensional space, offering a way to avoid issues like gimbal lock found in Euler angles.

Hilbert Space

A metric space is used to measure distances between points and perform general geometric operations. A metric space is a topological space, but not all topological spaces are metric spaces. Banach spaces and Normed spaces are spaces that combine the elements of both algebra and geometry. Norm generally refers to length; thus, these spaces yield stronger results.

Named after David Hilbert, a Hilbert space is a mathematical concept that essentially acts like a generalized version of our normal 3D space but can have infinitely many dimensions, allowing us to represent complex systems using tools from linear algebra, where we can measure "distances" and "angles" between elements within this space, using an idea called an "inner product"; it's particularly important in represent for describing the state of particles as wave functions.

Imagine the world you live in. You can describe any location with three coordinates: (x, y, z). In this 3D space, we can do some very intuitive things: We can draw an arrow from the origin (0, 0, 0) to any point. This arrow is a vector. We can calculate the length of any vector using the Pythagorean theorem. For a vector v = (x, y, z), its length (or norm) is ||v|| = $\sqrt{x^2 + y^2 + z^2}$. We can determine the angle between two vectors, v and w, using the dot product: v · w = ||v|| ||w||cos(θ). The dot product gives us a way to understand the geometry of the space – how vectors relate to each other.

This familiar 3D space is a beautiful, well-behaved mathematical structure. It's a vector space (we can add and scale vectors) that also has a way to measure lengths and angles (an inner product, which in this case is the dot product). A Hilbert space is a powerful generalization of this familiar three-dimensional world.

Three coordinates define our 3D space. A 4D space would have four. What if we needed an infinite number of coordinates to describe something? Imagine a "vector" that is an infinite sequence of numbers:

$$v = (x1, x2, x3, ...)$$

Even more abstractly, what if our "vector" isn't a list of numbers at all, but a function, like f(x) = sin(x)? A function can be thought of as a vector with an infinite number of "components"–one for every possible value of x. A Hilbert space allows vectors to exist in not just two or three dimensions, but potentially in an infinite number of dimensions. This is the most abstract yet crucial part. A space is "complete" if it has no "holes" or "missing points." Think about the set of all rational numbers (fractions). You can create a sequence of rational numbers that gets closer and closer to 2:

1,1.4,1.41,1.414,1.4142,...

The numbers in this sequence are getting closer and closer to *each other*. This is called a Cauchy sequence. However, the point they are converging to, 2, is *not* a rational number. So, the space of rational numbers has a "hole" where 2 should be. It is incomplete. If we use the real numbers instead, this problem vanishes. Any sequence of real numbers that gets arbitrarily close to itself (a Cauchy sequence) will always converge to a limit that is also a real number. The real number line is complete.

A Hilbert space demands this same property of completeness. Any Cauchy sequence of vectors in the space must converge to a vector that is also *within that space*. It guarantees that if we follow a path that leads somewhere, the destination actually exists in our space. A Hilbert space is a complete inner product space.

A vector space is a set where you can add elements together and multiply them by scalars (numbers), and the results stay within the set – a vector space equipped with an inner product. The inner product (written $\langle v,w \rangle$) is a generalization of the dot product. It's an operation that takes two vectors and produces a single number. It must satisfy certain rules that allow us to define:

1. ***Norm (Length):*** $\|v\| = \langle v,v \rangle$
2. ***Angle (& Orthogonality):*** 2 vectors are orthogonal if $\langle v,w \rangle = 0$.
3. Every Cauchy sequence of vectors in the space converges to a limit that is also in the space.

The reason Hilbert spaces are fundamental to modern science and engineering is that they provide an ideal mathematical framework for modeling phenomena that require infinite-dimensional representations.

The state of a quantum system is not a set of numbers but a vector in an infinite-dimensional Hilbert space. A wave function often represents this vector $\psi(x)$,

Measurements are related to inner products. The probability of a system in state $|\psi\rangle$ being measured in state $|\phi\rangle$ is given by $|\langle \phi|\psi \rangle|2$. Mathematical operators that act on the vectors in this space represent Observables (like position, energy, momentum). The entire mathematical structure of quantum mechanics is built on the foundation of Hilbert spaces.

In short, a Hilbert space takes the familiar geometric concepts of length and angle from our 3D world. It extends them to the far more complex and abstract space of infinite-dimensional vectors, such as wave-functions and signals.

Imaginary (and Complex) Numbers

To understand imaginary numbers and their role in quantum mechanics, it is helpful to shift from thinking of numbers as points on a line to thinking of them as movements on a map. The term "imaginary" is somewhat of a historical accident. In the 16th century, mathematicians encountered equations they couldn't solve, like "what number multiplied by itself equals -1?" Since any real number–positive or negative–becomes positive when squared (e.g., $-2 \times -2 = 4$), they called the solution to this problem "imaginary," symbolized by the letter i.

So, $i^2 = 0 - 1$ or more simply: $i = \sqrt{-1}$

However, it is not "fake." A better way to view it is as a 90° turn. If you are on a standard number line, multiplying a number by -1 is like an 180° flip (turning plus one into minus one). If you multiply by i twice, you get -1. So multiplying by i just once must be half of that flip–a 90° turn off the number line and into a new dimension. When we combine a "real" number with an "imaginary" number (like 3 + 2i), we get a complex number. This creates a two-dimensional "number plane." While real numbers can only move left or right, complex numbers can represent any point in a two-dimensional space, effectively allowing us to track both distance and direction (or rotation) simultaneously.

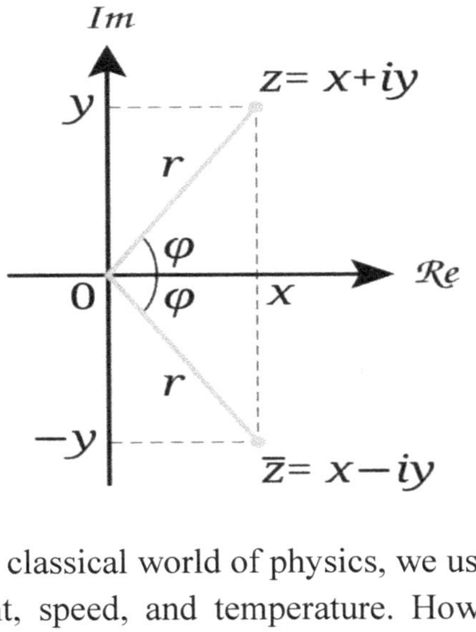

An illustration of the complex plane. Imaginary numbers are on the vertical coordinate axis.

In the classical world of physics, we use real numbers to describe quantities such as weight, speed, and temperature. However, the subatomic world is governed by waves. To describe the "state" of a particle, physicists use a mathematical tool called the wave function (ψ). In the Schrödinger equation the imaginary unit i appears right at the beginning. Imaginary numbers are essential for two main reasons:

1. Quantum particles behave like waves that are constantly oscillating or "spinning" in a mathematical sense. If you tried to describe a wave using only real numbers, you would need two separate equations – one for the height of the wave and one for its timing (phase). Because *i* represents a 90° rotation, complex numbers allow physicists to combine both the "height" and the "timing" of a quantum wave into a single, elegant equation. It captures the "rhythm" of the particle as it moves through time.

In our daily lives, if you have one chance out of ten to find a ball in a box, and someone adds another ball, your chances go up. Probabilities usually just add together. In the quantum world, however, particles can "interfere" with themselves. Two possibilities can add together to increase the probability, or they can cancel each other out entirely, leaving a probability of zero.

2. Real numbers (like 0.5 or 0.2) can't cancel each other out to zero unless one of them is negative, but you can't have a "negative probability" in the real world. In quantum mechanics, complex numbers are used to describe something called probability amplitudes. These amplitudes can point in different directions on the complex plane. If two amplitudes point in opposite directions, they cancel out perfectly. This "cancellation" explains why electrons stay in specific orbits around an atom and how light creates interference patterns.

Without imaginary numbers, the math of the subatomic world would be a cluttered mess of thousands of separate parts. The inclusion of i is what allows the universe to maintain its wave-like consistency and explains why matter is stable.

Lagrangian

Joseph-Louis Lagrange introduced Lagrangian mechanics in 1760. A Lagrangian is a mathematical function in physics that describes the state of a system by subtracting the potential energy from the kinetic energy, essentially representing the "difference in energy" that drives the motion of a system, allowing us to analyze its movement without directly applying forces; it's like a formula that captures the essence of a system's dynamics based on its position and velocity.

The typical Lagrangian formula is $L = T - V$, where L is the Lagrangian, T is the kinetic energy, and V is the potential energy. By finding the "stationary points" of the Lagrangian (using calculus of variations), you can derive the equations of motion for a system, which tell you how it will move over time.

In physics, a Lagrangian is a function that describes the state of a physical system in terms of its position coordinates and time derivatives. It is the difference between the system's kinetic and potential energy $L = KE - PE$. The theorem states that for any finite group, the order of any subgroup divides the group's order.

Laplacian

A Laplacian (or Laplace operator) is a second-order differential operator, denoted as ∇^2 or Δ. It is defined as the divergence of a function's gradient and is used in many areas of physics and mathematics. The Laplacian measures the extent to which a function's value at a specific point deviates from its average value in the immediate neighborhood. A zero Laplacian value indicates a state of equilibrium, such as a steady-state temperature distribution. It appears in fundamental equations describing physical phenomena, including heat diffusion, fluid flow, electromagnetism, and quantum mechanics. In three-dimensional Cartesian coordinates, it is the sum of the second partial derivatives of a function.

$$\nabla^2 f = \frac{\partial^2 f}{\partial x^2} + \frac{\partial^2 f}{\partial y^2} + \frac{\partial^2 f}{\partial z^2}$$

In mathematics, the Laplacian is an operator that measures how much a point is a local minimum or maximum. The Laplacian is the divergence of the gradient of a scalar function. It's represented by the symbol ∇^2, sometimes called "del squared." In Cartesian coordinates, the Laplacian is the sum of the second partial derivatives of a function with respect to each independent variable.

The Laplacian appears in the Schrödinger equation of quantum mechanics and in the equation governing electrostatics. It's also used in spectral clustering to partition data points into subsets. The Laplacian is similar to the second-derivative test in single-variable calculus. Laplace's equation $\nabla^2 u = 0$ states that the sum of the second-order partial derivatives of an unknown function equals zero. The Laplacian operator often represents this sum. It is used to model the gravitational potential in empty space outside of a mass.

In simple terms, the Laplacian is a mathematical operator that measures how a function's value at a point compares to the average value of its neighbors. It's a generalization of the second derivative to multiple variables, and its value can indicate whether a point is a local minimum (positive Laplacian) or a local maximum (negative Laplacian).

The Laplacian is a way to take a "second derivative" of a function with more than one input variable (e.g., x and y). Mathematically, it is the divergence of the gradient of a function. It appears in many fundamental equations, such as the heat, wave, and Schrödinger equations.

Leptons, Baryons, Mesons, Fermions, Hadrons, & Bosons

In particle physics, baryons and mesons are both types of hadrons (strongly interacting particles), with baryons being made up of three quarks while mesons consist of a quark and an antiquark; leptons are a separate class of elementary particles not affected by the strong force, and both baryons and leptons are classified as fermions (particles with half-integer spin). In contrast, mesons are considered bosons (particles with integer spin).

- **Baryons:** These are heavy particles composed of three quarks, including well-known examples like protons and neutrons.
- **Leptons:** These lighter particles do not interact via the strong force; they include electrons, muons, and neutrinos.
- **Bosons:** Particles with integer spin mediate forces, like the photon and the Higgs boson.
- **Fermions:** Have half-integer spin, which makes up matter, including all quarks and leptons.
- **Hadrons:** The collective term for particles that interact via the strong force, including both baryons and mesons.
- **Mesons:** These are composed of a quark and an antiquark and are considered bosons (e.g., Pions and Kaons).

Elementary Bosons

- **Higgs boson:** A scalar boson responsible for giving other particles mass.
- **Gauge bosons:** Particles that mediate fundamental forces.
  - **Photons:** Carry the electromagnetic force.
  - **W and Z bosons:** Mediate the weak nuclear force.
  - **Gluons:** Mediate the strong nuclear force, binding quarks together.
- **Gravitons:** A hypothetical particle that would mediate the gravitational force.

Composite bosons

- **Phonons:** Quasiparticles that represent a quantum of vibrational energy in a crystal lattice.
- **Cooper pairs:** Two electrons that pair up to behave as a single boson in some superconductors.

Lorentz Transformations

When people refer to "*theories of Lorentz*," they primarily mean the work of physicist Hendrik Antoon Lorentz, which paved the way for Einstein's Special Relativity (SR).

Lorentz didn't develop a single, unified theory in the modern sense. Instead, he developed a mathematical framework, the Lorentz Transformations, and an associated conceptual model, the Lorentz Ether Theory (LET). The Lorentz Transformations are a set of mathematical equations that describe how measurements of space and time change for two observers moving at a constant velocity relative to each other. These equations are the mathematical heart of Lorentz's model and Einstein's SR.

Before Lorentz, physicists used Galilean transformations, which worked well for Newtonian mechanics but failed for electromagnetism (Maxwell's Equations). Maxwell's Equations predicted that the speed of light (c) was constant, but Galilean transformations implied that (c) it should be different for observers in different reference frames. Lorentz derived his transformations to ensure that Maxwell's Equations (and thus the speed of light (c) remained invariant (the same) for all inertial observers. For an object moving along the x-axis, the transformations relate the spacetime coordinates (x, t) measured by one observer to the coordinates (x', t') measured by a second observer moving at velocity v:

$$x' = \gamma(x - vt)$$
$$t' = \frac{\gamma}{(t - \frac{vx}{c^2})}$$

Where γ (gamma) is the Lorentz factor:

$$\gamma = \frac{1}{\sqrt{1 = v^2 c^2}}$$

These transformations mathematically necessitate the relativistic effects we observe, such as Time Dilation: a change in the time coordinate (t) means that time slows for moving objects. The change in the space coordinate (x') requires that an object's length be shortened in its direction of motion.

Time dilation equation: $\Delta t' = \gamma \Delta t$. Length contraction equation: $L' = \frac{L}{\gamma}$

The Lorentz Ether Theory (LET) was Lorentz's physical interpretation of his mathematical transformations. Unlike Einstein's later interpretation, LET required the existence of a stationary, invisible medium called the luminiferous aether. LET proposed that Maxwell's Equations were only exactly valid in the frame of reference belonging to the stationary aether. All motion was absolute motion *relative to this aether*.

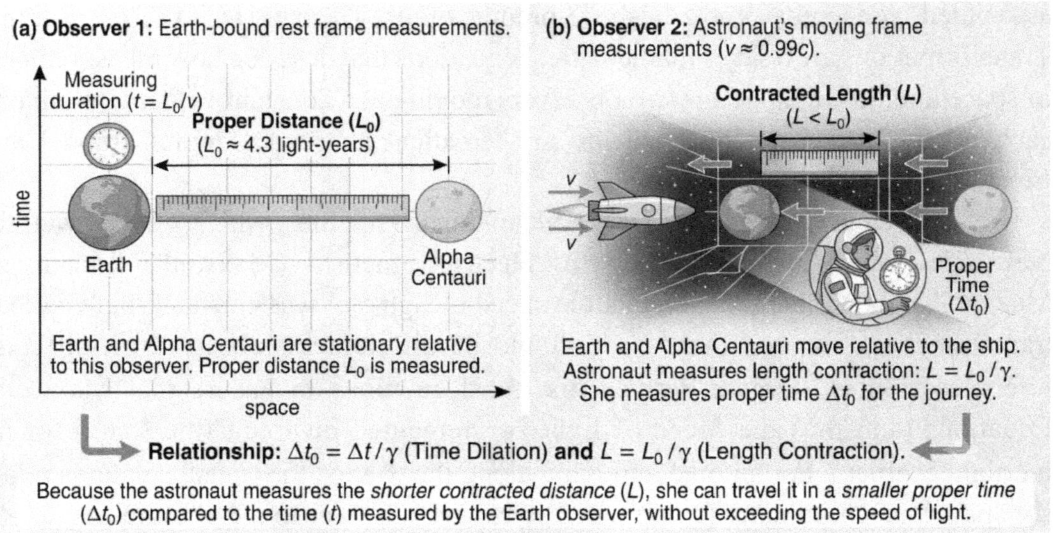

To explain why the speed of light *appeared* constant to all observers, Lorentz proposed that motion through the aether caused real, physical changes in objects: it physically slowed the rate of all internal processes (chemical, electrical, etc) within a clock. Motion through the aether physically compressed objects along their direction of travel.

LET was eventually superseded by Einstein's Special Relativity (SR) in 1905, primarily because SR eliminated the need for the unobservable aether. Einstein proposed two postulates (the laws of physics are the same for all observers, and the speed of light is constant for all observers) and derived the Lorentz Transformations from them, turning them into a statement about the nature of spacetime itself.

SR explained that time dilation and length contraction are not physical changes caused by motion through a medium, but fundamental consequences of measuring reality across different inertial frames. Lorentz provided the correct mathematics (Lorentz Transformations), but Einstein provided the correct, revolutionary physical interpretation that the transformations described the geometry of spacetime.

Minkowski Space

Minkowski space is a mathematical model of spacetime used in Albert Einstein's theory of special relativity. It's a four-dimensional space that combines the three dimensions of physical space with the time dimension into a single continuum. Think of it as an extension of the familiar three-dimensional Euclidean space (the space we experience every day) to include time.

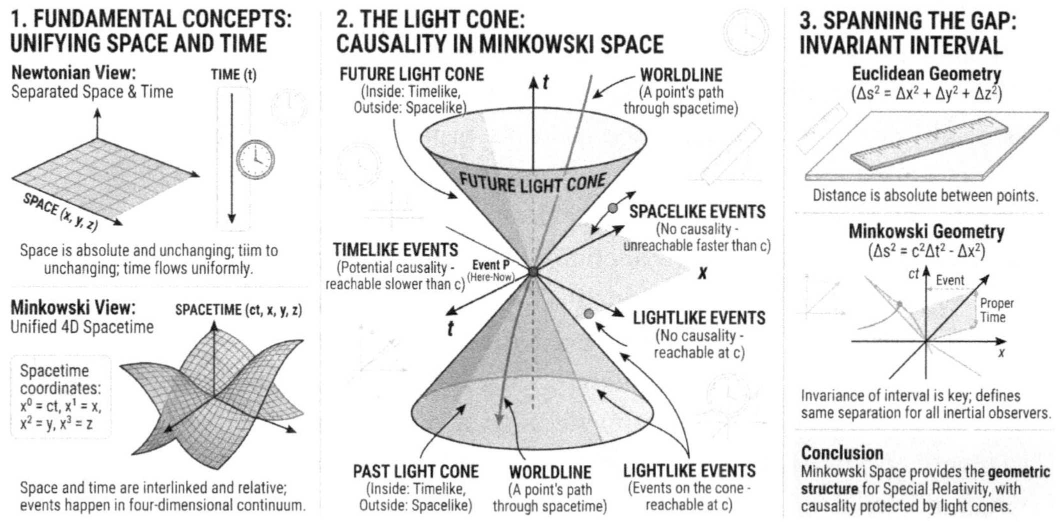

We live in a four-dimensionaluniverse of spacetime. There are three spatial dimensions of length, width, and height (or x, y, and z coordinates) that we use to locate objects in space, and the one dimension of time that measures the progression of events. So, instead of just specifying where something is (x, y, z), in Minkowski space, you specify *where and when* it is (x, y, z, t).

A point in Minkowski space is called an event and represents a specific location at a particular moment. The crucial difference between Minkowski space and ordinary Euclidean space lies in how we measure "distances" or "intervals" between points. In Euclidean space, the distance squared between two points is given by the Pythagorean theorem. In Minkowski space, the "distance" between two events is called the spacetime interval. It's defined with a crucial minus sign (or sometimes plus sign, depending on the convention used) associated with the time dimension. The "distance" in Minkowski space is not always positive and can be zero or even negative. This reflects the fact that, in relativity, space and time are intertwined in a specific way.

One of the most important aspects of Minkowski space is that the spacetime interval between two events is invariant under Lorentz transformations. Lorentz transformations relate the coordinates of two observers moving at constant relative velocity. This invariance is analogous to the fact that the distance between two points in Euclidean space remains invariant under rotations and translations of the coordinate system. It means that even though different observers might measure different spatial and temporal separations between two events, they will all agree on the value of the spacetime interval. This is a cornerstone of special relativity.

Minkowski space provides the natural mathematical framework for describing the geometry of spacetime in the absence of gravity (the domain of special relativity). It allows us to understand the relationship between space and time, formulate the laws of physics in a way consistent with the principle of relativity, and derive key relativistic effects such as time dilation and length contraction.

While Minkowski space describes flat spacetime (no gravity), it serves as the local tangent space to curved spacetime. In general relativity, the presence of mass and energy curves spacetime, and this is more complex than Minkowski space. However, at any given point in curved spacetime, we can approximate a small enough region as being a flat Minkowski space. Imagine replacing your regular 3D map with a 4D map in which time is also a dimension. The way you measure "distances" on this 4D map is a bit different because time and space are linked. This 4D map of Minkowski space uses a special method for measuring distances. It helps us understand how space and time behave according to special relativity.

Møller scattering

The name given to electron-electron scattering in quantum field theory, named after the Danish physicist Christian Møller, who derived it in 1932. The electron interaction that is idealized in Møller scattering forms the theoretical basis of many familiar phenomena such as the repulsion of electrons in the helium atom. While formerly many particle colliders were designed specifically for electron-electron collisions, electron-positron colliders have become more common. That said, Møller scattering remains a paradigmatic process within the theory of particle interactions.

Neutrinos – The Ghost Particle

The neutrino, often called the "ghost particle" for its elusive nature, has undergone one of the most remarkable journeys in scientific history – from a "desperate remedy" scribbled in a letter to the cornerstone of modern physics beyond the Standard Model. It's a nearly massless, electrically neutral fermion that interacts only via the weak nuclear force and gravity, allowing it to pass through light-years of lead without hitting a single atom.

Don't be alarmed, but some 100 trillion neutrinos pass completely harmlessly through your body every single second. Because they do not feel the electromagnetic force, they completely ignore the atoms making up your cells. They flow right through the empty space inside your body – and right through the entire planet Earth – as if it weren't even there. Statistically, the odds of a neutrino colliding with an atom in your body are so microscopic that only one single neutrino will actually hit a particle inside you over the course of your entire lifetime.

Its discovery altered our understanding of mass and the universe's symmetry. In 1930, Wolfgang Pauli addressed a letter to a group of physicists at a conference in Tübingen, proposing a particle he called the "neutron" (later renamed by Enrico Fermi). At the time, physicists were baffled by beta decay; the emitted electrons showed a continuous energy spectrum rather than a discrete set of values. This led some, including Niels Bohr, to suggest that energy conservation might only hold statistically. Pauli's "desperate remedy" was to hypothesize an undetected, neutral particle that carried away the missing energy, momentum, and spin. In 1934, Enrico Fermi formalized this idea in his theory of beta decay, which described a neutron transforming into a proton, an electron, and a neutrino:

$$n \rightarrow p + e^- + \bar{V}_e$$

Despite its solid theoretical basis, neutrinos went undetected for over two decades. In 1956, Clyde Cowan and Frederick Reines set out to "catch the ghost" at the Savannah River Plant in Georgia.[86] Their experiment, known as Project Poltergeist, used the intense flux of antineutrinos from a nuclear reactor. They employed a process called inverse beta decay, in which an antineutrino interacts with a proton in a large tank of water doped with cadmium chloride, producing a neutron and a positron. The detection of two simultaneous gamma rays from the positron's annihilation, followed microseconds later by a gamma ray from the neutron's capture by cadmium, provided the definitive proof of the neutrino's existence.

As particle physics advanced, it became clear that the neutrino was not a single entity but part of a family. In 1962, at Brookhaven, Leon Lederman, Melvin Schwartz, and Jack Steinberger fired a beam of high-energy pions at a target and produced a shower of particles that decayed into neutrinos. These neutrinos produced muons when they interacted with matter, rather than electrons, thereby proving the existence of the muon neutrino. Following the 1975 discovery of the tau lepton by Martin Perl, the tau neutrino (v_T) was directly observed in 2000 by the DONUT collaboration at Fermilab.

An amazing shift in neutrino physics occurred when scientists realized these particles could transform between flavors. Raymond Davis Jr. hinted at this as early as the late 1960s at the Homestake Mine. Davis used a 100,000-gallon tank of cleaning fluid to detect neutrinos from the sun, but he consistently found only about one-third of the predicted amount. This "Solar Neutrino Problem" suggested that either the solar model was wrong or the neutrinos were changing into types that Davis's detector could not see.

In 1998, the Super-Kamiokande experiment in Japan provided the first evidence of neutrino oscillations using atmospheric neutrinos, showing that the ratio of muon to electron neutrinos varied with the distance traveled through the Earth. Conclusive proof came in 2001 and 2002 from the Sudbury Neutrino Observatory (SNO) in Canada. SNO used heavy water to detect all three types of neutrinos, confirming that the total flux of neutrinos from the sun matched predictions exactly, but many had arrived as muon or tau flavors.

The ability to oscillate implies that neutrinos must have mass, since a massless particle traveling at the speed of light would experience no time dilation and so could not change state. The transformation is described by the PMNS (Pontecorvo–Maki–Nakagawa–Sakata) matrix, which relates the flavor eigenstates (v_e, v_μ, v_τ) to the mass eigenstates (v_1, v_2, v_3). The probability of a neutrino of flavor α oscillating into flavor β after traveling a distance L with energy E is given by the simplified two-flavor equation:

$$P(v_\alpha \to v_\beta) = sin^2(2\theta)sin^2\left(1.27\frac{\Delta m^2 L}{E}\right)$$

In this equation, θ represents the mixing angle, and Δm^2 is the difference in the squares of the masses of the two states. This discovery, recognized with the 2015 Nobel Prize, forced a revision of the Standard Model, which had previously assumed neutrinos were massless.

Modern Frontiers and New Theories

Neutrino research remains at the forefront of cosmology and particle physics. New findings in 2025 and 2026 have narrowed the search for physics beyond the Standard Model. For instance, the MicroBooNE experiment provided high-precision measurements that effectively ruled out the existence of a single "sterile" fourth neutrino, a particle once thought to explain certain anomalies in decay rates. This "paradigm shift" is pushing researchers toward more complex models of dark matter and universal symmetry.

Experiments such as KATRIN and CUORE continue to improve the upper limit on the neutrino's mass, confirming that it is at least a million times lighter than the electron. As experiments like DUNE (Deep Underground Neutrino Experiment) prepare for operation, the neutrino remains our most potent probe into the deep architecture of reality.

Sterile Neutrinos

A sterile neutrino is a hypothetical type of neutrino that does not interact via any of the fundamental interactions of the Standard Model except for gravity. While the three known neutrinos – electron, muon, and tau neutrinos – interact through the weak nuclear force and gravity, the sterile neutrino is "sterile" because it lacks a "weak charge," meaning it does not participate in the weak interaction at all.

In the Standard Model, neutrinos are fermions that come in three "flavors." These neutrinos are always observed to be "left-handed" (referring to their helicity or the way their spin relates to their motion). Because the weak force only acts on left-handed particles, the standard neutrinos are part of this interaction. Theoretical physics suggests there may be "right-handed" counterparts. If these right-handed neutrinos exist, they would not feel the weak force, making them sterile neutrinos.

The primary motivations for the sterile neutrino hypothesis come from experimental anomalies and the mystery of neutrino mass. Experiments have proven that neutrinos have a tiny but non-zero mass. The Standard Model does not inherently explain how they get this mass. Many theories, such as the See-Saw Mechanism, predict that for every light, left-handed neutrino, there is a very heavy, right-handed (sterile) neutrino. This mechanism "pushes" the mass of visible neutrinos to the tiny levels we observe.

Several experiments, including LSND (Liquid Scintillator Neutrino Detector) and MiniBooNE, observed more electron-neutrino-like events than predicted. One explanation is that standard neutrinos oscillate (transform) into a fourth, sterile flavor and then back again. Observations of neutrinos produced by nuclear reactors have consistently shown a slightly lower flux than predicted, suggesting some neutrinos might be "disappearing" into a sterile state.

If sterile neutrinos exist, they are a leading candidate for Dark Matter. Because they interact only through gravity and may have significant mass (depending on the model), they could provide the invisible gravitational "glue" that holds galaxies together. Specifically, sterile neutrinos in the keV (kilo-electronvolt) mass range are considered candidates for "Warm Dark Matter." Unlike "Cold Dark Matter," which moves slowly, warm dark matter would move fast enough to smooth out the distribution of matter on small scales, which might better match some astronomical observations.

Since we think dark matter accounts for roughly 27% of the cosmic mass budget, but the Standard Model of Particle Physics currently has nothing that fits the description. This is where the hypothetical "sterile" (or heavy) neutrino enters. It is one of the most elegant proposed solutions for the Dark Matter problem, primarily because it fixes a glaring asymmetry in quantum theory while perfectly matching the cosmic data.

To understand the sterile neutrino, we first have to look at the standard version. Neutrinos have zero electrical charge and do not feel the strong nuclear force. They only interact via gravity and the Weak Nuclear Force (which governs radioactive decay). For a long time, physicists thought standard neutrinos might be Dark Matter. However, because their mass is almost zero, they travel at near the speed of light.

In cosmological terms, they are "Hot Dark Matter." They move too fast to clump together. If Dark Matter were made of standard neutrinos, their speed would have smeared out the early universe, and galaxies could never have formed. Dark Matter must be "Cold" (or at least "Warm") – meaning the particles must be heavy and sluggish enough for gravity to herd them together.

In particle physics, every particle has a property called "spin," and they come in two orientations: left-handed and right-handed. Every quark and electron in the universe comes in both versions. Nature loves this symmetry.

However, we have only ever observed left-handed neutrinos. This asymmetry is a massive red flag. It implies our understanding may be incomplete. The "sterile neutrino" is the proposed missing half of the equation: the right-handed neutrino.

If this right-handed neutrino exists, why haven't we seen it in our particle accelerators? According to the math, the right-handed version of the neutrino is completely "unplugged" from the Weak Nuclear Force. It is entirely *sterile*. It does not interact with electromagnetism, the strong force, or the weak force. The only way it communicates with the rest of the universe is through the curvature of spacetime itself – Gravity.

This perfectly matches the profile of Dark Matter: an invisible entity that only reveals itself by exerting gravitational pull on surrounding galaxies.

How does a neutrino get heavy enough to be "Cold" Dark Matter? Theorists propose an architectural relationship called the See-Saw Mechanism. In the quantum realm, the left-handed (standard) and right-handed (sterile) neutrinos are mathematically entangled on a massive cosmic see-saw.

The heavier the right-handed sterile neutrino is, the lighter the left-handed standard neutrino is forced to be. Because we know standard neutrinos are incredibly light (almost massless), the See-Saw Mechanism dictates that their sterile partners must be incredibly heavy. If heavy sterile neutrinos exist, they are the ultimate background process. They were forged in the immense heat of the Big Bang, but because they are so heavy, they cooled down and slowed down very quickly. Being "sterile," they couldn't interact with the hot plasma of the early universe to emit light or form atoms. Instead, they drifted, letting their immense, collective gravity pool together to form the invisible gravitational wells that eventually pulled in ordinary matter to ignite the first stars.

One of the deepest theories about heavy neutrinos is that they might be Majorana particles. In most matter, a particle and its antiparticle are distinct (like an electron and a positron). A Majorana particle is its own antiparticle. If heavy neutrinos have this property, they could explain the Matter-Antimatter Asymmetry of the universe through a process called Leptogenesis:

1. In the incredibly hot early universe, heavy neutrinos were abundant.
2. Because they are their own antiparticles, they could decay in ways that slightly favored the production of matter over antimatter.
3. This bias in the early universe is why we have a world made of matter today, rather than a universe filled only with pure energy from total annihilation.

In summary, sterile neutrinos help with the Great Divide beautifully: a microscopic quantum anomaly (a missing right-handed spin) scaling up to become the scaffolding of the entire universe. Because sterile neutrinos do not interact via the weak force, they cannot be directly detected by standard neutrino detectors such as Super-Kamiokande or IceCube. Instead, scientists must look for "disappearance" or "appearance" signatures:

***Disappearance*:** Watching a beam of standard neutrinos to see if more vanish than expected.

Appearance: Watching for a standard neutrino flavor to appear in a beam where it shouldn't be, mediated by a sterile middle-step.

Because these particles are "heavy" and "sterile," we cannot see them directly. Instead, physicists look for *Neutrinoless Double Beta Decay*, a rare type of radioactive decay that would only be possible if the neutrino is its own antiparticle, a kink in Beta Decay spectra, a tiny "missing" amount of energy in standard decays that would indicate a heavy neutrino was produced, and Cosmological Signatures, patterns in the Cosmic Microwave Background (CMB) or galaxy distribution that would differ if heavy sterile neutrinos were acting as dark matter.

Noether's Theorem

For every continuous symmetry in a physical system,
there is a corresponding, fundamental conservation law.

Noether's Theorem is widely considered one of the most profound and elegant results in the history of theoretical physics. Developed by the mathematician Emmy Noether in 1915, it provides a fundamental bridge between two seemingly disparate concepts: symmetry and conservation laws. Before this theorem, conservation laws (such as the conservation of energy or momentum) were largely observed as empirical facts, but their underlying "why" remained a mystery. Noether's work proved that these laws are not arbitrary but are the direct mathematical consequences of the symmetries of the universe.

To understand the theorem, we must first define symmetry in a physical context. In physics, a symmetry is a transformation – such as moving an experiment to a different location, rotating it, or performing it at a later time – that leaves the underlying laws of physics unchanged. If the equations governing a system do not change when you apply one of these transformations, the system is said to be invariant. Noether's Theorem states that for every continuous symmetry of the laws of nature, there is a corresponding physical quantity that is conserved.

One of the most intuitive applications of the theorem is the relationship between space translation and momentum. Imagine a universe where the laws of physics are the same wherever you are. If you perform an experiment in one location and then move the entire setup three feet to the left, the results do not change because the laws are "translationally invariant." Noether showed that this specific symmetry of space leads directly to the conservation of linear momentum. If the laws of physics were different at every coordinate in space, momentum would not be conserved.

Similarly, the theorem explains the conservation of angular momentum in terms of rotational symmetry. If the laws of physics are isotropic – meaning they do not change if you rotate your coordinate system – then the system possesses rotational symmetry. This symmetry requires that the angular momentum remains constant over time. This explains why objects in space maintain their rotation and why a spinning figure skater accelerates when pulling their arms inward; they are participating in the geometric "logic" of the universe's rotational invariance.

Perhaps the most important application of Noether's Theorem is the link between time and energy. If the laws of physics are the same today as they were yesterday and will be tomorrow, the system has "time translation symmetry." Noether proved that invariance under time translation implies energy conservation. This provides the theoretical foundation for the First Law of Thermodynamics. It also implies that, in scenarios where the geometry of spacetime changes over time – such as in an expanding universe – energy conservation becomes a much more complex, and sometimes non-local, concept.

The theorem is mathematically expressed through the framework of Lagrangian mechanics and the Principle of Least Action. Noether's proof demonstrated that if the Lagrangian of a system remains unchanged (invariant) under a continuous transformation, there exists a specific mathematical "current" that remains constant.

Beyond classical mechanics, Noether's Theorem is the bedrock of the Standard Model of Particle Physics. It extends beyond the physical movement of objects into "internal symmetries." For instance, the conservation of electric charge results from a specific mathematical symmetry, called "gauge invariance," in the wave functions of particles. By linking the abstract symmetries of mathematics to the tangible conservation laws of the physical world, Noether's Theorem transformed physics from a study of disconnected forces into a unified study of the geometric architecture of reality.

Particle Spin Classifications

Bosons (Force Carriers and Scalars): Bosons follow Bose-Einstein statistics and can occupy the same quantum state simultaneously.

- ***Spin-0 (Scalar Bosons):*** Associated with fields that do not have a direction. If you rotate a spin-0 particle, it would look identical from every angle. It possesses no directional preference or "polarization" in its resting state. In quantum field theory, spin-0 particles are associated with fields that occupy every point in space with a simple numerical value rather than a direction.
  - ***Higgs Boson:*** The only known fundamental scalar particle.
  - ***Mesons (Pions, Kaons):*** Composite particles made of a quark and an antiquark that can result in a net spin of 0.
- ***Spin-1 (Vector Bosons):*** These are the force carriers for the fundamental interactions. These particles look the same after a full 360° rotation, similar to a standard arrow.
  - ***Photon:*** Carrier of the electromagnetic force.
  - ***Gluon:*** Carrier of the strong nuclear force.
  - ***W and Z Bosons:*** Carriers of the weak nuclear force.
- ***Spin-2 (Tensor Bosons):*** These are associated with the geometry of spacetime.
  - ***Graviton:*** A theoretical particle that carries the force of gravity.

Fermions (Matter Particles): Fermions follow the Pauli Exclusion Principle, meaning two identical fermions cannot occupy the same quantum state at the same time. This property is what allows matter to have volume and structure.

- **Spin-$\frac{1}{2}$:** All known fundamental matter particles possess this spin. A spin-$\frac{1}{2}$ particle must be rotated 720° (two full turns) to return to its original quantum state. (See Spinors in Chapter 9: The Amazing Electron).
  - ***Quarks:*** The building blocks of protons and neutrons (Up, Down, Charm, Strange, Top, Bottom).

  - *Leptons:* This group includes the Electron, Muon, Tau, and their three corresponding Neutrinos.
- **Spin-$\frac{3}{2}$**: These particles must be rotated 480° (one and one-third turns) to return to their original state.
  - *The Gravitino:* In theories of supersymmetry, the gravitino is the predicted spin-$\frac{3}{2}$ partner of the graviton.
  - *Delta Baryons:* You can find spin-$\frac{3}{2}$ states in composite particles made of three quarks where the spins are aligned ($\frac{1}{2}+\frac{1}{2}+\frac{1}{2}$).

Perturbation Theory

Perturbation theory is a fundamental mathematical framework in theoretical physics used to find approximate solutions to problems that cannot be solved exactly. It is often described as the "physics of almost" – it allows scientists to start with a system they understand perfectly and calculate how it changes when a small, complicating factor is added.

In classical and quantum mechanics, most systems are too complex for a "closed-form" solution (an exact equation). Perturbation theory sidesteps this by splitting the total energy of a system (the Hamiltonian, H) into two parts: $H = H_0 + \lambda V$.

- **H_0** ***(The Unperturbed System):*** A simple, idealized version of the problem that we already know how to solve exactly (like a perfectly spherical atom or a simple harmonic oscillator).
- **V** ***(The Perturbation):*** The "complication" or external force (like an electric field or a slight bump in a potential well).
- λ ***(The Coupling Constant):*** A small parameter that represents the "strength" of the perturbation. For the theory to work, λ must be small enough that the system's behavior doesn't change fundamentally.

The beauty of perturbation theory is that it allows for systematic improvement. You express the unknown energy (E) and the state of the particle (ψ) as a power series:

$$E = E^{(0)} + \lambda E^{(1)} + \lambda^2 E^{(2)} + \ldots |\psi\rangle = |\psi^{(0)}\rangle + \lambda|\psi^{(1)}\rangle + \lambda^2|\psi^{(2)}\rangle + \ldots$$

- ***First-Order Correction*** ($\mathrm{E}^{(1)}$)*:* This is often just the average value of the perturbation. It tells you the immediate "shift" in energy caused by the external force.

- ***Second-Order Correction*** ($\mathrm{E}^{(2)}$)*:* This accounts for how the perturbation allows the system to "leak" into other possible states. It is almost always a downward shift in energy for the ground state.

Depending on the problem, physicists use one of two categories:

- ***Time-Independent Perturbation Theory:*** Used for static problems, such as calculating how the spectral lines of an atom split when placed in a magnetic field (the Zeeman Effect) or an electric field (the Stark Effect).

- ***Time-Dependent Perturbation Theory:*** Used to calculate transition rates–the probability that a system will jump from one state to another over time. This leads to Fermi's Golden Rule, which is essential for understanding how atoms absorb or emit photons.

Perturbation theory is not just an "extra" tool; it is the primary way we understand the subatomic world. In Quantum Field Theory, perturbation theory is visualized through Feynman Diagrams. Each line and vertex in a diagram represents a term in a perturbation series.

- A simple diagram with one interaction vertex represents a "first-order" term.
- A diagram with a "loop" represents a higher-order (often second-order) correction.

In Quantum Electrodynamics, the "small parameter" λ is the Fine-Structure Constant ($\approx 1/137$). Because $1/137$ is so small, each higher power in the perturbation series (e.g.,$(1/137)^2$, $(1/137)^3$) becomes rapidly smaller. This allows

A major hurdle in theoretical physics was that higher-order perturbation terms often resulted in "infinities." The process of Renormalization was developed to manage these infinities, essentially "hiding" them within the definitions of mass and charge to produce finite, usable results.

Perturbation theory only works if the interaction is weak. If the coupling constant λ is large (greater than 1), the series "diverges"–meaning each correction gets larger rather than smaller, making the whole method useless.

The Strong Force (QCD): At low energies, the force holding quarks together is so strong that perturbation theory fails. Physicists must instead use Lattice QCD – simulating the universe on a grid – or other non-perturbative methods.

- ***Bound States:*** Perturbation theory is great at describing how particles deflect off each other, but it cannot easily describe how they stick together to form a permanent bond (like a proton or a neutron).
- ***Solitons and Instantons:*** These are "lumpy" solutions to equations that cannot be built up by adding small corrections to empty space; they exist fundamentally as non-perturbative structures.

Planck Units

Planck units represent physical constants that are theoretical limits at the quantum level. Quantum fluctuations exceeding these limits could tear apart the fabric of spacetime and reality, creating rifts into other worlds or dimensions. At Planck scales, the strength of gravity becomes comparable with the other forces, and the fundamental forces are unified.

Max Planck created the scale in 1906. It is constructed out of these universal constants that have a numeric value of 1 when expressed in these units. At its heart, Planck's constant (h) is the fundamental constant of quantum mechanics that defines the smallest possible unit of energy exchange in the universe.

In the early 1900s, Planck was studying the light and heat emitted by hot objects (a phenomenon known as black-body radiation). Classical physics theories failed miserably to predict the observed colors. To solve this puzzle, Planck proposed a radical idea: what if energy couldn't be exchanged in any arbitrary amount? What if it could only be transferred in discrete little packets, which he called "quanta"?

Planck proposed that the energy (E) of these packets was directly proportional to the frequency (f) of the light. The "constant of proportionality" connecting them was a new number he introduced, now known as Planck's constant (h).

The relationship is captured in one of physics' most famous equations: $E = hf$

- E is the energy of a single quantum (a single particle of light, a photon).
- f is the frequency of the light wave.
- h is Planck's constant.

Gravitational constant	$G = 6.673(10) \times 10^{-11}\ m^3\ kg^{-1}\ s^{-2}$
Planck's constant	$h = 6.62607015 \times 10^{-34}\ J \cdot Hz^{-1}$
Reduced Planck's Constant	$\hbar$ (h-bar) $= \frac{h}{2\pi} = 1.054571596(82) \times 10^{-34}\ kg\ m^2\ s^{-1}$
Boltzmann constant	$k = 1.380649 \times 10^{-23}\ J \cdot K^{-1}$

The Planck Constants

Unit	Scale	Comment
Planck length	$1.616 \times 10^{-35} m$	If a particle or dot about 0.1 mm in size were magnified to be as large as the universe, then inside that universe-

		sized dot, the Planck length would be roughly the size of a 0.1 mm dot. It would take more Planck lengths to span a grain of sand than it would take grains of sand to span the observable universe.
Planck mass	$2.176 \times 10^{-8} kg$	An object of such mass would be a quantum black hole created at Planck time, with a Schwarzschild diameter of Planck length. Each time an electron (for example) travels the reduced Compton wavelength, they counter-strike. The electron is in a mass state only a fraction of the time. This is why the Planck mass can be so enormous compared to the electron's rest mass and still account for the electron and other subatomic particles.
Planck time	$5.391 \times 10^{-44} s$	In the Big Bang, the Planck epoch, or Planck era, is the earliest stage, lasting approximately 10^{-43} seconds.
Planck temperature	$1.417 \times 10^{32} K$ (kelvin)	It's a billion billion times the highest natural temperatures in the current universe, found in gamma-ray bursts and quasars. This is *'absolute hot'*, conceived as the opposite of absolute zero. Everything 5×10^{-44} seconds after the Big Bang.
Planck area	$10^{70} m^2$	According to the Bekenstein bound, the entropy of a black hole is proportional to the number of Planck areas that it would take to cover the black hole's event horizon.
Planck volume	$10^{105} m^3$	A quantum black hole is contained within a Planck volume. There are about 10^{186} Planck volumes in the universe.
Planck energy	$10^9 J$ (joules)	A quantum black hole must have Planck mass and Planck energy such that its escape velocity exceeds (c).
Planck energy density	$10^{-29} g/cm^3$	Analogous to Planck's law, describing the spectral density of electromagnetic radiation emitted by a black body.
Planck charge	10^{-18} C (coulombs)	The electric potential energy of one Planck charge on the surface of a sphere that is one Planck length in diameter.
Planck force	10^{44} N (newtons)	The amount of force required to accelerate one Planck mass by one Planck acceleration.

Planck density	10^{96} kg/m^3	Equivalent to the mass of the universe packed into the volume of a single atomic nucleus.
Planck pressure	10^{113} Pa (pascals)	Equal to one Planck force in one Planck area. It's the gravitational force of attraction between two equal-sized universes, all concentrated on 1/4 of the Planck Area.
Planck acceleration	10^{51} m/s^2	The acceleration due to gravity at the surface of a Planck mass or quantum black hole.
Planck frequency	10^{43} /s	Upper bound for the frequency (vibrations per second) of an electromagnetic wave.

Planck's constant is an incredibly tiny number: $h \approx 6.626 \times 10^{-34} J \cdot s$, which is why we don't notice quantum effects in our everyday lives. The "steps" of energy are so incredibly small that the world around us appears smooth and continuous. It's like looking at a high-resolution digital photo – from a distance, it seems smooth, but if you zoom in, you can see the individual pixels. Planck's constant sets the size of those "pixels" for reality itself.

While Planck's constant (h) relates energy to frequency, physicists quickly found that in many quantum mechanical equations–especially those involving rotation, waves, or angular momentum – the term $\frac{h}{2\pi}$ appeared over and over again. To simplify their equations and make them neater, they defined a new constant, the reduced Planck constant, or, more commonly, h-bar ($\hbar$). The definition is simple:

$$\hbar = 2\pi h$$

The factor 2π comes from the relationship between frequency and angular frequency. Frequency (f) is measured in cycles per second (Hertz). It asks, "How many full cycles of a wave pass by in one second?" Angular frequency (ω) is measured in radians per second. It asks, "How many radians of a wave pass by in one second?" Since one complete cycle of a wave is equal to 2π radians, the relationship is: $\omega = 2\pi f$. By using $\hbar$, the fundamental energy equation can be rewritten in terms of angular frequency, which is often more natural for describing waves and rotations:

- Original equation: E = hf
- Substitute $f = \frac{\omega}{(2\pi)} : E = h(2\pi\omega)$
- Rearrange and use the definition of $\hbar$: $E = (2\pi h)\omega = \hbar\omega$

So, the equation $E = \hbar\omega$ is the same statement as $E = hf$, just expressed in a different rotational language. While it may seem like just a mathematical convenience, $\hbar$ is considered more fundamental in the context of quantum mechanics. It is the fundamental unit of quantum angular momentum.

One of quantum mechanics' shocking discoveries (proven by the Stern-Gerlach experiment) is that angular momentum is quantized. An object can't just have any amount of spin; it must be an integer or half-integer multiple of $\hbar$. For example, the spin of an electron is $\frac{1}{2}\hbar$. This is why $\hbar$ appears constantly in key equations of quantum mechanics, like the Schrödinger equation and Heisenberg's uncertainty principle.

Planck Length

The Planck length (lP) is considered the smallest possible unit of length in the universe, calculated using fundamental constants like the speed of light and gravity, and is so tiny (around 1.6×10^{-35}meters) that it's essentially impossible to measure anything smaller with our current understanding of physics; think of it as the "quantum" of distance, where the concept of space itself might start to break down at that scale.

The Planck length is extremely small; it's many times smaller than an atom. Physicists use it to explore extreme scenarios, such as the very early universe, where quantum effects and gravity become highly intertwined. The Planck length is calculated using the speed of light, Planck's constant, and the gravitational constant. The Planck length is special because it is the scale at which quantum mechanics and general relativity both become significant and *must* be unified. At this length, gravity is expected to become a dominant force, and our current laws of physics are predicted to break down. It is the scale at which the concept of spacetime becomes invalid, and the universe may be dominated by quantum fluctuations, creating a "quantum foam".

The Planck length is the scale at which both quantum and gravitational effects are equally strong. A theory of quantum gravity is needed to describe what happens at this scale. At this length scale, the laws of general relativity, which describe gravity and the geometry of spacetime, fail. Any attempt to measure a distance smaller than the Planck length would require so much energy that it would create a black hole, obscuring the measurement.

Some theories suggest that at the Planck length, spacetime itself is not smooth but is a "quantum foam" of fluctuations. The Planck length is derived from fundamental constants, including the speed of light (c), the gravitational constant (G), and the Reduced Planck's Constant $(\hbar)$. This means it is a natural unit of length that is independent of human-made standards. Note: (G) is an empirical constant, meaning it is determined strictly through measurement rather than derived from a theoretical formula.

Some theories propose that spacetime is not continuous at this scale but rather consists of discrete, Planck-sized units, similar to pixels on a screen. Extra dimensions could lower the Planck scale, potentially making quantum gravity effects observable at lower energies than previously expected.

Planck Time

Planck time (t_p) is the smallest possible unit of time that can be theoretically measured. It's the time it takes for light to travel a distance equal to a Planck length, which is an incredibly tiny distance; it's considered the shortest measurable time interval, roughly equivalent to 5.39×10^{-44} seconds, $(10^{-43}$ seconds$)$ and is used to describe extremely early moments in the universe like just after the Big Bang.

A Planck time is so small that it's practically impossible to measure directly with current technology. This time unit is calculated using the speed of light, Planck's constant, and the gravitational constant. Current physical theories may be unable to describe events on timescales shorter than the Planck time, suggesting the need for a new theory to explain such extreme conditions.

At this incredibly small scale, our current understanding of physics, which relies on classical gravity, is expected to break down, making it a key focus for theories of quantum gravity and the Big Bang. Planck time is the smallest unit of time that can be measured in principle. A second is vastly longer than a Planck time; there are more Planck times in one second than there are seconds in the age of the universe. It represents the earliest moment in the universe – the Planck epoch – when the four fundamental forces of nature were believed to have been unified before gravity separated from the others. It is derived from the fundamental constants of nature $(G, c$ and $\hbar)$.

The formula for Planck time is $t_p = \sqrt{\frac{\hbar G}{c^5}}$

It's not a precise value; it is a characteristic time unit that sets the scale of the smallest possible interactions in the universe. It is not currently measurable, and although it is theoretically the smallest unit, the smallest time interval ever measured is the zeptosecond (10^{-21} seconds). Some physicists speculate that the true minimum time increment may differ, but the Planck time remains the foundation for understanding the universe at its most fundamental level.

Planck Density

The Planck density (ρp) is the unit of density in the system of natural units known as Planck units. It represents the incredibly extreme density of matter or energy at which the known laws of physics – specifically General Relativity (GR) and Quantum Mechanics – are expected to break down. It is derived by combining the three fundamental constants of nature: the speed of light (c), the Gravitational Constant (G). and the Reduced Planck's Constant $(\hbar)$

The Planck density is defined as the Planck mass (m_p) divided by the cube of the Planck length (l_p). The formula for Planck density is:

$$\rho P = \frac{m_p}{l_p^3} = \frac{c^5}{\hbar G^2}$$

The value is massive:

$$\rho_p = \frac{c^5}{\hbar G^2} \approx 5.1 \times 10^{96} \text{ kilograms per cubic meter (kg/m}^3\text{)}$$

To put the magnitude of the Planck density into perspective: the nucleus of an atom has a density of about $10^{17} kg/m^3$. The Planck density is 10^{79} times denser than that. The Planck density marks the physical conditions that existed during the very first instant of the universe (the Planck Epoch). It is crucial because it defines the limit of classical physics. Any region of spacetime that reaches this density, such as the singularity inside a black hole or the Big Bang itself, requires a theory of quantum gravity to describe it accurately.

Planck Epoch

The Planck Epoch is the period from the Big Bang (t = 0) up to 10^{-43} seconds. During this period, the entire observable universe was confined to a volume with a density on the order of the Planck density. At this density and temperature, gravitational effects become so strong and occur at such small quantum scales that General Relativity alone cannot accurately describe them. This is the regime where the effects of quantum gravity are necessary. If General Relativity is extrapolated backward to the Big Bang, it predicts an infinite density (a singularity). The Planck density marks the point before the singularity at which current physics fails, indicating the limit of our understanding.

The singularity is a prediction of General Relativity. However, it is an artifact of the theory's incompleteness. Modern quantum gravity theories aim to eliminate this singularity by introducing quantum effects at the Planck scale.

The quantization of space is what prevents matter from achieving infinite density inside a black hole: Since there is a minimum size limit (the Planck volume) and a maximum mass limit for any given volume (the Planck density), mass can no longer be squeezed into an infinitely small space. As the matter within the black hole collapses, it compresses spacetime down to its minimum, quantized size. At this point, quantum mechanical effects take over, generating a massive, short-range repulsive force. Instead of collapsing into an infinite point (the singularity), the matter "bounces" due to this quantum pressure. This effectively means the black hole is not a dead end but a tunnel or a bridge that immediately expands into another region of spacetime (perhaps another universe, or simply another part of our own universe, but moving forward in time). This model is sometimes called the "Quantum Tunneling" or "Black Hole Bounce" scenario.

In Loop Quantum Gravity (LQG), the density of matter inside a black hole never reaches infinity. It reaches the Planck density and reverses its collapse due to quantum mechanics. A region of maximum but finite density replaces the singularity.

Riemannian Geometry

Riemannian Geometry is a branch of differential geometry that studies Riemannian manifolds. Imagine a space that locally looks like Euclidean space (our familiar flat space). A curved surface (like a sphere or a saddle) is a 2D manifold. Our universe, according to general relativity, is a four-dimensional manifold (three spatial dimensions plus one time dimension). Riemannian geometry concerns smooth, curved spaces of any dimension.

The core of Riemannian geometry is the metric tensor. This is a mathematical object that tells you how to measure distances and angles within the manifold. In flat Euclidean space, the metric is simple (think of the Pythagorean theorem). In curved spaces, the metric becomes more complex and can vary from point to point. It encodes the curvature of the space. The metric tensor allows you to calculate the length of a curve, the angle between two vectors, the area of a surface, and the volume of a region within the curved space.

Riemannian geometry provides tools to quantify the curvature of a manifold. There are different types of curvature tensors (such as the Riemann curvature tensor, the Ricci tensor, and the scalar curvature) that capture how the space deviates from flatness. Positive curvature (like on a sphere) means that initially parallel lines will converge. Negative curvature (like on a saddle) means that initially parallel lines will diverge. Zero curvature means the space is flat (Euclidean).

A key concept in Riemannian geometry is that the "straightest possible paths" between two points on a curved manifold are geodesics. In flat space, geodesics are straight lines. On a sphere, they are great circles (like the equator or lines of longitude). Secondly, when moving a vector along a curve while keeping it "parallel" to its initial orientation in curved space, its final orientation will generally depend on the path taken. And finally, Riemannian geometry extends the concepts of calculus (differentiation and integration) to curved spaces.

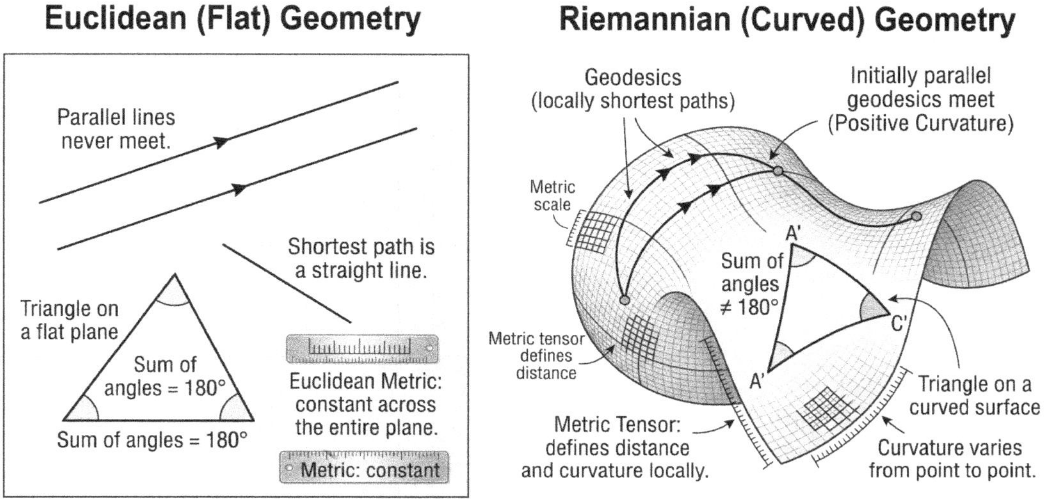

Riemannian geometry is the mathematical foundation of Einstein's theory of General Relativity. Gravity is not a force but a manifestation of the curvature of spacetime. Objects move along geodesics in this curved spacetime.

Riemannian geometry, as the language of general relativity, provides our current best description of spacetime geometry at macroscopic scales. However, it is expected to break down at the Planck scale due to quantum fluctuations.

Some theories suggest that spacetime and its geometry, as described by Riemannian geometry, might not be fundamental but rather emerge from some underlying quantum degrees of freedom at the Planck scale.

While Riemannian geometry is incredibly successful at describing gravity at macroscopic scales, a Planck-scale formalism aims to provide a more fundamental description that incorporates quantum effects. This likely involves either modifying, generalizing, or emerging from a structure that is very different from the smooth, continuous geometry of Riemannian manifolds. The exact nature of the Planck-scale formalism and its relationship to Riemannian geometry are among the biggest open questions in modern physics. Developing a successful theory of quantum gravity that can merge both theories remains a major goal of theoretical research.

Scalars and Vectors

Scalars: In physics, a scalar is a physical quantity that is fully described by its magnitude (a numerical value with a unit) alone, without any direction. Examples of scalars include mass, temperature, time, distance, and energy. Unlike vectors, which require both magnitude and direction for a complete description, scalars are simply numbers that can be added, subtracted, and multiplied using standard algebraic rules.

A scalar quantity is defined by magnitude, its size or amount, such as 50 minutes or 10 kilograms. Scalars do not have a direction. Scalars can perform algebraic operations and can be added, subtracted, and multiplied like ordinary numbers. For example, 50 minutes minus 10 minutes equals 40 minutes.

Examples:

- ***Mass:*** A 5 kg bag of sugar.
- ***Temperature:*** A 20°C room.
- ***Time:*** A 30-minute class period.
- ***Distance:*** The 100-meter length of a road.
- ***Energy:*** A 200-calorie serving of food.

The main distinction between a scalar and a vector is the presence of direction. A scalar has magnitude only (e.g., speed), whereas a vector has both magnitude and direction (e.g., velocity). For instance, a car's speed is a scalar (e.g., 60 km/h), whereas its velocity is a vector because it includes both the magnitude (e.g., 60 km/h) and the direction (e.g., 60 km/h north).

Vectors: In mathematics and physics, a vector is a quantity that possesses both magnitude (size or length) and direction. It can be visualized as a directed line segment, with its length representing the magnitude and its arrow indicating the vector's direction. Common examples include velocity, force, and displacement, which have both a speed/strength and a direction.

A vector has magnitude (length or size), direction (orientation in space), and can be decomposed into components, often represented as ordered collections of numbers (e.g., (x, y) in 2D or (x, y, z) in 3D). A vector begins at the "tail" and ends at the "head," which is marked by the arrow.

Examples:

- A car traveling at 50 mph due north is a vector (velocity), while just 50 mph is a scalar (speed).
- Pushing a box with a certain strength (force) in a specific direction.

- The change in position from the starting point to the ending point, regardless of the path taken (displacement).

Vector mathematical operations include addition, multiplication, dot product, and cross product.

- Combining vectors to find a resultant vector, such as adding two forces acting on an object.
- Multiplying a vector by a scalar (a real number) changes its magnitude.
- An operation between two vectors that results in a scalar, used to find the projection of one vector onto another or to check for orthogonality.
- An operation between two vectors that results in a new vector perpendicular to the plane of the original two vectors.

Sigma - Statistical Significance

In experiments, Sigma (σ) is a unit of measurement for statistical significance. It tells you how likely it is that an observed result (like a "bump" in a data graph) is a real discovery rather than just a random fluke. Mathematically, σ represents one standard deviation. If your data follows a normal distribution (the "bell curve"), the sigma level indicates how far your result is from the "expected" average.

The Bell Curve and Probability

Sigma's power comes from the 68-95-99.7 rule. In a random set of data:

- 1σ 68% of the data falls into this category. A1σ result is very common and likely just noise.
- 2σ 95% of the data falls here. Very often, the threshold for "statistically significant" applies to social sciences or medicine.

3σ 99.7% of data falls here. This is a "hint" of something new in physics.

Sigma Level	Confidence	Probability of a Fluke	Significance
1σ	68%	1 in 3	Background noise
2σ	95%	1 in 20	"Interesting" (Medical standard)
3σ	99.7%	1 in 370	Evidence / A "Hint"
5σ	99.99994%	1 in 3,500,000	**Official Discovery**

Tensors

A tensor is a mathematical object that generalizes scalars, vectors, and matrices to higher dimensions. In machine learning, a multidimensional array is often referred to as a tensor; its core definition is that it maintains its fundamental properties regardless of the coordinate system used to describe it.

Imagine a physical quantity, like temperature. A thermometer might display 20°C or 68°F, but the underlying physical temperature is the same no matter which unit system you use. The temperature is a "scalar," which is a rank-0 tensor. The value itself changes, but the physical property it represents remains the same.

- **Rank-0 Tensor:** Scalar (0 indices, e.g., T) – Magnitude only.
- **Rank-1 Tensor:** Vector (1 index, e.g., V^i) – Magnitude and direction.
- **Rank-2 Tensor:** Matrix/Linear Map (2 indices, e.g., $g_{\mu\nu}$ or $T^{\mu\nu}$) – Relates vectors to vectors or maps directions to directions.

Tensors take this idea further for more complex quantities involving magnitude and multiple directions. The components of a tensor will change when you change your coordinate system. Still, they will do so in a predictable way that ensures the tensor represents the same underlying physical or geometric object.

A tensor's rank refers to the number of dimensions in its array of components. A rank-0 tensor (scalar) is a single number, like temperature or mass, that has only magnitude. A rank-1 tensor (vector) is a list of numbers that represents a quantity with both magnitude and a single direction, such as a force or velocity. A rank-2 tensor (matrix) is a two-dimensional grid of numbers that describes a relationship between two vectors. A classic example is the stress tensor in physics, which relates a force vector to an area vector. Higher-rank tensors can have any number of dimensions and are represented as multi-dimensional arrays. A three-dimensional tensor could be visualized as a cube of numbers, and higher-rank tensors are nearly impossible to visualize directly.

The term "tensor" is used differently depending on the context. Tensors are essential for describing physical laws in a coordinate-independent manner. In general relativity, Einstein used tensors to describe the curvature of spacetime. Continuum mechanics describes the stress-strain relationship in materials using a rank-2 tensor. In electromagnetism, the electric susceptibility and permeability of materials are described using tensors.

The best way to understand tensors is to think of them as generalizations of things you already know:

A scalar is just a single number. It has magnitude but no direction–examples include temperature, mass, and speed. Mathematically, we can think of a scalar as a 0th-order tensor. Vectors have both magnitude and direction. Think of an arrow in space – examples: velocity, force, displacement. Mathematically, a vector is a 1st-order tensor. You can represent a vector as a list of numbers (components), like (x, y, z) in 3D space.

A matrix is a rectangular array of numbers. You may have seen them in linear algebra. Matrices can represent transformations, vector relationships, and other relationships. Mathematically, a matrix is a 2nd-order tensor. It has two indices to specify each element in the grid. Tensors are not *just* matrices. They can be in any order (rank). They can be 3rd-, 4th-, etc. order. Imagine a 3D array of numbers, or even higher-dimensional structures – those can be tensors.

What truly defines a tensor and distinguishes it from just a collection of numbers is how its components transform when you change your coordinate system. This can be illustrated with vectors and then generalized. Imagine a vector in 2D space. If you rotate your coordinate axes, the *components* of the vector (the x and y values) will change. However, the vector itself – the arrow in space – remains the same. The *rule* of how components change is very specific and defines a vector.

Tensors have similar, but more complex, transformation rules. When you switch from one coordinate system to another, the components of a tensor will transform according to a specific mathematical prescription.

Think of it like this: the tensor is the underlying physical or mathematical entity, such as "velocity," "stress," or "curvature." The components of a tensor are just the numbers we use to *represent* the tensor in a particular coordinate system. Just like coordinates depend on your choice of axes, tensor components depend on your coordinate system. Why is this "transformation rule" important? Because in physics and geometry, we want to describe things in a way that's coordinate-independent. Physical laws should be the same no matter how we choose to orient our coordinate axes, or whether we use Cartesian, spherical, or some other coordinate system.

Tensors provide a mathematical language for expressing these coordinate-independent relationships. If you write an equation using tensors, and the equation is valid in one coordinate system, it will automatically be true in *any* coordinate system. This is a powerful property!

You'll often see tensors written with indices.

- ***Scalar:*** Just a symbol, like s.
- ***Vector:*** Often written with one index, like v^i or v_i. The index i can

represent components along the x-, y-, and z-directions (i=1, 2, 3).

- ***Matrix (2nd-order tensor):*** Written with two indices, like T^{ij}, T_{ij}, or T^{i}_{j}. The indices j and i could represent rows and columns in a matrix.
- ***Higher-order tensors:*** Written with more indices, like R^{ijk}, S_{ijkl}, etc.

The positions of the indices also have meaning in terms of their transformation properties (covariant vs. contravariant tensors). However, for a basic understanding, simply recognizing the multiple indices is a good start. You can do various mathematical operations with tensors, such as add or subtract tensors of the same order, component by component. You can multiply a tensor by a scalar, multiplying each component by that scalar. The Tensor Product (Outer Product) method allows the creation of a higher-order tensor from two existing tensors. Contraction is a method for reducing the order of a tensor by summing over its indices.

To truly master tensors, you need to study linear algebra, differential geometry, and tensor calculus.

For those with a physics background

- *A Student's Guide to Vectors and Tensors* by Daniel A. Fleisch: A highly recommended, accessible, and solid introduction that isn't "watered down".
- *Spacetime and Geometry: An Introduction to General Relativity* by Sean Carroll: A great starting point for learning tensors from a physics perspective, with introductory chapters that cover the basics.

For online and visual learning

- *Profound Physics* website (profoundphysics.com): "*The Ricci Tensor: A Complete Guide With Examples.*" It details how the tensor functions as a tool for tracking volume changes and gravitational tides, using straightforward analogies.
- *Pavel Grinfeld's YouTube channel:* A comprehensive playlist of over 100 videos that complements his book, providing explanations of tensor calculus.
- *Tensors.net tutorials:* Provides accessible tutorials for those with a basic knowledge of linear algebra, including code snippets for numerical methods.

For a quick start

- *An Introduction to Tensors for Students of Physics and Engineering* by NASA: A concise and accessible document for a quick start.
- *Quick Introduction to Tensor Analysis* by Ruslan Sharipov: A self-contained PDF that can be read without prior prerequisites.

General Relativity Equations

Einstein's Field Equations (1915)
Equation: $G_{\mu\nu} + \Lambda g_{\mu\nu} = \frac{8\pi G}{c^4} T_{\mu\nu}$
Discoverer: **Albert Einstein**
Significance: The core of general relativity, describing how the geometry of spacetime (as expressed by the Einstein tensor $G_{\mu\nu}$) is curved by the presence of matter and energy $T_{\mu\nu}$.

The Geodesic Equation (1915)
Equation: $\dfrac{d^2x^\mu}{d\tau^2} + \Gamma^\mu_{\alpha\beta}\dfrac{dx^\alpha}{d\tau}\dfrac{dx^\beta}{d\tau} = 0$
Discoverer: **Albert Einstein**
Significance: Describes the motion of a particle in curved spacetime. It shows that in the absence of external forces, particles follow the "straightest possible" path (a geodesic) through the 4D manifold.

The Schwarzschild Radius Equation (1915)
Equation: $R_s = \dfrac{2GM}{c^2}$
Discoverer: **Karl Schwarzschild**
Significance: The equation determines the size of the event horizon for a non-rotating mass, marking the threshold where the pull of gravity becomes so intense that not even light can escape.

The Schwarzschild Metric (1916)
Equation: $ds^2 = -\left(1 - \dfrac{r_s}{r}\right) dt^2 + \left(1 - \dfrac{r_s}{r}\right)^{-1} dr^2 + r^2 d\Omega^2$
Discoverer: **Karl Schwarzschild**
Significance: The first exact solution to Einstein's Field Equations. It describes the spacetime geometry outside a non-rotating, spherical mass and provides the first mathematical basis for black holes.

The Friedmann Equations (1922)
The First Friedmann Equation: $H^2 = \left(\dfrac{\dot{a}}{a}\right)^2 = \dfrac{8\pi G}{3}\rho - \dfrac{kc^2}{a^2} + \dfrac{\Lambda c^2}{3}$
The Acceleration Equation: $\dfrac{\ddot{a}}{a} = -\dfrac{4\pi G}{3}\left(\rho + \dfrac{3p}{c^2}\right) + \dfrac{\Lambda c^2}{3}$
Discoverer: **Alexander Friedmann**

Significance: Derived from Einstein's Field Equations, the first equation describes the expansion rate (H) and relates the kinetic energy of expansion to the total mass-energy density. The second equation describes the acceleration (or deceleration) of the universe's expansion, driven by the pressure of the cosmic fluids and the cosmological constant.

Quantum Theory Equations

The Planck Relation (1900)
Equation: $E = hf$
Discoverer: **Max Planck**
Significance: This equation states that energy is quantized, meaning it is exchanged in discrete packets called "quanta." The energy (E) of a photon is directly proportional to its frequency (f) by Planck's constant (h).

Einstein's Photoelectric Equation (1905)
Equation: $K_{\text{max}} = h(\nu - \nu_0)$
Discoverer: **Albert Einstein**
Significance: This describes how light behaves as particles (photons). It shows that the maximum kinetic energy (K_{max}) of an ejected electron depends on the energy of the incident light minus the work function (ϕ) required to remove the electron from the material.

De Broglie's Hypothesis (1924)
Equation: $\lambda = \frac{h}{p}$
Discoverer: **Louis de Broglie**
Significance: Introduces wave-particle duality for all matter. Calculates the wavelength (λ) of a particle based on its momentum (p) and Planck's constant (h).

The Schrödinger Equation (1925/1926)
Equation: $i\hbar\frac{\partial}{\partial t}\Psi(\mathbf{r},t) = \hat{H}\Psi(\mathbf{r},t)$
Discoverer: **Erwin Schrödinger**
Significance: The fundamental equation of motion for quantum systems. It describes how the wave function (ψ), which contains the probability information of a system, evolves over time under the influence of the Hamiltonian operator $(\hat{H})$.

The Born Rule (1926)
Equation: $P(\mathbf{r},t) = |\Psi(\mathbf{r},t)|^2 = \Psi^*(\mathbf{r},t)\Psi(\mathbf{r},t)$

Discoverer: **Max Born**

Significance: Provides the physical interpretation of Schrödinger's wave function. It states that the probability density (P) of finding a particle at a specific point equals the squared magnitude of its wave function at that point.

Heisenberg's Uncertainty Principle (1927)

Equation: $\Delta x \Delta p \geq \frac{\hbar}{2}$

Discoverer: **Werner Heisenberg**

Significance: Sets a fundamental limit on the precision with which certain pairs of physical properties, such as position (x) and momentum (p), can be known simultaneously.

The Dirac Equation (1928)

Equation: $(i\gamma^{\mu}\partial_{\mu} - m)\psi = 0$

Discoverer: **Paul Dirac**

Significance: Reconciled quantum mechanics with special relativity for fermions (like electrons). It predicted the existence of spin and famously led to the discovery of antimatter (the positron).

Famous Laws

Law	Description
Kepler's First Law (Law of Ellipses)	Planets move around the Sun in elliptical orbits, with the Sun at one focus.
Kepler's Second Law (Law of Equal Areas)	A line segment joining a planet and the Sun sweeps out equal areas during equal intervals of time.
Kepler's Third Law (Law of Harmonies)	The square of a planet's orbital period is proportional to the cube of the semi-major axis of its orbit.
Newton's First Law of Motion	An object remains at rest or in uniform motion in a straight line unless acted upon by an external force.
Newton's Second Law of Motion	The acceleration of an object is directly proportional to the net force acting on it and inversely proportional to its mass.
Newton's Third Law of Motion	For every action, there is an equal and opposite reaction.
Newton's Law of Universal Gravitation	Every particle of matter attracts every other particle with a force that depends on their masses and the distance between them.

Coulomb's Law	The electrical force between two charged objects is directly proportional to the product of their charges and inversely proportional to the square of the distance between them.
Ampère's Circuital Law	The magnetic field induced around a closed loop relates to the electric current passing through the loop.
Ohm's Law	The electric current through a conductor between two points is directly proportional to the voltage across the two points.
Faraday's Law of Induction	A changing magnetic field induces an electromotive force in a conductor.
First Law of Thermodynamics	Energy cannot be created or destroyed, only transformed from one form to another. (Rudolf Clausius / William Rankine)
Second Law of Thermodynamics	The total entropy of an isolated system always increases over time. (Rudolf Clausius)
Hubble's Law	Galaxies move away from Earth at velocities proportional to their distance.

Greek Symbols: Physics

- **α (alpha):** Angular acceleration, alpha particle, or fine-structure constant.
- **β (beta):** Beta particle or velocity in terms of the speed of light.
- **γ (gamma):** Gamma ray, photon, or the Lorentz factor in relativity.
- **Δ (Delta – uppercase):** Change in a quantity (e.g., Δt for a time interval).
- **ε (epsilon):** Permittivity or strain.
- **η (eta):** Efficiency or viscosity.
- **θ (theta):** Angle or angular displacement.
- **κ (kappa):** Spring constant or dielectric constant.
- **λ (lambda):** Wavelength or linear density.
- **μ (mu):** Coefficient of friction, magnetic permeability, or prefix "micro".
- **ν (nu):** Frequency or kinematic viscosity.
- **π (pi):** Mathematical constant (≈ 3.14159...).
- **ρ (rho):** Density or resistivity.
- **Σ (Sigma – uppercase):** Summation operator.

- **σ (sigma – lowercase):** Stress or electrical conductivity.
- **τ (tau):** Torque, time constant, or shear stress.
- **φ (phi):** Magnetic or electric flux, or an angle/phase angle.
- **ψ (psi):** Wave function in quantum mechanics.
- **Ω (Omega – uppercase):** The unit for electrical resistance (Ohm).
- **ω (omega – lowercase):** Angular velocity or angular frequency.

The Greek Alphabet

α	alpha	Α	a
β	beta	Β	b
γ	gamma	Γ	g
δ	delta	Δ	d
ε	epsilon	Ε	e
ζ	zêta	Ζ	z
η	êta	Η	êê
θ	thêta	Θ	th
ι	iota	Ι	i
κ	kappa	Κ	k
λ	lambda	Λ	l
μ	mu	Μ	m
ν	nu	Ν	n
ξ	xi	Ξ	ks
ο	omikron	Ο	o
π	pi	Π	p
ρ	rho	Ρ	r
σ, ς	sigma	Σ	s
τ	tau	Τ	t
υ	upsilon	Υ	u
φ	phi	Φ	f
χ	chi	Χ	ch
ψ	psi	Ψ	ps
ω	omega	Ω	ôô

- **Sigma (σ, ς):** There are two forms for the letter Sigma. When written at the end of a word, it is like this: ς. If it occurs elsewhere, it is written as σ.
- **Upsilon (υ):** In the above table, we suggest that you pronounce this letter like "u" in "put." The preferred pronunciation is similar to the German "ü" sound, as in "Brücke," or the French "u" sound, as in "tu."
- **Xi (χ):** This is the same sound as "ch" in "Bach," which does not sound like "ch" in "chair."

Commonly Used Symbols

Calculus & Differential Operators

∂ : Differential (e.g., ∂x)

$\int$**:** Integral (single, double, triple)

$\oint$: Contour integral, Closed surface integral

∇: Gradient (nabla)

$\nabla\cdot$: Divergence

$\nabla\times$: Curl

∇^2**:** Laplacian

Σ: Summation

Vectors & Special Notations

$\rightarrow$: Vector (e.g., ($\rightarrow$v), approaches

^: Hat (unit vector, operator, e.g., $\hat{x}$, $\widehat{H}$

x̄: Mean, average⟨x⟩: Time average, expectation value

Δ: Increment, change (e.g., ΔE)

ẋ, ẍ: Time derivatives (velocity, acceleration)

|ψ:⟩ Ket (Quantum state vector. The fundamental symbol for a quantum state vector, representing the system's description.)

⟨ψ|: Bra (Dual vector. The complex conjugate transpose of a ket, used with kets to form inner products (e.g., ⟨φ|ψ⟩)

⟨ψ|φ⟩: Inner product (Probability amplitude)

Notes

Copyright

All efforts have been made to ensure that no text or images violate copyright. While this book has a copyright, you may feel free to use any of the materials herein, though I would appreciate a mention.

Artificial Intelligence (AI) Usage

Google's *Gemini*™ was invaluable in finding appropriate information, summarizing data, improving the overall flow, and providing illustrations. I have reviewed the data to try and avoid any errors – so blame me, not the AI. I also modified, added, and subtracted as needed to present the information appropriately. What I did not do, is give the AI a list of topics, and ask it to write the book. The topics, the ordering, how I wanted to present the data, and many long hours of thought and study went into it.

Recommended Reading

Here are a few of the books that I found educational, enlightening, amusing, or inspirational. All of them are readable by the layperson, but feel free to find your own, or use the Bibliography. I highly recommend Alan Lightman and Carlo Rovelli for their elegant writing skills.

- *Quantum Entanglement* by Jed Brody, MIT Press
- *Reality Is Not What It Seems: The Journey to Quantum Gravity* by Carlo Rovelli
- *The Order of Time* by Carlo Rovelli
- *Quantum Physics for Beginners* by Carl J. Pratt
- *Quantum Physics For Beginners, Into the Light* by John Stoddard
- *Quantum Physics for Hippies* by Dr. Lukas Neumeier & Dr. James Douglas
- *General Relativity: The Essentials* by Carlo Rovelli
- *Spacetime and Geometry* by Sean Carroll
- *Relativity For Beginners, The Special And The General Theory* by John Stoddard
- *The Elegant Universe: Superstrings, Hidden Dimensions, and the Quest for the Ultimate Theory* by Brian Greene
- *Existential Physics* by Hossenfelder, Sabine
- *Ghost Particle* by Alan Chodos and James Riordon.
- *The Big Picture* by Sean Carroll
- *Einstein's Dreams* by Alan Lightman
- *Great Ideas in Physics* by Alan Lightman
- *Six Easy Pieces* by Richard P. Feynman, Robert B. Leighton, and Matthew Sands

Glossary

Absolute Zero: The theoretical lowest temperature possible (-273.15°C or 0 Kelvin), at which all classical thermal motion of atoms ceases.

Anisotropy: Refers to the property of a material or system that exhibits different physical or mechanical characteristics depending on the direction of measurement. In contrast to isotropy (where properties are uniform in all directions), an anisotropic item's behavior changes depending on how it is oriented.

Annihilation: The process that occurs when a subatomic particle collides with its corresponding antiparticle (e.g., an electron and a positron). Upon collision, the particles disappear, and their combined mass is entirely converted into pure energy in the form of photons, neutrinos, or other lighter particles.

Antimatter: Matter composed of antiparticles, which have the same mass as ordinary particles but opposite physical charges (such as the positron, the antimatter counterpart of the electron).

Anyons: In quantum physics, an anyon is a type of quasiparticle that exists only in two-dimensional systems. Unlike standard 3D particles (which are strictly fermions or bosons), anyons have fractional quantum statistics, exhibiting properties that fall anywhere between those of standard particles.

Asymptotic: An asymptotic definition describes the long-term, large-scale behavior of a mathematical function, algorithm, or system as its variables or inputs approach infinity. It shows how values converge on a specific limit or how computational complexity grows, without focusing on minor, localized details.

Asymptotic Safety: Asymptotic safety is a quantum field theory framework, primarily used for quantum gravity, that proposes a theory can remain valid at arbitrarily high energies by approaching a non-trivial fixed point in its Renormalization Group (RG) flow.

***Baryons*:** A type of subatomic particle, such as a proton or neutron, made of three quarks held together by the strong force.

***Big Bang*:** The prevailing cosmological model describing the early development and expansion of the universe from a state of extremely high density and temperature.

***Black Hole*:** A region of spacetime where the gravitational pull is so intense that nothing, not even light, can escape its influence.

Bosons: Fundamental particles that carry forces, such as the photon (electromagnetism) or the gluon (strong force). Bosons can occupy the same quantum state simultaneously.

Bragg's Law: It describes how waves interfere when reflected off a crystal lattice. The formula ($n\lambda = 2d \sin\theta$) explains the conditions required for X-rays to bounce off the atomic planes of a crystal and produce constructive interference (diffraction). This allowed scientists to figure out the exact spacing and 3D structures of atoms within a crystal. It was developed by W. L. Bragg 1912.

Calabi-Yau Manifolds: A special type of complex geometric space that satisfies specific mathematical equations regarding its curvature. Named after mathematicians Eugenio Calabi and Shing-Tung Yau, they are fundamental in algebraic geometry and theoretical physics, particularly superstring theory.

Cherenkov Radiation: The electromagnetic radiation emitted when a charged particle passes through a dielectric medium at a speed greater than the phase velocity of light in that medium.

Compatibilism: The philosophical belief that free will and determinism are not mutually exclusive, meaning it is possible to believe in both without logical inconsistency. It defines free will as acting according to one's desires without external impediment, allowing for moral responsibility even if actions are part of a causal chain.

Cosmic Microwave Background: Also referred to as the CMB or CMBR or Relic radiation. It is the thermal electromagnetic radiation left over from the Big Bang, filling the entire observable universe.

Cosmological Constant: A value introduced by Einstein to represent the energy density of empty space, now often associated with dark energy and the accelerating expansion of the universe.

Dark Energy: A theoretical form of energy that permeates all of space and is believed to be responsible for the accelerating expansion of the universe.

Dark Matter: A form of matter that does not emit or reflect electromagnetic radiation, detected only by its gravitational effects on visible matter and the large-scale structure of the universe.

Decoherence: The process by which a quantum system loses its "quantumness" (superposition and entanglement) through interaction with its environment, leading to classical behavior.

Electron: A stable subatomic particle with a negative electric charge, acting as a primary carrier of electricity and the fundamental building block of chemical bonds.

Entanglement: A quantum phenomenon where two or more particles become linked such that the state of one particle instantly influences the state of the others, regardless of the distance between them.

Epistemic: An adjective meaning relating to knowledge, understanding, or the methods used to gain knowledge. It stems from the Greek word epistēmē ("knowledge") and is used to describe how we know what we know, often focusing on the nature, scope, and limitations of knowledge.

Equivalence Principle: A foundational concept in physics stating that the effects of a uniform gravitational field are completely indistinguishable from the effects of uniform acceleration.

Event Horizon: The "point of no return" surrounding a black hole, beyond which the escape velocity exceeds the speed of light.

Fermions: A fundamental particle of matter, such as a quark or lepton. Unlike bosons, fermions obey the Pauli Exclusion Principle, meaning no two fermions can occupy the same quantum state.

Frame-dragging: An effect in general relativity where a rotating massive object twists the geometry of spacetime around itself. In classical physics, a rotating mass only exerts a gravitational pull toward its center. Einstein's field equations show that mass in motion alters spacetime differently than mass at rest. When an object rotates, it drags the surrounding coordinate system along with its spin.

Functional Renormalization Group: An advanced theoretical framework in physics used to study quantum and statistical systems with many interacting particles, especially in complex, strongly correlated materials. It allows physicists to scale from fundamental microscopic laws (high-energy/short-distance) to observable macroscopic phenomena (low-energy/long-distance).

Gauge Theory: A type of field theory in which the laws of physics remain unchanged (invariant) under certain local transformations, necessitating the existence of force-carrying particles.

Gauss's Laws: Carl Friedrich Gauss was a German mathematician and physicist. Gauss's Laws form a core pillar of Maxwell's Equations, which govern how electricity and magnetism interact. When physicists write out the mathematical equations for a quantum particle's probability cloud, they use Gaussian wave packets. Gauss was one of the very first to realize that geometry didn't have to be flat. His work laid the mathematical foundation that Albert Einstein later used to develop General Relativity.

***General Relativity*:** Albert Einstein's theory of gravitation, which describes gravity not as a force, but as the curvature of spacetime caused by the presence of mass and energy.

***Gluons*:** The gauge boson that carries the strong nuclear force, "gluing" quarks together to form protons, neutrons, and other hadrons.

***Gravity*:** The fundamental interaction that causes objects with mass or energy to be attracted to one another.

Heisenberg's Uncertainty Principle: A fundamental limit in quantum mechanics stating that certain pairs of physical properties, such as position and momentum, cannot be known simultaneously with absolute precision.

Hermitian: A special type of linear operator that equals its own conjugate transpose: $\hat{A}^\dagger = \hat{A}$. In matrix notation, taking the conjugate transpose († or "dagger") means swapping rows with columns (transposition) and taking the complex conjugate of every entry (i $\rightarrow -i$). For a matrix A, this is written as $\mathrm{A}^\dagger = (A^*)^T = A$.)

Higgs Boson: The elementary particle associated with the Higgs field, which provides mass to other fundamental particles through their interaction with it.

It From Qubit: "It from qubit" is a modern extension of physicist John Wheeler's "It from bit" concept, which posits that all physical reality – particles, forces, and spacetime – is fundamentally information-theoretic in origin. Instead of the universe being built from classical 0s and 1s, "it from qubit" suggests reality emerges from quantum bits (qubits) and their complex entanglements.

Kinetic Energy: A form of energy an object possesses due to its motion, measured in Joules (J). It's equal to half the mass of an object multiplied by the square of its velocity.

Landau Levels: Landau levels are the discrete, quantized energy states of a charged particle (like an electron) moving in a uniform magnetic field. Discovered by Soviet physicist Lev Landau in 1930, they force electrons into circular "cyclotron" orbits, causing the continuous energy spectrum to split into distinct, highly degenerate energy bands.

Length Contraction: A concept in Albert Einstein's theory of special relativity. It is the phenomenon where a moving object's length appears shorter along its direction of motion than its actual length at rest.

Lie Algebra: A vector space equipped with a specialized binary multiplication operation called the Lie bracket, denoted as [X, Y]. It serves as the local, linear counterpart to a Lie group. Instead of analyzing curved, nonlinear transformation spaces directly, a Lie algebra examines the tangent vector space at the identity element of a continuous symmetry group. This converts problems involving differential geometry into standard linear algebra.

Lie Group: A Lie Group is a foundational mathematical concept that links abstract algebra (group theory) with smooth geometry (differential calculus). It provides the primary mathematical framework for describing continuous symmetries in differential geometry, quantum mechanics, and general relativity.

***Light-year*:** The distance light travels in a vacuum in one year, approximately 5.88 trillion miles (9.46 trillion kilometers).

Mach's principle: A philosophical hypothesis stating that the inertia of any physical object is not an intrinsic property, but rather determined by the large-scale distribution and motion of all other matter in the universe. Coined by Albert Einstein after the 19th-century Austrian physicist and philosopher Ernst Mach, this concept seeks to resolve the problem of absolute motion in physics.

Magnetic Moment: In physics, the magnetic moment is a vector quantity that dictates a magnet's strength and its orientation in an external magnetic field. It determines the amount of torque a magnet will experience, pointing from a magnet's south pole to its north pole.

Møller Scattering: The elastic scattering process of two electrons in quantum field theory, named after Danish physicist Christian Møller, who derived the cross-section in 1932. It is a foundational QED process involving the electromagnetic exchange of a photon between two interacting electrons.

Momentum: Describes an object's resistance to stopping. It is defined as "mass in motion" – a measure of how much force is required to stop an object based on its size and speed.

Moore's Law: Moore's Law is the observation that the number of transistors on a microchip doubles roughly every two years, while the cost of computers is halved. Coined by Intel co-founder Gordon Moore in 1965, this exponential growth became the foundational guiding principle for the entire technology industry.

Mott insulators: A Mott insulator is a class of materials that conventional band theory predicts should be electrical conductors, but which actually act as insulators. This behavior occurs because traditional band theory treats electrons as independent particles, whereas Mott insulators are dominated by strong electron-electron interactions (Coulomb repulsion). Named after physicist Sir Nevill Mott, these materials are a cornerstone of strongly correlated electron physics.

Neutrino: An extremely light, neutral subatomic particle that interacts only via the weak nuclear force and gravity, allowing it to pass through ordinary matter almost entirely unimpeded.

Newton's Laws of Motion: 1. An object at rest will remain at rest, and an object in motion will remain in motion at a constant velocity in a straight line, unless acted upon by an outside net force. 2. The acceleration of an object depends directly on the net force acting upon it and inversely on its mass. 3. For every action, there is an equal and opposite reaction.

Noether's Theorem: A mathematical theorem stating that every continuous symmetry in nature corresponds to a specific conservation law (e.g., time symmetry leads to the conservation of energy).

Ontic: An adjective that relates to entities, facts, or the concrete properties of things that actually exist in reality, as opposed to theoretical concepts, abstract ideas, or modes of being. In Quantum Mechanics, it's used to classify whether a concept (like the quantum state) represents a real, physical entity rather than just a subjective "epistemic" state of knowledge.

Particle-wave Duality: A foundational concept in quantum mechanics stating that all fundamental entities of the universe – such as light (photons) and matter (electrons) – exhibit both wave-like and particle-like properties, depending on how they are observed or measured.

Pauli Exclusion Principle: The Pauli Exclusion Principle states that no two identical fermions (such as electrons) in a single atom or system can simultaneously occupy the exact same quantum state. In practical terms for chemistry, this means no two electrons in an atom can have the same four quantum numbers.

Phenomenology: The application of theoretical physics to experimental data by deriving quantitative predictions from known theories. It is related to the philosophical notion of the same name in that these predictions describe anticipated behaviors for the phenomena in reality. Phenomenology stands in contrast to the scientific method's experimentation, which aims to test a scientific hypothesis rather than make predictions.

Phonon: A phonon is a quantum or discrete unit of vibrational mechanical energy within a crystal lattice, functioning as a "particle" of sound or heat. As collective excitations of oscillating atoms, they travel through solids to transport thermal and acoustic energy. Phonons are essential to understanding heat capacity, thermal conductivity, and sound propagation.

Photon: The fundamental particle of light and the gauge boson responsible for the electromagnetic force. It is massless and travels at the speed of light.

Planck Scale: The scale of length, time, and energy at which the effects of quantum gravity are expected to become dominant; it marks the limit of our current physical theories.

Poincaré group: The Poincaré group (also known as the inhomogeneous Lorentz group) is the full 10-dimensional Lie group of spacetime symmetries in special relativity. It consists of all isometric transformations that leave the spacetime interval between two events in flat Minkowski spacetime invariant:

$$ds^2 = -c^2dt^2 + dx^2 + dy^2 + dz^2$$

Mathematically, the Poincaré group is the semidirect product of four-dimensional spacetime translations $\mathbb{R}^{1,3}$ and the Lorentz group $O(1,3)$:

Poincaré group $= \mathbb{R}^{1,3} \rtimes O(1,3)$

While the Lorentz group only covers spatial rotations and velocity shifts (boosts), the Poincaré group extends these operations by adding origin shifts in space and time.

Polariton: Polaritons are bosonic quasiparticles formed by the strong coupling of electromagnetic waves (photons) with dipolar excitations in matter, such as excitons, phonons, or plasmons. Often described as half-light, half-matter hybrids, they enable the fast, wavelike flow of light combined with strong, matter-like interactions.

Quantum Mechanics: The branch of physics that describes the behavior of matter and energy at the atomic and subatomic scales, where actions are quantized and probabilistic.

Quark: An elementary particle and a fundamental constituent of matter. Quarks combine to form composite particles called hadrons, the most stable of which are protons and neutrons.

Quaternion Group: The quaternion group (often denoted as Q_8) is a fundamental, non-commutative (non-Abelian) mathematical group containing exactly eight elements. It is widely used in abstract algebra, quantum mechanics, and 3D computer graphics (such as game engines and aerospace orientation).

Rayleigh-Jeans Law: A classical physics formula that attempts to describe the density of electromagnetic radiation emitted by a blackbody at a given temperature. Developed by Lord Rayleigh and Sir James Jeans in the early 20th century, it states that the energy output of a blackbody is directly proportional to its absolute temperature and inversely proportional to the fourth power of the wavelength. While it successfully describes radiation at long wavelengths (such as radio waves and infrared light), it fails spectacularly at short wavelengths (like ultraviolet light), a conceptual breakdown famously known as the ultraviolet catastrophe.

Relativity Postulate: Or the Principle of Relativity, dictates that the laws of physics are identical in all inertial (non-accelerating) frames of reference. This fundamental assumption of Albert Einstein's special theory of relativity means no observer can perform an experiment to determine if they are absolutely "stationary" or "moving" at a constant velocity.

Renormalization Group: The renormalization group is a mathematical toolkit that physicists use to study how a system changes when you view it at different scales of distance or energy – averaging out microscopic details, and helping physicists predict big-picture phenomena such as how water turns to steam or how magnets lose their strength. It allows scientists to examine a complex system with billions of microscopic moving parts and systematically zoom out to see the large-scale behavior. Imagine looking at a digital photograph.

Ricci Tensor: In differential geometry and general relativity, the Ricci curvature tensor (often just called the Ricci tensor) is a mathematical object that measures how a curved space differs from ordinary, flat space.

Rydberg Atom: An atom with one or more electrons excited to a very high principal quantum number (n). Because the outermost electron orbits exceptionally far from the nucleus, the atom becomes physically massive, highly sensitive to electric and magnetic fields, and useful for advanced quantum computing and sensors.

Simultaneity: In Albert Einstein's theory of Special Relativity, simultaneity is not absolute; it is relative. Because the speed of light is constant, whether two spatially separated events happen at the same time depends entirely on the observer's frame of reference (how fast they are moving).

Singularity**:** A point of infinite density and gravity, predicted by General Relativity to exist at the center of a black hole and at the origin of the universe.

Spacetime**:** The four-dimensional mathematical model that fuses the three dimensions of space and the one dimension of time into a single continuum.

SQUIDs: Superconducting Quantum Interference Devices are the most sensitive magnetic field sensors ever created. They detect incredibly faint magnetic fields, even those billions of times smaller than the Earth's magnetic field, making them invaluable in medical imaging, quantum computing, and materials science.

Sterile Neutrino: Sterile neutrinos are hypothetical, fourth types of elementary particles that do not interact via the weak force, unlike the three known active neutrino flavors. They are called "sterile" because they interact only through gravity and may blend with active neutrinos via oscillation, making them a leading candidate for dark matter and for explaining anomalies in anomaly detection.

***Strong Nuclear Force*:** The most powerful of the four fundamental forces, responsible for binding quarks into protons and neutrons, and holding atomic nuclei together.

Superposition: The principle that a quantum system can exist in multiple states or locations simultaneously until it is measured or interacts with its environment.

Time Dilation: A physical phenomenon where time passes at different rates for observers depending on their relative motion or proximity to a gravitational field. It is a fundamental consequence of Einstein's theory of relativity and proves that time is not absolute.

Torque: In physics, torque is the rotational equivalent of linear force. It is a measure of the turning or twisting force that causes an object to rotate around an axis or pivot point. It is also known as the "moment of force".

Van der Waals Interactions: Named after Nobel Prize-winning physicist Johannes Diderik van der Waals (1837–1923), these interactions are the weak, short-range electric forces of attraction that exist between neutral atoms and molecules. Unlike ionic or covalent bonds, they do not involve the sharing or transferring of electrons, but rather the subtle, temporary shifting of electrical charges.

***Weak Nuclear Force*:** The fundamental force responsible for radioactive decay and neutrino interactions; it operates only over very short subatomic distances.

Worldline: The path that an object traces in four-dimensional spacetime.

***Wormhole*:** A hypothetical "bridge" or tunnel through spacetime that could create a shortcut between two distant points in the universe.

Zeno: Zeno of Elea (c. 490–430 BC) was an ancient Greek philosopher and disciple of Parmenides, known for creating paradoxes that argued motion and change are illusions. His famous paradoxes, such as Achilles and the Tortoise, suggest that traversing any distance requires an infinite number of steps, theoretically making movement impossible.

Bibliography

Books

Quantum Mechanics

- *The Principles of Quantum Mechanics* by P. A. M. Dirac
- *Unlocking the Mysteries of the Universe: Your Essential Guide to Quantum Physics for Beginners* by Diego Marchesi
- *On the Trail of Blackbody Radiation* by Don S. Lemons
- *Reality Is Not What It Seems: The Journey to Quantum Gravity* by Carlo Rovelli
- *Helgoland: Making Sense of the Quantum Revolution* by Carlo Rovelli
- *Experimental Search for Quantum Gravity* by Sabine Hossenfelder (Editor)
- *Something Deeply Hidden* by Sean Carroll
- *Supergravity* by Daniel Z. Freedman and Antoine Van Proeyen
- *The Biggest Ideas in the Universe: Space, Time, and Motion* by Sean Carroll
- *Quantum Entanglement* by Jed Brody, MIT Press
- *Quantum Physics for Beginners* by Carl J. Pratt
- *Quanta and Fields: The Biggest Ideas in the Universe* by Sean Carroll
- *Quantum Physics For Beginners, Into the Light* by John Stoddard
- *What Is Real?* by Adam Becker
- *The Age of Entanglement* by Louisa Gilder
- *Quantum Mechanics: The Theoretical Minimum* by Leonard Susskind
- *No-Nonsense Quantum Field Theory* by Jakob Schwichtenberg
- *Quantum Field Theory, as Simply as Possible* by Anthony Zee
- *Chronicles of Quantum Chromodynamics* by Richard Gray
- *Quantum Physics For Dummies* by Steven Holzner
- *Supersymmetry and Supergravity* by Julius Wess and Jonathan Bagger.
- *Modern Supersymmetry: Dynamics and Duality* by John Terning.
- *A Supersymmetry Primer* by Stephen P. Martin.

Quantum Oddities

- *How the Hippies Saved Physics* by David Kaiser
- *Quantum Physics for Hippies* by Dr. Lukas Neumeier & Dr. James Douglas
- *God and Quantum Physics* by C. Allan Boyles
- *Quantum Enigma: Physics Encounters Consciousness* by Fred Kuttner and Bruce Rosenblum.
- *Mindful Universe: Quantum Mechanics and the Participating Observer* by Henry P. Stapp.
- *The Self-Aware Universe* by Amit Goswami

- *Quantum Physics of Consciousness* by Fred Kuttner and Bruce Rosenblum.

General Relativity

- *General Relativity: The Essentials* by Carlo Rovelli
- *Spacetime and Geometry* by Sean Carroll
- *Relativity: The Special And The General Theory* by Albert Einstein
- *Relativity For Beginners, The Special And The General Theory* by John Stoddard
- *General Relativity* by Leonard Susskind and André Cabannes
- *Faster than Light* by Robert J Nemiroff
- *White Holes* by Carlo Rovelli
- *Six Not-So-Easy Pieces: Einstein's Relativity, Symmetry, and Spacetime* by Richard P. Feynman, Robert B. Leighton, and Matthew Sands
- *From Eternity to Here:* by Sean Carroll
- *The Big Bang and the Fine-tuned Universe (In a Nutshell)* by Robert A Alston

General Physics & Related Topics

- *Lost in Math: How Beauty Leads Physics Astray* by Hossenfelder, Sabine
- *Six Easy Pieces* by Richard P. Feynman, Robert B. Leighton, and Matthew Sands
- *Special Relativity and Classical Field Theory: The Theoretical Minimum* by Leonard Susskind and Art Friedman
- *Seven Brief Lessons on Physics* by Carlo Rovelli
- *The Order of Time* by Carlo Rovelli
- *The Beginning of Infinity* by David Deutsch
- *The Fabric of Reality:* by David Deutsch
- *The Emperor's New Mind* by Roger Penrose
- *Ghost Particle* by Alan Chodos and James Riordon.
- *The Big Picture* by Sean Carroll
- *The God Equation* by Michio Kaku
- *Existential Physics* by Hossenfelder, Sabine
- *Einstein's Dreams* by Alan Lightman
- *Great Ideas in Physics* by Alan Lightman
- *General Relativity for Babies* by Chris Ferrie
- *Einstein and Religion* by Max Jammer
- *Spinoza* by Stuart Hampshire

YouTube Channels

- *Science with Sabine*–Sabine Hossenfelder
- *The Biggest Ideas in the Universe*–Sean Carroll
- *WhatDaMath*–Anton Petrov
- *Complex Questions Explained Simply*–Arvin Ash
- *PBS Space Time*–Dr. Matt O'Dowd
- *Astrum*–Alex McColgan
- *Into The Impossible* – Dr. Brian Keating
- *The Science Asylum*–Nick Lucid
- *Ri*–The Royal Institution
- *How the Universe Works*
- *Float Head Physics*–Mahesh Shenoy

Podcasts

General Quantum Physics

- *Physics World Weekly Podcast:* Includes topics on quantum physics.
- *Sean Carroll's Mindscapes:* Interviews with science experts.
- *The Quantum Spin:* A podcast about the world of quantum physics.
- *Clear as Quantum:* Focuses on quantum science and technologies.
- *Ologies:* This podcast has an episode covering quantum physics.
- *The Titanium Physicists:* For those who want a deeper dive into the subject.
- *Math & Physics Podcast:* Discusses both mathematics and physics.
- *Quanta Magazine Podcast (Quanta Podcast):* Covers a range of topics in physics, including quantum mechanics.

Quantum Computing

- *The Quantum Divide: A* focus on quantum technology and networking.
- *IEEE Quantum:* Features discussions with scientists and engineers about quantum research.
- *Nonlocal:* A Quantum Computing Podcast: Provides a look behind the scenes of quantum computing research.

General Relativity (Astronomy and the Cosmos)

- *Robinson's Podcast: A* masterclass with philosopher Tim Maudlin on general relativity, covering black holes, spacetime curvature, and more.
- *The Last Theory:* Explores deriving GR from Wolfram Physics.
- *Sean Carroll:* Frequently discusses relativity on podcasts like Lex Fridman's, offering expert insights.
- *Theory of Relativity: Simplified*: Breaks down GR, time dilation, $E=mc^2$, and gravitational waves for newcomers.
- *Astronomy Cast:* Offers accessible explanations, such as the classic rubber-sheet analogy for gravity warping spacetime.
- *Universe Today:* Has podcasts dedicated to explaining Einstein's theory.
- *Ask a Spaceman:* Features a scientist answering questions about space, which often touches on quantum topics.

Websites

https://profoundphysics.com/: Ville Hirvonen, Abir Ghosh, and Cameron Bunney.

About the Author

Les Short spent thirty years at IBM as a software engineer and project manager. His professional journey was driven by his technical background and a degree in mathematics and computer science. These provided the analytical framework for his interest in the laws of nature.

As an enthusiast of astronomy and physics, he has spent many years exploring the complexities of general relativity and quantum mechanics. His retirement allowed him to transition from a student of the cosmos to an author, with a mission to translate the most challenging concepts of modern physics into a language accessible to anyone with a spark of curiosity.

Today, he resides in North Carolina, with his wife, and their two dogs, Koba and JoJo. When he is not writing about the nature of reality, he enjoys the coastal lifestyle, whether walking on the beach, flying kites, or continuing his astronomical observations. This work marks his second publication, following a book of his personal journey to happiness and renewal.

Les Short

2026, North Carolina, USA

Website: lesshort.com

If you have any comments, please contact me at: lesshor2@gmail.com

Alphabetical Index

C

D

E

Q

R

S

T

X

Y

Z

† Reference Section

1 Gell-Mann, M & Zweig, G. (1964). A Schematic Model of Baryons and Mesons, An SU(3) Model for Strong Interaction Symmetry and its Breaking. *CalTech Authors, CERN preprint.* https://authors.library.caltech.edu/records/d4jws-7eh11 https://cds.cern.ch/record/570209/files/CERN_Report_TH-412_corrected.pdf

2 Rubbia, C., van der Meer, S. (1983). CERN. Discovery of the W and Z bosons. *Physics Letters B.* https://www.sciencediract.com/journal-physics-letters-b

3 CERN (European Organization for Nuclear Research). (2012). The Higgs Boson Discovery. Large Hadron Collider (LHC). https://atlas.cern/Discover/Physics/Higgs

4 CERN. (2026). https://home.cern/cms-looks-deep-inside-quarks/

5 Wu, C., Lee, T., & Yang, C. (1956). Question of Parity Conservation in Weak Interactions. *Physical Review. 106, 1371 (1957).* https://doi.org/10.1103/PhysRev.104.254

6 Hasert, F. J., et al. (1973). Observation of neutrino-like interactions without muon or electron in the Gargamelle neutrino experiment. *Physics Letters B*, *46*(1), 138–140. [Insert DOI if available]

7 Arnison, G., et al. (1983). Experimental observation of isolated large transverse energy electrons with associated missing energy at $\sqrt{s} = 540$GeV. *Physics Letters B*, *122*(1), 103–116. [Insert DOI if available]

8 Bloom, E. D., Coward, D. H., DeStaebler, H., Drees, J., Miller, G., Mo, L. W., Taylor, R. E., Breidenbach, M., Friedman, J. I., Hartmann, G. C., & Kendall, H. W. (1969). High-energy inelastic *e–p* scattering at 6° and 10°. *Physical Review Letters*, *23*(16), 930–934.

9 Schlamminger, S., Chao, L., Lee V., Shakarji C., Newell D., Stirling J., Cochrane, R., & Speake. C. (2026). Redetermination of the gravitational constant with the BIPM torsion balance at NIST. *The Metrologia Journal.* https://www.nist.gov/publications/redetermination-gravitational-constant-bipm-torsion-balance-nist

10 Witten, E. (1995). String Theory Dynamics In Various Dimensions. Cornell University Arxiv. https://arxiv.org/abs/hep-th/9503124

11 Chao, Y.-X., Ge, P., Hua, Z.-X., Jia, C., Wang, X., Liang, X., Yue, Z., Lu, R., Tey, M. K., & You, L. (2026). Probing false vacuum decay and bubble nucleation in a Rydberg atom array. *Physical Review Letters, 136(12), Article 120407.* https://doi.org/10.1103/PhysRevLett.136.120407

12 Wigner, E. P. (1961). Remarks on the mind-body question. In I. J. Good (Ed.), *The scientist speculates: An anthology of partly-baked ideas* (pp. 284–302). Heinemann.

13 Everett, H., III. (1957). On the foundations of quantum mechanics *[Doctoral dissertation, Princeton University].* Princeton University Finding Aids.

Everett, H., III. (1957). "Relative state" formulation of quantum mechanics. *Reviews of Modern Physics*, *29*(3), 454–462.

14 Tegmark, M. (1998). The interpretation of quantum mechanics: Many worlds or many words? *Fortschritte der Physik*, *46*(6–8), 855–862. https://doi.org/10.1002/(SICI)1521-3978(199811)46:6/8 *Preprint.* arXiv. (1997).

15 Bohm, D. (1952). A suggested interpretation of the quantum theory in terms of "hidden" variables. I. *Physical Review, 85(2), 166–179.* https://doi.org/10.1103/PhysRev.85.166

16 Englert, B.-G., Scully, M. O., Süssmann, G., & Walther, H. (1992). Surrealistic Bohm trajectories. *Zeitschrift für Naturforschung A*, *47*(12), 1175–1186. https://doi.org/10.1515/zna-1992-1201

17 Sharoglazova, V., Puplauskis, M., Mattschas, C., Toebes, C., & Klärs, J. (2025). Energy –speed relationship of quantum particles challenges Bohmian mechanics. *Nature*, *643*(8070), 67–72. https://doi.org/10.1038/s41586-025-09099-4

18 Ghirardi, G. C., Grassi, R., & Rimini, A. (1990). A continuous-spontaneous-collapse model for quantum mechanics. *Physical Review A*, *42*(3), 1057–1064. https://doi.org/10.1103/PhysRevA.42.1057

18 Drezet, A., Lazarovici, D., & Nabet, B. M. (2026). Bohmian mechanics remains unchallenged by tunnelling experiment. *Nature*, *653*(8116), E13–E15. https://doi.org/10.1038/s41586-026-10450-6

19 Hameroff, S., & Penrose, R. (1996). Orchestrated reduction of quantum coherence in brain microtubules: A model for consciousness. *Mathematics and Computers in Simulation*, *40*(3-4), 453-480. https://doi.org/10.1016/0378-4754(96)80476-9

Hameroff, S., & Penrose, R. (2014). Consciousness in the universe: A review of the 'Orch OR' theory. *Physics of Life Reviews*, *11*(1), 39-78.

20 Ghirardi, G. C., Rimini, A., & Weber, T. (1986). Unified dynamics for microscopic and macroscopic systems. *Physical Review D*, *34*(2), 470–491. https://doi.org/10.1103/PhysRevD.34.470

21 Cramer, J. G. (1986). The transactional interpretation of quantum mechanics. *Reviews of Modern Physics*, *58*(3), 647–687.

22 Caves, C. M., Fuchs, C. A., & Schack, R. (2001). Quantum probabilities as Bayesian probabilities. *Phys. Rev. A 65, 022305 (2002).* Cited by: 655

23 Fritzsch, H., Gell-Mann, M., & Leutwyler, H. (1973). Advantages of the color octet gluon picture. *Physics Letters B*, *47*(4), 365–368. Cited by: 3590

24 Tomonaga, S. (1946). On a relativistically invariant formulation of the quantum theory of wave fields. *Progress of Theoretical Physics*, *1*(2), 27–42.

Schwinger, J. (1948). Quantum electrodynamics. I. A covariant formulation. *Physical Review*, *74*(10), 1439–1461. https://doi.org/10.1103/physrev.74.1439

Feynman, R. P. (1949). Space-time approach to quantum electrodynamics. *Physical Review*, *76*(6), 769–789.

25 Gross, D. J., & Wilczek, F. (1973). Ultraviolet behavior of non-abelian gauge

theories. *Physical Review Letters*, *30*(26), 1343–1346.
Politzer, H. D. (1973). Reliable perturbative results for strong interactions? *Physical Review Letters*, *30*(26), 1346–1349.

26 Chern, S.-S., & Simons, J. (1974). Characteristic forms and geometric invariants. *Annals of Mathematics, 99(1), 48–69.* https://doi.org/10.2307/1971013 Cited by: 1498
Schwarz, A. S. (1979). The partition function of a degenerate functional. Communications in Mathematical Physics, 67(1), 1–16. https://doi.org/10.1007/bf01223197 Cited by: 335

27 Yukawa, H. (1935). On the interaction of elementary particles. I. *Proceedings of the Physico-Mathematical Society of Japan, 17, 48–57.*

28 Yang, C. N., & Mills, R. L. (1954). Conservation of isotopic spin and isotopic gauge invariance. *Physical Review, 96(1), 191–195.* https://doi.org/10.1103/PhysRev.96.191

29 Schwinger, J. (1966). Particles and sources. *Physical Review, 152(4), 1219–1226.*
Schwinger, J. (1970). Particles, sources, and fields (Vol. 1). Addison-Wesley.

30 Scully, M. O., & Drühl, K. (1982). Quantum eraser: A proposed photon correlation experiment concerning observation and "delayed choice" in quantum mechanics. *Physical Review A*, *25*, 2208. https://doi.org/10.1103/PhysRevA.25.2208

31 Kim, Y.-H., Yu, R., Kulik, S. P., Shih, Y. H., & Scully, M. O. (2000). A delayed choice quantum eraser. *Physical Review Letters, 84(1), 1–5.* https://doi.org/10.1103/PhysRevLett.84.1

32 Bell, J. (1964). On the Einstein Podolsky Rosen Paradox. *Physique Fizika Journal.* https://arxiv.org/html/2408.04483v1

33 Clauser, J. F., Horne, M. A., Shimony, A., & Holt, R. A. (1969). Proposed experiment to test local hidden-variable theories. *Physical Review Letters, 23(15), 880–884.* https://doi.org/10.1103/PhysRevLett.23.880

34 Freedman S.J. & Clauser, J.F. (1972). Experimental Test of Local Hidden-Variable Theories. *Physical Review Letters, Volume 28, Issue 14, Pages 938–941* https://journals.aps.org/prl/abstract/10.1103/PhysRevLett.28.938

35 Aspect, A., Dalibard, J & Roger, G. (1982) Experimental Test of Bell's Inequalities Using Time-Varying Analyzers. *Physical Review Letters.* https://journals.aps.org/prl/abstract/10.1103/PhysRevLett.49.1804

36 Clarke, J., Devoret, M.H., & and Martinis, J.M. (1985). Measurements of Macroscopic Quantum Tunneling out of the Zero-Voltage State of a Current-Biased Josephson *Junction. Phys. Rev. Lett. 55, 1908* https://physics.aps.org/articles/v18/170

37 Pedalino, S., Ramírez-Galindo, B.E., Ferstl, R., Hornberger, K., Arndt, M., & Gerlich, S. (2026). Probing quantum mechanics with nanoparticle matter-wave interferometry. *The Nature Journal.* https://www.nature.com/articles/s41586-025-09917-9

38 Pusey, M. F., Barrett, J., & Rudolph, T. (2012). On the reality of the quantum state. *Nature Physics, 8(6), 475–479.* _Cited by: 1218

39 Hall, E. H. (1879). On a new action of the magnet on electric currents. *American*

Journal of Mathematics, 2(3), 287–292. https://doi.org/10.2307/2369245

von Klitzing, K., Dorda, G., & Pepper, M. (1980). New method for high-accuracy determination of the fine-structure constant based on quantized Hall resistance. *Physical Review Letters, 45(6), 494–497.* https://doi.org/10.1103/PhysRevLett.45.494

40 Palmer, T. (2026). Rational quantum mechanics: Testing quantum theory with quantum computers. *Proceedings of the National Academy of Sciences, 123(12), Article e2523350123.* https://doi.org/10.1073/pnas.2523350123. Cited by: 3

41 Aliev, H., Kim, H., & Choi, S. (2020). A Scalable and Secure Group Key Management Method for Secure V2V Communication. Sensors, 20(21), 6137.

42 Zhao, C., Hao-Qi, S., Wang, X., Wei-Lai, Y., Rui-Hong, G., Rui-Hong, G., Yi-Kun, W., & Yi-Kun, W. (2024). Analysis of Beam Walk in Inter-Satellite Laser Link: Implications for Differential Wavefront Sensing in Gravitational Wave Detection. Applied Sciences, 14(13), 5526.

43 Kretschmann, E. (1917). On the physical meaning of relativity principles: A. Einstein's new and his original theory of relativity (R. Rynasiewicz, Trans.). Annalen der Physik, 53, 575–614.

44 Penrose, R. (1965). Gravitational Collapse and Space-Time Singularities. Physical Review Letters, 14(3), 57–59. _Cited by: 4666

45 Abbott, B. P., Abbott, R., Abbott, T. D., et al. (2016). Observation of Gravitational Waves from a Binary Black Hole Merger. Physical Review Letters, 116(6), 061102. _Cited by: 21861

46 Hawking, S. W. (1975). Particle creation by black holes. *Communications in Mathematical Physics, 43(3), 199–220.* https://doi.org/10.1007/bf01608497 Cited by: 20766

47 Susskind, L., Thorlacius, L., & Uglum, J. (1993). The stretched horizon and black hole complementarity. *Physical Review D, 48(8), 3743–3761.* https://doi.org/10.1103/physrevd.48.3743 Cited by: 1606

48 Almheiri, A., Engelhardt, N., Marolf, D., & Maxfield, H. (2019). The entropy of bulk quantum fields and the entanglement wedge of an evaporating black hole. *Journal of High Energy Physics, 2019(12), Article 63.* https://doi.org/10.1007/jhep12(2019)063 Cited by: 1362

Penington, G. (2020). Entanglement wedge reconstruction and the information paradox. *Journal of High Energy Physics, 2020(9), Article 2.* https://doi.org/10.1007/jhep09(2020)002 Cited by: 1442

49 Latham Boyle, L. & Neil Turok, N. (2024). Black Mirrors: CPT-Symmetric Alternatives to Black Holes. *The European Physical Journal* C (Vol. 85, Article 305). https://arxiv.org/abs/2412.09558

50 Gaztañaga, E., Kumar, K. S., & Marto, J. (2026). A new understanding of Einstein-Rosen bridges. *Classical and Quantum Gravity, 43, Article 015023.* https://doi.org/10.1088/1361-6382/ae3044 Cited by: 19

51 Kerr, R. P. (1963). Gravitational field of a spinning mass as an example of algebraically special metrics. Physical Review Letters, 11(5), 237–238. https://doi.org/10.1103/PhysRevLett.11.237

52 Rovelli, C., & Vidotto, F. (2014). Planck stars. International Journal of Modern Physics D, 23(12), 1442026. https://doi.org/10.1142/s0218271814420267

53 Wiltshire, D. L. (2009). Average observational quantities in the timescape cosmology. *Physical Review D, 80(12), Article 123512.* https://doi.org/10.1103/PhysRevD.80.123512 Cited by: 171

54 Arnowitt, R., Deser, S., & Misner, C. W. (1962). The dynamics of general relativity. In L. Witten (Ed.), *Gravitation: An introduction to current research (pp. 227–265). John Wiley & Sons. 6)022.* https://doi.org/10.1088/1475-7516/2016/02/022 Cited by: 91

55 Rovelli, C., & Smolin, L. (1990). Loop space representation of quantum general relativity. *Nuclear Physics B, 331(1), 80–152.* https://doi.org/10.1016/0550-3213(90)90019-a Cited by: 1215

Rovelli, C., & Smolin, L. (1995). Spin networks and quantum gravity. Physical Review D, 52(10), 5743–5759. https://doi.org/10.1103/PhysRevD.52.5743

56 Ashtekar, A., Olmedo, J., & Singh, P. (2023). Regular black holes from loop quantum gravity. *Springer Series in Astrophysics and Cosmology, 235–282.* https://doi.org/10.1007/978-981-99-1596-5_7 Cited by: 92

57 Mannheim, P. D. (2012). Making the case for conformal gravity. *Foundations of Physics*, *42*(3), 388–420.

58 Mannheim, P. D., & Kazanas, D. (1989). Exact vacuum solution to conformal Weyl gravity and galactic rotation curves. *The Astrophysical Journal*, *342*, 635–648. https://doi.org/10.1086/167623

59 Mannheim, P. D. (1990). Conformal cosmology with no cosmological constant. *General Relativity and Gravitation*, *22*(3), 289–298. https://doi.org/10.1007/BF00756202

60 Mannheim, P. D. (2006). Alternatives to dark matter and dark energy. *Progress in Particle and Nuclear Physics*, *56*(2), 340–445. https://doi.org/10.1016/j.ppnp.2005.08.001

61 Bender, C. M., & Mannheim, P. D. (2008). No-ghost theorem for the fourth-order derivative Pais-Uhlenbeck oscillator model. *Physical Review Letters*, *100*(11), Article 110402.

62 Hoyle, F., & Narlikar, J. V. (1964). A new theory of gravitation. Proceedings of the Royal Society of London. Series A. Mathematical and Physical Sciences, 282(1389), 191–207.

Hoyle, F., & Narlikar, J. V. (1966). Conformal Rees–Chala gravitation theory. Proceedings of the Royal Society of London. Series A. Mathematical and Physical Sciences, 294(1437), 138–148.

63 Shaposhnikov, M., & Zenhausern, D. (2009). Quantum scale invariance, cosmic inflation and dark energy. Physics Letters B, 671(4-5), 162–166. https://doi.org/10.1016/j.physletb.2008.11.054

Shaposhnikov, M., & Zenhausern, D. (2009). Scale invariance, the Higgs mechanism, and calculation of cosmological constants. Physics Letters B, 671(4-5), 187–192. https://doi.org/10.1016/j.physletb.2008.11.055

64 Engle, J., Noui, K., Perez, A., & Pranzetti, D. (2010). Black hole entropy from an SU(2) invariant formulation of Type I isolated horizons. Physical Review D, 82(4), Article 044050.

65 Penrose, R. (1967). Twistor algebra. Journal of Mathematical Physics, 8(2), 345–366.

Penrose, R. (1977). The twistor programme. Reports on Mathematical Physics, 12(1), 65 –76. https://doi.org/10.1016/0034-4877(77)90047-7

66 Modesto, L. (2012). Super-renormalizable quantum gravity. Physical Review D, 86(4), Article 044005. https://doi.org/10.1103/PhysRevD.86.044005

Modesto, L., & Rachwał, L. (2014). Super-renormalizable & finite gravitational theories. Nuclear Physics B, 889, 228–248.

67 Weinberg, S. (1978). Critical phenomena for field theorists. In A. Zichichi (Ed.), Understanding the Fundamental Constituents of Matter (pp. 1–52). Plenum Press.

Weinberg, S. (1979). Ultraviolet divergences in quantum theories of gravitation. In S. W. Hawking & W. Israel (Eds.), General Relativity: An Einstein Centenary Survey (pp. 790–831). Cambridge University Press.

68 Ambjørn, J., & Loll, R. (1998). Non-perturbative Lorentzian quantum gravity, causality and topology change. *Nuclear Physics B, 536(1–2), 407–434.* https://doi.org/10.1016/s0550-3213(98)00692-0 Cited by: 504

Ambjørn, J., Jurkiewicz, J., & Loll, R. (2001). Dynamically triangulating Lorentzian quantum gravity. *Nuclear Physics B, 610(1–2), 347–382.* _Cited by: 413

69 Engle, J., Pereira, R., & Rovelli, C. (2007). The loop-quantum-gravity vertex-amplitude. *Physical Review Letters*, *99*(16), Article 161301. Cited by: 355

70 Milgrom, M. (1983). A modification of the Newtonian dynamics as a possible alternative to the hidden mass hypothesis. *The Astrophysical Journal*, *270*, 365–370. https://doi.org/10.1086/161130 Cited by: 5975

Milgrom, M. (1983). A modification of the Newtonian dynamics: Implications for galaxies. *The Astrophysical Journal*, *270*, 371–383. https://doi.org/10.1086/161131

71 Gallardo, P. A., Pardo, K., Philcox, O. H. E., Battaglia, N., Battistelli, E. S., Bean, R., & Atacama Cosmology Telescope Collaboration. (2026). The Atacama Cosmology Telescope: A test of the gravitational force law on cosmological scales using the kinematic Sunyaev-Zeldovich effect. *Physical Review Letters*. Cited by: 3

72 Belenkiy, A. (2012). Alexander Friedmann and the origins of modern cosmology. *Physics Today*, *65*, 38–43. https://doi.org/10.1063/pt.3.1750 Cited by: 49

Pfeifer, C., Voicu, N., Friedl-Szász, A., & Popovici-Popescu, E. (2025). From kinetic gases to an exponentially expanding universe - The Finsler-Friedmann equation. *JCAP10(2025)050*. https://doi.org/10.1088/1475-7516/2025/10/050 Cited by: 7

73 Cadoni, M., Murgia, R., Pitzalis, M., & Sanna, A. P. (2024). Quasi-local masses and cosmological coupling of black holes and mimickers. *Journal of Cosmology and*

Astroparticle Physics, *2024*(3), Article 026. https://doi.org/10.1088/1475-7516/2024/03/026

74 DeWitt, B. S. (1967). Quantum theory of gravity. I. The canonical theory. *Physical Review*, *160*(5), 1113–1148.

Wheeler, J. A. (1968). Superspace and the nature of quantum geometrodynamics. In C. M. DeWitt & J. A. Wheeler (Eds.), *Battelle Rencontres: 1967 Lectures in Mathematics and Physics* (pp. 242–307).W. A. Benjamin.

75 Bartini, G. (2026). Emergent entropic time within a partitioned Bose-Einstein condensate. *Journal of Quantum Cosmology*, *14*(2), 112–125.

76 Hartman, T. E. (1962). Tunneling of a wave packet. *Journal of Applied Physics*, *33*, 3427–3433. Cited by: 1183

77 Mi, X., Ippoliti, M., Quintana, C., et al. (2021). Observation of time-crystalline eigenstate order on a quantum processor. *Nature*, *601*, 531–536. Cited by: 493

78 Kongkhambut, P., Skulte, J., Mathey, L., Cosme, J.G., Hemmerich, A., & Keßler. H. University of Hamburg. (2022). Observation of a continuous time crystal. *Science Journal.* https://www.science.org/stoken/author-tokens/ST-697/full

79 Böhmer, T, Jan P. Gabriel, J.P., Costigliola, L. Niklas, J., Kociok, N., Hecksher, T., Jeppe C., Dyre, J.C., & Blochowicz, T. (2024). Time reversibility during the ageing of materials. *The Nature Physics Journal.* https://forskning.ruc.dk/en/publications/time-reversibility-during-the-ageing-of-materials/

80 Guff, T, Shastry, C.U., Rocco, A. University of Surrey (2025). Emergence of opposing arrows of time in open quantum systems. *Scientific Reports 3658 (2025).*

81 Gunther, K. (2025). Three-Dimensional Time: A Mathematical Framework for Fundamental Physics. *Reports in Advances of Physical Sciences (Vol. 9, Article 2550004).* https://www.worldscientific.com/doi/10.1142/S2424942425500045

82 Cooper, L. N. (1956). Bound electron pairs in a degenerate Fermi gas. *Physical Review*, *104*(4), 1189–1190.

Pęcak, D., & Sowiński, T. (2022). Unconventional pairing in few-fermion systems at finite temperature. *Scientific Reports*, *12*. https://doi.org/10.1038/s41598-022-22411-w

Subedi, M. S. (2017). Superconductivity and Cooper Pairs. *Himalayan Physics*, *6*, 104– 107.

83 Kong, L. (2025). Cooper-pair density modulation state in an iron-based superconductor. *Nature*, *640*(8057). https://doi.org/10.1038/s41586-025-08703-x

84 Ranni, A., Brange, F., Mannila, E. T., Flindt, C., & Maisi, V. F. (2021). Real-time observation of Cooper pair splitting showing strong non-local correlations. *Nature Communications*, *12*(1), 6358. https://doi.org/10.1038/s41467-021-26627-8 Cited by: 49

85 STAR collaboration at the U.S. Department of Energy's Brookhaven National Laboratory (2011). Detection of the antimatter partner of the helium nucleus: antihelium-4. *Nature 473, pages 353–356 (2011) (DOI: 10.1038/nature10079).*

86 Cowan, C. L., Reines, F., Harrison, F. B., Kruse, H. W., & McGuire, A. D. (1956). Detection of the Free Neutrino: a Confirmation. *Science, 124(3212),* 103–104. https://doi.org/10.1126/science.124.3212.103

www.ingramcontent.com/pod-product-compliance
Lightning Source LLC
LaVergne TN
LVHW060542061126
851857LV00017B/1871
* 9 7 9 8 9 9 7 0 2 2 9 0 7 *